AXIOMATIC QUALITY

AXIOMATIC QUALITY

Integrating Axiomatic Design with Six-Sigma, Reliability, and Quality Engineering

BASEM SAID EL-HAIK

With Foreword by Nam P. Suh

Published by John Wiley & Sons, Inc., Hoboken, New Jersey.
Published simultaneously in Canada.

For general information on our other products and services please contact our Customer Care Department within the U.S. at 877-762-2974, outside the U.S. at 317-572-3993 or fax 317-572-4002.

Wiley also publishes its books in a variety of electronic formats. Some content that appears in print, however, may not be available in electronic format.

Library of Congress Cataloging-in-Publication Data:

El-Haik, Basem.
Axiomatic quality: integrating axiomatic design with six-sigma, reliability, and quality engineering / by Basem Said El-Haik.
p. cm.
"Wiley-Interscience publication."
Includes bibliographical references and index.
ISBN 0-471-68273-X (cloth : alk. paper)
1. Process control. 2. Quality control. 3. Six sigma (Quality control standard). 4. Production management. I. Title.

TS156.8.E44 2005
658.5′62–dc22

2004059408

Printed in the United States of America.

10 9 8 7 6 5 4 3 2 1

To my parents, wife, and children
for their continuous support

Download the Axiomatic Design Software

“Acclaro DFSS Light®” from

ftp://ftp.wiley.com/public/sci_tech_med/axiomatic_quality/

See Section 2.2.2

CONTENTS

FOREWORD

To be successful, a company must have products that are multisuperior—that is, superior in design, technology, quality, reliability, and cost. This is true whether the products are materials, machinery, software systems, manufacturing processes and systems, complex systems such as the Space Shuttle, or a variety of services provided by industrial firms or other businesses and organizations to its customers.

Traditionally, these product requirements have been satisfied through a recursive and repetitious process of *design/build/test/fix*, characterized by trial and error. This paradigm of product development is expensive, uncompetitive, unpredictable—and ultimately prone to failures. Despite this, the idea that companies might approach product development in a systematic and scientific basis has been resisted by far too many businesses, notwithstanding the enormous benefits that such an approach can bring to them and their customers.

As businesses seek innovative ways to magnify their bottom lines, how could such a faulty tradition of product development be allowed to endure? This is in part due to their lack of understanding of design theories and principles that have the power to guide and discipline the thought processes of engineers and managers and lead to more efficient and effective product development. In many companies, engineers learn and repeat the experientially acquired practices used in the company but do not have opportunities and incentives to acquire new knowledge.

Fortunately, many companies have recently realized that their profit and growth are closely tied to the quality of their products, which has led to the current *design for six-sigma* (DFSS) movement. Although DFSS has helped many companies, including General Electric, its reputation has become greater than its capabilities and is now used to represent an overall quality movement. In the past, DFSS approaches were appropriately used to address the problems caused primarily by coupled designs of products and manufacturing processes. But now, *six-sigma* stands for *designing it right* and *making it right* to satisfy customer needs to produce superior products at low cost. In reality, to achieve the ultimate goal of DFSS, companies must *first* design the product right, since even an inordinate amount of investment in subsequent manufacturing processes cannot overcome the errors committed during the design stage of product development.

When a product is a coupled design due to the reliance on the traditional trial-and-error development approaches, the product lacks robustness, requiring that every manufacturing process be executed precisely, which is difficult to do. Furthermore, coupled designs often cost more to manufacture and operate, in addition to being unreliable. Even when a product appears to perform well when it is new, products with coupled functional requirements do not have the long-term reliability and stability.

This is where the introduction of axiomatic design is critical. The idea that well-designed products maintain the independence of functional requirements—one of the two axioms of axiomatic design—is integral to designing superior products, developing complex systems, and achieving technology innovation. The design of products and systems must be done logically and systematically so that even in the event that a product is poorly made, it will still perform well, satisfying the functional requirements and constraints. Such a disciplined approach leads to technology breakthroughs.

By combining axiomatic design with traditional six-sigma methodologies, this book by Dr. Basem El-Haik is a welcome addition to the growing body of literature on the importance of the design decisions and the quality of products. This work is a result of his many years of experience and leadership in quality efforts at Ford and now at Textron. He combines his professional background in statistics, axiomatic design, and six-sigma to create an integrated approach to the manufacture of quality products. This book also provides a guideline for how a company can implement the goals of six-sigma and axiomatic design across its entire organization.

Every company should encourage their engineers to read this book and apply the basic principles of axiomatic design and six-sigma to their daily tasks. They will be pleased with the bottom-line results of such an endeavor. Doing it right the first time—rather than spending most of their resources to correct mistakes made at the design stage through testing of prototypes—will pay off handsomely in terms of profit, technology innovation, efficiency, and reputation. The faulty paradigm of product development and manufacturing wastes resources and creates risks for companies and organizations. Complex products cannot be developed using such tired paradigms unless cost over-run, poor performance, missed schedule, and lack of technological innovation are to be accepted or explained away. Since changing the behavior of people is an extremely daunting task, strong leadership in these organizations to implement a new paradigm will be necessary, but it is one that can result in multisuperior results.

My congratulations to Basem for writing such a fine book.

NAM P. SUH

Sudbury, MA
January 18, 2005

PREFACE

Today's product solutions of current engineering and design activities in many manufacturing companies and design houses are generally suffering from much vulnerability, such as complexity, coupled functional performance, and modest use of the modularity principle, among others. Functional requirements coupling is a common conceptual design vulnerability that generates hidden and unnecessary developmental effort and, later, operational costs in the hands of the customer. Coupling exhibits itself by the degree of lack of control that can be exerted by both the design team and the customer. Complexity is operational design vulnerability that can be attributed to the variability exhibited in the design's functional requirements and is caused by sources of variation, the noise factors. Complexity has far-reaching implications on a design entity beyond the intuitive "sum of parts" meaning. Complexity vulnerability, similar to coupling, has a conceptual origin and is established in the design entity as a result of failure of obedience to design principles, in particular, those promoted to axioms. Unfortunately, the effect of conceptual design vulnerability is rarely considered. The symptoms generated by such vulnerabilities can be worsened further by ignorance on the part of design teams and the lack of design process diligence that emphasizes upstream quality and reliability thinking. Indeed, the lack of diligence of design processes and procedures coupled with the wrong impression of perceiving design as an art rather than a technical field has contributed largely to the creation of vulnerable products and services.

Many vulnerable engineered products are the result of the limited scope of traditional design practices. This may be attributed primarily to the lack of a comprehensive process to achieve robustness at both the conceptual and operational levels. Additional causes of design vulnerability are inherent in the nature of the current design processes themselves, being algorithmic focus with limited attention to early design phases. Fredrikson (1994) claimed that as much as 80% of the total life cycle cost is determined during the concept development stages. An equivalent percentage is equally applicable to the quality, and reliability levels of the design entity are also committed at these stages. Both assessments are obedient to the Pareto principle. It is interesting to notice that most treatment effort in terms of methods, techniques, and approaches is spent on fixing operational performance vulnerabilities, with little or no attention paid to addressing conceptual types of vulnerabilities. Coupling vulnerability will always produce operational vulnerabilities; the reverse, however, is false. Therefore, conceptual treatments should eliminate or reduce conceptual operational vulnerabilities, hence the *axiomatic quality process*, the subject of this book.

The axiomatic quality process has three ingredients, representing the common conceptual intersection of design for six-sigma, axiomatic design, and robust design, as depicted in Figure P.1. The success of six-sigma and design for six-sigma as a company wide initiative and axiomatic design as a prescriptive design methodology has generated enormous interest in the engineering world. In addition, the axiomatic quality process gains extra strength by building on the current deployment experience of robust design to address residual operational vulnerabilities that were not addressed by conceptual means.

Starting with the voice of the customer, axiomatic quality focuses on establishing a comprehensive design process that utilizes ingredients from comparative tools: quality engineering, axiomatic design, theory of inventive problem solving, deterministic optimization, and in the absence of quantitative data, fuzzy set theory. Selection of the axiomatic design approach as the design method to be used in this book is based on its inherent possession of several possible complementary interfaces to many synthesis and analysis design tools that can be molded into one comprehensive process.

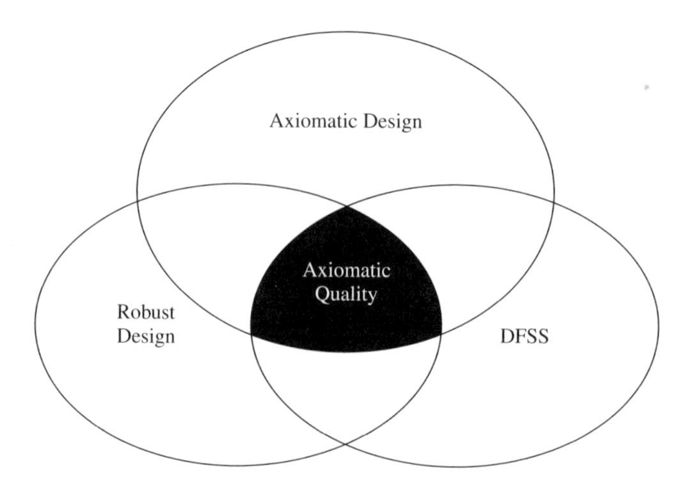

Figure P.1 Axiomatic quality ingredients.

Unlike other quality improvement movements, where the focus is primarily on the quality of the product or service to external customers, axiomatic quality focuses on the whole quality of product and service design. Axiomatic quality is aligned very closely with the spirit of design for six-sigma as outlined in Yang and El-Haik (2003). The axiomatic quality process puts a lot of focus on design and tries to do things right the first time. The ultimate axiomatic quality goal is *do the right things* and to *do things right all the time*. Doing the right things means achieving absolute excellence in design, be it a product, a manufacturing process, a service process, or a business process. Doing the right things all the time means that not only should we have superior design, but the actual product or process that we build according to our design should always deliver what it is supposed to deliver. Doing things right all the time means high consistency and extremely low variation in performance. Nowadays, high consistency is necessary not only for product performance and image but is also a matter of survival.

Axiomatic design, the theory of inventive problem solving (TRIZ), the transfer function, and score cards are really powerful methods to create superior designs, that is, to do the right things. Axiomatic quality also brings another class of powerful methods, Taguchi's methods, into its toolbox. The fundamental objective of the Taguchi methods is to create a superior product or process that can perform highly consistently despite external disturbances and uncertainties, called *noise factors*, thus making it possible to do things right all the time.

The implementation of axiomatic quality will take more effort and training than that of other methods, but it will be more in line with design principles and axioms. This book's mission is to give readers a complete picture of the axiomatic quality process.

Objectives of the Book

1. To provide clear, in-depth coverage of all the important philosophical, theoretical, implementation, and technical aspects of axiomatic quality.
2. To discuss and illustrate very clearly the entire axiomatic quality deployment and execution process.
3. To discuss and illustrate clearly all methods used in the axiomatic quality process.
4. To discuss the theory and background of each method's part in the axiomatic quality process clearly, together with examples.
5. To help develop practical skills in applying axiomatic quality in industrial contexts.

Background Needed

The background required to study this book is some familiarity with simple statistical principles, such as normal distribution, mean, variance, and simple data analysis techniques. A good mathematical background is also needed.

Summary of Chapter Contents

Chapter 1 is an introduction to the axiomatic quality process. The chapter begins by presenting the case for the process, introduces the process ingredients (i.e., axiomatic design, six-sigma and design for six-sigma, and robust design), and then proceeds to discuss axiomatic quality within the big picture of product life cycle, including development. The following chapters build on this introductory chapter by going deeper in concepts and wider in applications.

Chapter 2 is basically a high-level introduction to axiomatic design method as developed by Nam P. Suh of MIT. Several concepts of the method are highlighted, such as design domains, axioms, design mappings, and hierarchy, as well as design vulnerabilities. In a collective sense, axiomatic design method is a prescriptive design method that enables teams to design for conceptual and operational robustness by following design axioms and applying theories and corollaries that are derived from them. Axiomatic design method entertains two axioms, the independence axiom and the information axiom, which together with the set of theories and corollaries deduced from them constitute an axiomatic system.

In Chapter 3 the independence axiom is demystified from both the practical and theoretical perspectives. In this chapter the mathematical formulation of the independence axiom is stressed, to lay down the background deemed necessary for the following axiomatic quality process concepts and derivations. We explore coupling vulnerability, a conceptual vulnerability that is created in the design entity due to the violation of the independence axiom. It features two measures of coupling. In addition, the zigzagging method, the axiomatic design method used for design mapping, is presented in several applications.

In Chapter 4 the information axiom is presented from the perspective of traditional axiomatic design method as well as from a new perspective developed by the author. The objective of this chapter is to present the implication of operational vulnerability created in a design entity due to violation of the information axiom and to develop measures to assess design complexity and information content as axiomatic measures. Such measures are used later in additional axiomatic quality process developments.

In Chapter 5 we present Taguchi's quality engineering procedure, with concentration on its phased deployment as well as its links to axiomatic design and axiomatic quality in general. Both conceptual and theoretical linkages are discussed within a mathematical framework that ties robustness measures to axiomatic measures. We strengthen the axiomatic quality process by providing the necessary building blocks from several perspectives, adding to the core of axiomatic quality process.

In Chapter 6 we present the big picture of the axiomatic quality process. The process extends over three phases: the customer attributes-to-functional requirements mapping phase, the conceptual design for capability (CDFC) phase, and the optimization phase. Functional requirements (FRs) are actionable engineering characteristics that are cascaded from the voice of the customer over two stages of quality function deployment. The array of FRs at a given mapping within a design hierarchical level is the vehicle used for both the CDFC and optimization phases of the axiomatic quality process. The core steps of the customer attributes-to-functional requirements mapping and CDFC phases are discussed in this chapter, which features a section on axiomatic quality process deployment. In Chapters 7 and 8 we discuss in more detail the CDFC phase steps that were mentioned briefly in Chapter 6.

In Chapter 7 we discuss the concept selection problem, a core CDFC phase step. The problem can be formulated and solved using an integer programming optimization. The objective function is pieced from measures such as complexity, customer satisfaction, and cost. Design complexity is measured by information content using Chapter 5 complexity derivations, which in turn takes the probability of success as arguments. Both crisp and fuzzy formulations are presented, depending on the type of data available to the design team.

Chapter 8 contains the balance of the CDFC phase not fully explored in Chapter 6 due to their taxing development. The CDFC phase is an integrating framework of axiomatic design and quality engineering methods such that quality and reliability can be designed up front with no or minimal conceptual design vulnerabilities. It is shown that consistent deployment of the independence axiom is critical to reducing significant design changes and improving quality in later phases of the development process. Emphasis on an understanding of the relationships between design modules (components and subassemblies) and hierarchical design mapping is stressed via the concept of physical structure. This chapter features a section on TRIZ and its use within the axiomatic quality process.

Chapter 9 covers the axiomatic quality optimization phase, including mean adjustment and variability optimization techniques. For a given design mapping, the idea is to have all design FR means on targets desired by the customer with minimal variability around them. The variability optimization techniques take the six-sigma target as an objective. This chapter is supplemented by mathematical formulations that quantify robustness measures at six-sigma quality levels of a design mapping. The optimization phase has a parameter component and a tolerance component that is at its core both deterministic and empirical. In the deterministic case, the objective function is pieced from cost incurred by the customer and by cost incurred by the manufacturing company to keep the functional requirements variation at six-sigma targets. Additionally, the formulation is supplemented by constraints to achieve FR independence according to the independence axiom.

In Chapter 10 we present a low-pass filter axiomatic quality case study and in Chapter 11 we discuss reliability engineering and its relation to the axiomatic quality process. A new definition of reliability that takes into account axiomatic design, the design method used to conceive the design entity, is introduced: axiomatic reliability. This will help design teams get an initial assessment of design reliability prior to testing or mass production. Axiomatic reliability is part of the axiomatic quality optimization phase, but was assigned its own chapter due to its depth of developmental effort.

What Distinguishes This Book from Others

This book's main distinguishing features are its completeness and comprehensiveness. Most important topics in the axiomatic quality process are discussed clearly and in depth. The organizational, implementational, theoretical, and practical aspects of both the process and its integrated methods are covered very carefully and in complete detail. This is the only book that focuses on axiomatic quality and reliability. It can be used either as a complete reference book on the axiomatic quality process or as a complete training material for design teams.

Additionally, this book features a copy of Acclaro DFSS Light®[1], a limited version of Acclaro Designer® software provided by Axiomatic Design Solutions, Inc., via free download from the Wiley ftp site

ftp://ftp.wiley.com/public/sci_tech_med/axiomatic_quality/

Acclaro DFSS Light® is a JAVA-based software package that implements axiomatic design processes as presented in Chapter 2 (Section 2.2.2).

Acknowledgments

In preparing this book I received advice and encouragement from several people, including James Wasiloff of DaimlerChrysler Company and Larry Smith of Ford Motor Company, as well as many others. I also wish to thank George Telecki and Rachel Witmer of John Wiley & Sons, Inc., and Matt Pallaver of Axiomatic Design Solutions, Inc.

Contacting the Author

Your comments and suggestions regarding this book will be greatly appreciated and I will give serious consideration to them for inclusion in a future edition. I also conduct public and in-house six-sigma, axiomatic design, quality engineering, and design for six-sigma workshops and provide consulting services. I can be reached by e-mail at *basemhaik@hotmail.com.*

BASEM SAID EL-HAIK

[1]Acclaro DFSS Light® is one of the software products of Axiomatic Design Solutions, Inc. It is protected under both copyright and pending patents. Acclaro is a registered trademark of Axiomatic Design Solutions, Inc. It is about a 30 megabyte download installation file. It requires Windows 95, 2000, NT, XP, or later revisions with 256k of memory.

CHAPTER 1

INTRODUCTION TO THE AXIOMATIC QUALITY PROCESS

1.1 WHY AXIOMATIC QUALITY?

Attention has begun to shift from the improvement of design quality in downstream development stages to early upstream stages. This shift is motivated by the fact that the design decisions made during early stages of a product development cycle have the largest impact on total life-cycle cost and the quality of the system. It has been claimed that as much as 80% of the total life-cycle cost is determined during the concept development stage (Fredrikson, 1994). Research in the design and manufacturing arenas, including product development, is currently receiving an increased focus on addressing industry efforts to shorten lead times, cut development and manufacturing costs, lower total life-cycle cost, and improve the quality of the end products and systems. It is the author's experience that at least 80% of design quality is also determined in the early design phases.

In the context of this book, the term *quality* can be defined as the degree to which the design vulnerabilities do *not* adversely affect product performance. This definition, as well as most of the developments of this book, are equally applicable to the service design arena because the design principles, in particular those promoted to *axioms*,[1] are universal. In the context of the axiomatic quality

[1]Fundamental knowledge that cannot be tested, yet is generally accepted as true, is treated as an axiom.

Axiomatic Quality: Integrating Axiomatic Design with Six-Sigma, Reliability, and Quality Engineering, by Basem Said El-Haik
ISBN 0-471-68273-X

process, the major design vulnerabilities are be categorized as:

- *Conceptual vulnerabilities,* established due to the violation of design principles.
- *Operational vulnerabilities,* created as a result of factors beyond the designer's control, called *noise factors*. These factors, in general, are responsible for causing a product's functional characteristic or process to deviate from target values. Controlling noise factors is very costly or difficult, if not impossible. Operational vulnerability is usually addressed by robust design (Taguchi et al., 1989).

Conceptual vulnerabilities will always result in operational vulnerabilities. However, the reverse is not true. That is, it is possible for a healthy concept that is in full obedience to design principles to be operationally vulnerable. Conceptual vulnerabilities are usually overlooked during product development, due to a lack of understanding of the principles of design, the absence of a compatible systemic approach to finding ideal solutions, the pressure of deadlines, and budget constraints. These vulnerabilities are usually addressed by traditional quality methods. These methods can be characterized as after-the-fact practices since they use lagging information relative to developmental activities such as bench tests and field data. Unfortunately, these practices drive development toward endless design–test–fix–retest cycles, creating what is broadly known in the manufacturing industry as a *fire fighting* operational mode. Companies that follow these practices usually suffer from high development costs, longer time to market, lower quality levels, and marginal competitive edge. In addition, firefighting actions to improve conceptual vulnerabilities are not only costly but also difficult to implement, as pressure to achieve design milestones builds during the development cycle. Therefore, it should be a goal to implement quality thinking in the conceptual stages of a development cycle. This goal can be achieved when systematic design theories are integrated with quality concepts and methods up front. Specifically, this book is geared toward developing an integration framework or process for quality in design by borrowing from quality engineering (Taguchi, 1986) and the axiomatic design principles of Suh (1990). In the context of this book, this framework is referred to as *axiomatic quality*. The objective axiomatic quality process is to address design vulnerabilities, both conceptual and operational, by providing tools, processes, and formulations for their quantification, then elimination or reduction.

Operational vulnerabilities necessitate the pursuit of variability reduction as an objective. Variability reduction has been a popular field of study. The method of robust design advanced by Taguchi (1986), Taguchi and Wu (1980), and Taguchi et al. (1989), time-domain control theory (Dorf, 2000), and tolerance design and the tolerancing technique[2] are just some of the proposals in this area. Tolerance research includes many areas that deal with the assignment of tolerances to design

[2]Some developments in this arena, including M-space theory, offset-solids theory, and virtual boundaries, have been discussed by Srinivasan and Wood (1992), Vasseur et al. (1993), Wood et al. (1993), and Zhang and Huq (1995), who linked process variation with inspection methods.

parameters and process variables, assessment and control of manufacturing processes, metrological matters, as well as geometric and cost modeling. In this book, tolerances play an important role in the axiomatic quality process in addressing operational vulnerability.

1.2 GOALS AND SCOPE OF THE BOOK

This book contributes to engineering design practice by devising tools and concepts within the axiomatic quality process. Such a process materializes by integrating design and quality methods early in the design process while dealing with the peculiarities of developmental activities and avoiding the pitfalls in current practices. In particular, this book focuses on establishing required connections among the various quality and design tools, identifying opportunities, and formulating methods to deal with conceptual and operational vulnerability elimination or reduction.

The major goals of this book can be summarized as follows:

- To reduce conceptual vulnerabilities of a design by integrating axiomatic design and quality methods in a conceptual framework that we call the *conceptual design for capability* (CDFC) *phase*. This phase addresses coupling and, conceptual design vulnerability.
- To reduce operational vulnerabilities of a design in terms of axiomatic measures, quality losses, and control costs of the *functional requirements* (FRs) delivered by the system. This is the core of the optimization phase.

The axiomatic quality process objectives can be achieved when design parameters and process variables can be adjusted (1) to reduce coupling design vulnerability and (2) to reduce complexity, an operational vulnerability, by minimizing the variability of the FRs and achieving goals 1 and 2 at the lowest possible cost. The framework of the axiomatic quality process is enforced further by mathematical relationships derived to bridge the gap between engineering design and quality methods. Such relationships will be used to formulate design vulnerability treatment techniques. The ultimate goal for all requirements is to have six times the standard deviation contained between the target and each side of the specification limits. This target is called the *six-sigma target* (or 6σ), where the Greek letter sigma represents the standard deviation, a statistical variability measure.

The purpose of the next sections is to present in brief the scientific components of the axiomatic quality process: the axiomatic design method, the design theory adopted in this book; the six-sigma philosophy; and Taguchi's quality engineering.

1.3 AXIOMATIC DESIGN

The axiomatic quality process hinges on axiomatic design (Chapters 2 and 3), a prescriptive[3] engineering design method. Systematic research in engineering

[3]*Prescriptive design* describes how design should be processed. Axiomatic design is an example of prescriptive design methodologies. *Descriptive design* methods, such as design for assembly, are descriptive of best practices and are algorithmic in nature.

design began in Germany during the 1850s. Recent contributions in the field of engineering design include axiomatic design (Suh, 1984, 1990, 1995, 1996, 1997, 2001), product design and development (Ulrich and Eppinger, 1995), the mechanical design process (Ulman, 1992), Pugh's total design (Pugh, 1991, 1996), and TRIZ (Altshuller, 1988, 1990; Arciszewsky, 1988; Rantanen, 1988). These contributions demonstrate that research in engineering design is an active field that has spread from Germany to most industrialized nations around the world. To date, most research in engineering design theory has focused on design methods. As a result, a number of design methods are now being taught and practiced in both industry and academia. However, most of these methods overlook the need to integrate quality methods in the concept stage. Therefore, the assurance that only healthy concepts are conceived, optimized, and validated with no (or minimal) vulnerabilities cannot be guaranteed.

Axiomatic design is a design theory that constitutes knowledge of basic and fundamental design elements. In this context, a scientific theory is defined as a theory comprising fundamental knowledge areas in the form of perceptions and understandings of various entities and the relationship among these fundamental areas. These perceptions and relations are combined by the theorist to produce consequences that can be, but are not necessarily, predictions of observations. Fundamental knowledge areas include mathematical expressions and categorizations of phenomena or objects, models, and so on, and are more abstract than observations of real-world data. Such knowledge and relations between knowledge elements constitute a theoretical system. A theoretical system may be one of two types: axioms or hypotheses, depending on how the fundamental knowledge areas are treated. Fundamental knowledge that is generally accepted as true, yet cannot be tested, is treated as an *axiom*. If the fundamental knowledge areas are being tested, they are treated as *hypotheses* (Nordlund, 1996). In this regard, axiomatic design is a scientific design method but with the premise of a theoretical system based on two axioms.

Motivated by the absence of scientific design principles, Suh (1984, 1990, 1995, 1996, 1997, 2001) proposed the use of axioms as the scientific foundation of design. The following are two axioms that a design needs to satisfy:

The Independence Axiom Maintain the independence of the functional requirements.

The Information Axiom Minimize the information content in a design.

In the context of this book, the independence axiom will be used to address the conceptual vulnerabilities, and the information axiom will be tasked with operational design vulnerabilities. Operational vulnerability is usually minimized and cannot be totally eliminated. Reducing the variability of design functional requirements and adjusting their mean performance to desired targets are two steps to achieving such minimization. Such activities will also result in reducing design information content, a measure of design complexity according to the

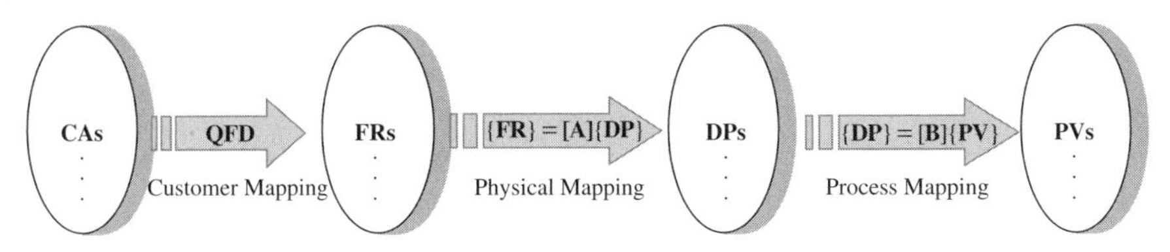

Figure 1.1 Design mapping process.

information axiom. Information content is related to the probability of successful manufacture of a design as intended by a customer.

The design process involves three mappings among four domains (Figure 1.1). The first mapping involves the mapping between *customer attributes* (CAs) and the FRs. This mapping is very important, as it yields a definition of the high-level minimum set of functional requirements needed to accomplish the design. This definition can be accomplished by the use of *quality function deployment* (QFD). Once the minimum set of FRs is defined, *physical mapping* may be started. This mapping involves the FR domain and the *design parameter* (DP) codomain. It represents product development activities and can be depicted by design matrixes; hence, the term *mapping* is used. This mapping is conducted over the design hierarchy as the high-level set of FRs, defined earlier, is cascaded down to the lowest hierarchical level. Design matrices reveal *coupling*, a conceptual vulnerability discussed in Chapter 2, and provide a means to track the chain of effects of design changes as they propagate across a design structure.

Process mapping is the last mapping of axiomatic design and involves the DP domain and the *process variable* (PV) codomain. This mapping can be represented formally by matrixes as well and provides the process elements needed to translate DPs to PVs in the manufacturing and production domains. A conceptual design structure called the *physical structure* is generally used as a graphical representation of design mappings.

The mapping equation $\mathbf{FR} = f(\mathbf{DP})$, or in matrix notation, $\{\mathbf{FR}\}_{m\times 1} = [\mathbf{A}]_{m\times p}$ $\{\mathbf{DP}\}_{p\times 1}$, is used to reflect the relationship between the domain, array $\{\mathbf{FR}\}$, and the codomain, array $\{\mathbf{DP}\}$, in the physical mapping, where the array $\{\mathbf{FR}\}_{m\times 1}$ is a vector with m requirements, $\{\mathbf{DP}\}_{p\times 1}$ is the vector of design parameters with p characteristics, and $\mathbf{A}$ is the design matrix. According to the independence axiom, the ideal case is to have one-to-one mapping so that a specific DP can be adjusted to satisfy its corresponding FR without affecting other requirements. However, perfect deployment of the design axioms may not be feasible, due to technological and cost limitations. Under these circumstances, different degrees of conceptual vulnerabilities are established in the measures (criteria) related to the unsatisfied axiom. For example, a degree of *coupling* may be created because of violation of the independence axiom, and this design may function adequately for some time in the use environment; however, a conceptually weak system may have limited opportunity for continuous success even with the aggressive implementation of an operational vulnerability improvement phase.

When matrix **A** is a square-diagonal matrix, the design is called *uncoupled* (i.e., each FR can be adjusted or changed independent of the other FRs). An *uncoupled design* is a one-to-one mapping. Another design that obeys the independence axiom, although with a known design sequence, is called *decoupled.* In a *decoupled design*, matrix **A** is a *lower* or *upper triangular matrix*. The decoupled design may be treated as an uncoupled design when the DPs are adjusted in some sequence conveyed by the matrix. Uncoupled and decoupled design entities possess conceptual robustness (i.e., the DPs can be changed to affect specific requirements without affecting other FRs unintentionally). A coupled design definitely results in a design matrix with a number of requirements, m, greater than the number of DPs, p. Square design matrices ($m = p$) may be classified as coupled design when the off-diagonal matrix elements are nonzeros. Graphically, the three design classifications are depicted in Figure 1.2 for the 2×2 design matrix case. Notice that we denote the nonzero mapping relationship in the respective design matrices by X; a zero denotes the absence of such a relationship.

Consider the uncoupled design in Figure 1.2*a*. An uncoupled design possesses the path independence property; that is, the design team could set the design to level 1 as a start point and move to setting 2 by changing DP_1 first (moving east to the right of the page, or parallel to DP_1) and then changing DP_2 (moving toward the top of the page, parallel to DP_2). Due to the path independence property of the uncoupled design, the team could start from setting 1 to setting 2 by changing DP_2 first (moving toward the top of the page, or parallel to DP_2) and then changing DP_1 (moving east or parallel to DP_1). The paths are equivalent; that is, they accomplish the same result. Notice also that FR independence is depicted as orthogonal coordinates as well as perpendicular DP axes that parallel the respective FR in the diagonal matrix.

Path independence is characterized, mathematically, by a diagonal design matrix (uncoupled design). Path independence is a very desirable property of

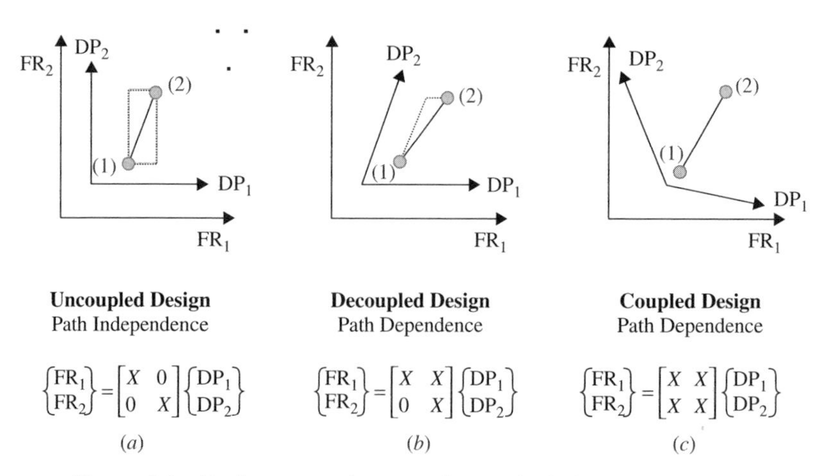

Figure 1.2 Design categories according to the independence axiom.

an uncoupled design and implies full control of the design team and ultimately the customer (user) over the design. It also implies a high level of design quality and reliability since interaction effects between the FRs are minimized. In addition, a failure in one (FR, DP) combination of an uncoupled design matrix is not reflected in mappings within the same design hierarchical level of interest.

For the decoupled design, the path independence property is somehow fractured. As depicted in Figure 1.2*b*, decoupled design matrices have design settings sequences that need to be followed for the functional requirements to maintain their independence. This sequence is revealed by the matrix as follows: First, we need to set FR_2 using DP_2, fix DP_2, then set FR_1 by leveraging DP_1. Starting from setting 1, we need to set FR_2 at setting 2 by changing DP_2, and then change DP_1 to the desired level of FR_1.

The discussion above is a testimony to the fact that uncoupled and decoupled designs have conceptual robustness; that is, coupling can be resolved with proper selection of the DPs, path sequence application, and use of design theorems (Chapter 2). The coupled design matrix in Figure 1.2*c* indicates the loss of path independence due to off-diagonal design matrix entries (on both sides), and the design team has no easy way to improve the controllability, reliability, and quality of their design. The design team is left with compromise practices (e.g., optimization) among the FRs as the only option since a component of individual DPs can be projected on all orthogonal directions of the FRs. The uncoupling or decoupling step of a coupled design is a conceptual activity that follows the design mapping and will be explored later.

An example of design coupling is presented in Figure 1.3, where two possible arrangements of the generic water faucet (Swenson and Nordlund, 1996) are

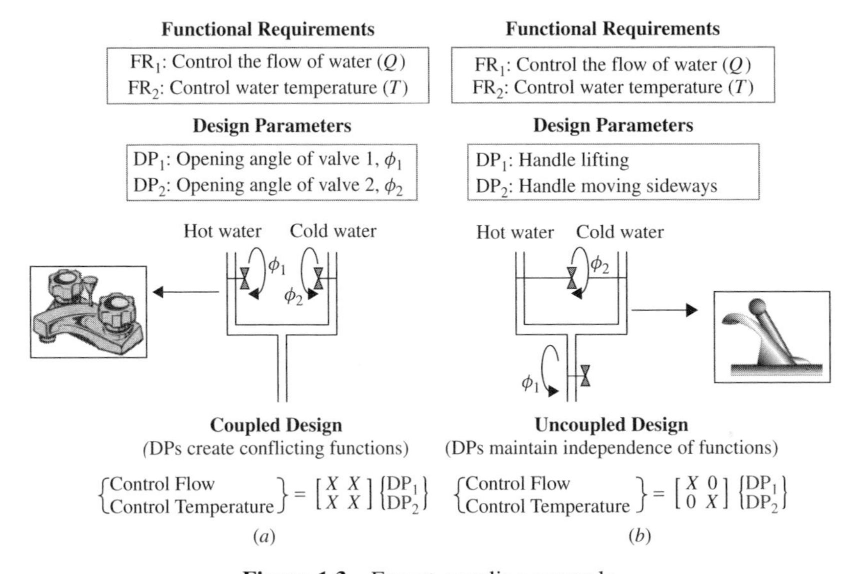

Figure 1.3 Faucet coupling example.

displayed (see Section 3.4 for more details). There are two functional requirements: water flow and water temperature. The faucet in Figure 1.3*a* has two design parameters, the water valves (knobs) (i.e., one for each water line). When the hot-water valve is turned, both flow and temperature are affected. The same happens when the cold-water valve is turned. That is, the functional requirements are not independent, and a coupled design matrix below the schematic reflects such fact. From a consumer perspective, optimization of the temperature will require reoptimization of the flow rate until a satisfactory compromise among the FRs, as a function of the DP settings, is obtained over several iterations.

Figure 1.3*b* exhibits an alternative design with one handle system delivering the FRs—however, with a new set of design parameters. In this design, flow is adjusted by lifting the handle; moving the handle sideways will adjust the temperature. In this alternative, adjusting the flow does not affect the temperature, and vice versa. This design is better since the functional requirements maintain their independence following the independence axiom. An uncoupled design will give the customer path independence to set either requirement without affecting the other. Note also that in the uncoupled design case, design changes to improve an FR can be done independently as well, a valuable design attribute.

The importance of design mapping has many perspectives. Chief among them is the identification of coupling among the functional requirements, due to the physical mapping process with design parameters in the codomain. Knowledge of coupling is important because it provides the design team with clues from which to find solutions, make adjustments, or design changes in proper sequence, and to maintain their effects over the long term with minimal negative consequences.

The design matrixes are obtained in a hierarchy and result from employment of the *zigzagging method of mapping*, depicted in Figure 1.4. The zigzagging

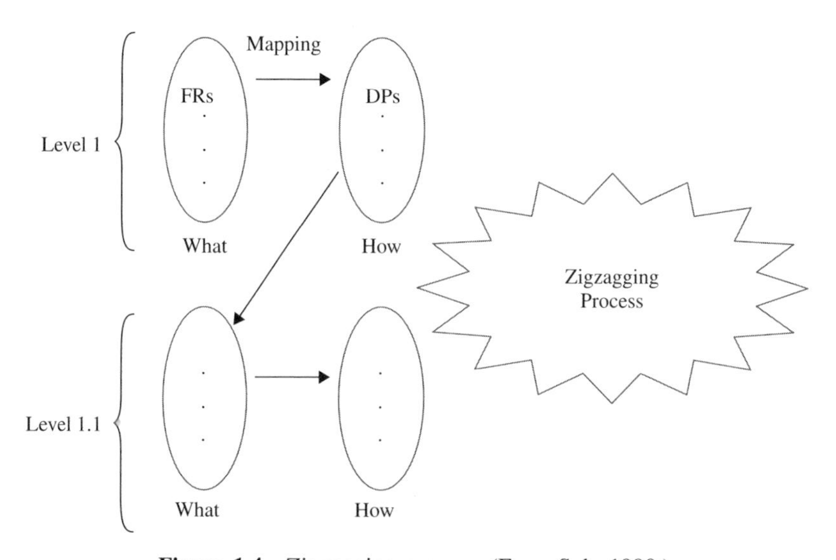

Figure 1.4 Zigzagging process. (From Suh, 1990.)

process requires a solution-neutral environment, where the DPs are chosen after the FRs are defined, and not vice versa. When the FRs are defined, we have to *zig* to the physical domain, and after proper DP selection, we have to *zag* back to the functional domain for further decomposition or cascading, although at a lower hierarchical level. This process is in contrast to traditional cascading processes, which utilize only one domain at a time, treating the design as a sum of functions or sum of parts.

At lower levels of hierarchy, entries of design matrixes can be obtained mathematically from basic physical and engineering quantities, enabling the definition and detailing of transfer functions, an operational vulnerability treatment vehicle. In some cases, these relationships are not readily available, and some effort needs to be paid to obtaining them empirically or via modeling [e.g., computer-aided engineering (CAE)]. Lower levels represent the roots of the hierarchical structure, where robust design and six-sigma concepts can be applied with some degree of ease. These concepts are summarized next.

1.4 SIX-SIGMA AND DESIGN FOR SIX-SIGMA PHILOSOPHY

The six-sigma methodology provides manufacturing and design companies with tools to improve the capability of their business processes. For six-sigma, a process is the basic unit for improvement. A process could be a product or service that a company provides to customers. A process could also be an internal process within a company, such as a billing process, a production process, and so on (Tadikamalia, 1994). In six-sigma, the purpose of process improvement is to shift the mean performance to target and to decrease variation. Improvements in both performance mean and variation will result in defect reduction, improvement in profit margins, better employee morale, higher quality levels, and eventually, business excellence.

Six-sigma is the fastest-growing business management system in industry today. Six-sigma is credited with saving billions of dollars for companies over the past 10 years. Developed by Motorola in the mid-1980s, the methodology became well known after Jack Welch at General Electric made it a central focus of his business strategy in 1995.

The name *six-sigma* comes from statistical terminology, where sigma, or σ, represents standard deviation. For a normal distribution, the probability of falling within a $\pm 6\sigma$ range around the mean is 0.9999966 (with a 1.5σ mean shift). In a production process, the *six-sigma standard* indicates that the process will produce defectives at the rate of 3.4 per million units. Clearly, six-sigma indicates a degree of extremely high consistency and extremely low variability. In statistical terms, the purpose of six-sigma is to reduce variation to achieve very small process standard deviations of the FRs.

The six-sigma strategy involves the use of statistical tools within a structured methodology for gaining the knowledge needed to achieve better, faster, and more cost-effective solutions: repeated, disciplined application of the master strategy on project after project, where the projects are selected based on the need to resolve key business issues that directly affect the bottom line. The achievement of financial improvement results in increased profit margins and return on investment. Project leaders are called *black belts*. The six-sigma initiative in a company is designed to change the culture through breakthrough improvement by focusing on out-of the-box thinking in order to achieve aggressive and stretched goals (see, e.g., Harry, 1994, 1998; Breyfogle, 1999).

A system is classified as *defective* if a desired requirement, denoted by y, is outside the customer's upper specification limit (USL) or lower specification limit (LSL). The half-tolerance range of y, denoted by t or Δy, equals half of the difference of the specification range (i.e., the design range). In addition to both limits, customers specify a target value, T, for y, which is typically the midpoint between USL and LSL. Lets say that the internal customer's specification of a shaft diameter is 50 ± 1 mm. In this example, $T = 50$ mm, LSL $= 49$ mm, USL $= 51$ mm, and the half tolerance $= 1$ mm. There are many sources of variability in manufacturing processes, such as machine feed rate, speed, vibration, material properties, and setup variables, that will make the central limit theorem hold for a typical design parameter average. Therefore, diameter can be treated as a normally distributed random variable. A sample is taken and the average, $\overline{Y}$, and sample standard deviation, S, are used to estimate the process average, μ, and the process standard deviation, σ, respectively, from the sample. With this approximation, the terms *sigma* and *standard deviation* are used interchangeably (Tadikamalia, 1994).

Diameter as a design parameter is called *centered* (i.e., the average equals the target); otherwise, it is *off-center*. The sigma level is defined by a *z-score*, or *sigma score*, given by

$$
\begin{aligned}
z_{\mathrm{USL}} &= \frac{\mathrm{USL} - \overline{Y}}{\sigma} \\
z_{\mathrm{LSL}} &= \frac{\overline{Y} - \mathrm{LSL}}{\sigma}
\end{aligned}
\qquad (1.1)
$$

When the process is centered in the design range and $\sigma = \Delta y/6, \Delta y/5, \Delta y/4, \Delta y/3, \ldots$, the z-scores are 6, 5, 4, 3, ..., respectively. As Table 1.1 shows, there

TABLE 1.1 Sigma Levels (Normal Distribution)

Long-Term Yield (%)	Short-Term Sigma	Defects per 1,000,000	Defects per 10,000	Defects per 100
99.9996599	6	3.40	0.034	0.00034
99.9767327	5	232.67	2.327	0.02327
99.3790320	4	6,209.68	62.097	0.62097
93.3192771	3	66,807.23	668.072	6.68072
69.1462467	2	308,537.53	3085.375	30.85375
30.8537533	1	691,462.47	6914.625	69.14625
6.6807229	0	933,192.77	9,331.928	93.31928

are approximately 67,000 defects per million at the 3σ level and 6200 defects per million at the 4σ level, assuming a 1.5σ shift in the average over time. If $S = \Delta y/6$ and assuming that the diameter is off-center by as much as 1.5σ, the maximum number of defects is 3.4 defects (or shafts) per million. Table 1.1 is derived from the normal distribution table.

When manufacturing companies embark on six-sigma quality programs, what is their objective? Is it to reduce the process variance so that the half tolerance of the product characteristic is equal to six times the standard deviation? From the technical viewpoint, it might make sense to talk in terms of the process variance, but from the managerial or customer viewpoint, quality standards can be described in terms of defects per million.

Usually, an adjustment to move the average closer to the target value is relatively easier than changing the system to reduce the variance. It is generally true with programs such as six-sigma that reducing process variance involves extensive efforts, which may include use of statistical techniques and in some cases capital investments in better technology. Alternatively, adjusting the process to the target value requires much less effort but may not result in significant improvement. Certainly, companies want to reduce process variance in the most cost-effective way. Variance reduction and adjustment of the mean to target are both activities usually taken to reduce operational vulnerabilities. However, such activities should be carried out in the light of the information axiom, which also calls for design simplicity by minimizing the information content needed for manufacturing.

The six-sigma philosophy molds statistical and practical thinking into an algorithm for problem solving over the phases *define, measure, analyze, improve*, and *control* (DMAIC). To date, most of the successes of six-sigma have been gained in deploying the philosophy in manufacturing problem solving, where short-term results can be achieved. However, the biggest potential can be gained only when manufacturing companies adopt the philosophy up front, in early design stages. This will transform a company from the firefighting mode to the prevention mode of operation.

1.4.1 Introduction to Design For Six-Sigma

Design for six-sigma (DFSS) is a disciplined methodology that embeds customer expectations into the design, applies the transfer function approach to ensure that customer expectations are met, predicts design performance prior to the pilot phase, builds into the design performance measurement systems with score cards to ensure effective ongoing process management, leverages a common language for design, and uses tollgate reviews to ensure accountability (Yang and El-Haik, 2003). In DFSS, emphasis is placed on design FRs; identification, optimization, and verification using the transfer function; and score card vehicles. The *transfer function*, in its simplest form, is a mathematical relationship (mapping) between an FR and the critical influential factors (called X's or DPs). Score cards help predict risks to the achievement of FRs by monitoring and recording their DP mean shifts and variability performance.

DFSS is a disciplined and rigorous approach to service, process, and product design by ensuring that new designs meet customer requirements prior to launch. It is a design approach that ensures complete understanding of development steps, capabilities, and performance measurements by using score cards and tollgate reviews to ensure accountability of design stakeholders, black belts, project champions, deployment champions, and the rest of the organization.

Unlike the six-sigma methodology, the phases or steps of DFSS are not universally defined, as evidenced by the many customized training curricula available in the marketplace. Often, deploying companies will implement the version of DFSS used by the vendor chosen to assist in the deployment. On the other hand, a company may customize a DFSS program to suit its business, industry, and culture, creating its own version. Therefore, DFSS is more of an approach, unfortunately, with little consensus across companies.

DFSS may be used to design or redesign a product or service. The expected process sigma level for a DFSS product or service is at least 4.5σ, but can be 6σ or higher, depending on the designed entity. The production of such a low defect level from product or service launch means that customer expectations and needs must be understood completely before a design can be operationalized. That is, quality is defined by the customer.

Deployment of the axiomatic quality process can be an initiative on its own or linked to DFSS deployment. Yang and El-Haik (2003) presented some aspects of a product DFSS implementation scheme.

1.5 ROBUSTNESS ENGINEERING: TAGUCHI'S QUALITY ENGINEERING

In the context of this book, the terms *quality* and *robustness* are used interchangeably. *Robustness* is defined as reducing the variation in FRs of a system and having them on target as defined by the customer (Taguchi and Wu, 1980; Taguchi, 1986; Phadke, 1989; Taguchi et al., 1989, 1999). Operational vulnerabilities have been the subject of robust design (Taguchi, 1986) through methods such as parameter design and tolerance design. The principal idea behind robust design is that statistical testing of a product should be carried out at the design stage or *off-line stage*. To make a product *robust* against the effects of sources of variation in the manufacturing and use environments, the design problem is viewed from the point of view of quality and cost (Taguchi and Wu, 1980; Taguchi, 1986; Taguchi et al., 1989, 1999; Nair, 1992).

Quality is measured through quantifying statistical variability through measures such as standard deviation or mean-squared error. The main performance criterion is to achieve the FR target on average while minimizing the variability around this target. *Robustness* means that a system performs its intended functions under all operating conditions (different causes of variations) throughout its intended life. Undesirable and uncontrollable factors that cause the FR under consideration to deviate from the target value are called *noise factors*. Noise factors affect quality adversely, and ignoring them will result in a system not optimized for the conditions of use. Eliminating noise factors may be expensive, so, instead, we seek to reduce the effect of noise factors on FR performance.

Robust design is a disciplined engineering process that seeks to find the best expression of a system design. "Best" is defined carefully to mean that the design is the lowest-cost solution to the specification, which itself is based on identified customer needs. Taguchi has included design quality as one more dimension of cost. High-quality systems minimize these costs by performing consistently at targets specified by the customer. Taguchi's philosophy of robust design is aimed at reducing loss due to variation of performance from the target value based on a portfolio of concepts and measures, such as *quality loss function* (QLF), *signal-to-noise* (SN) *ratio*, optimization, and experimental design. *Quality loss* is the loss experienced by customers and society and is a function of how far performance deviates from the target. The QLF relates quality to cost and is considered a better evaluation system than the traditional binary treatment of quality (i.e., within/outside specifications). The QLF of an FR has two components: mean (μ_{FR}) deviation from targeted performance value (T) and variability (σ^2_{FR}). It can be approximated by a quadratic polynomial of the functional requirement.

Consider two settings or means of a design parameter, setting 1 (DP*) and setting 2 (DP**), having the same variance and probability density function (statistical distribution), depicted in Figure 1.5. Consider also the given curve of a hypothetical transfer function (a mathematical form of design mapping), which

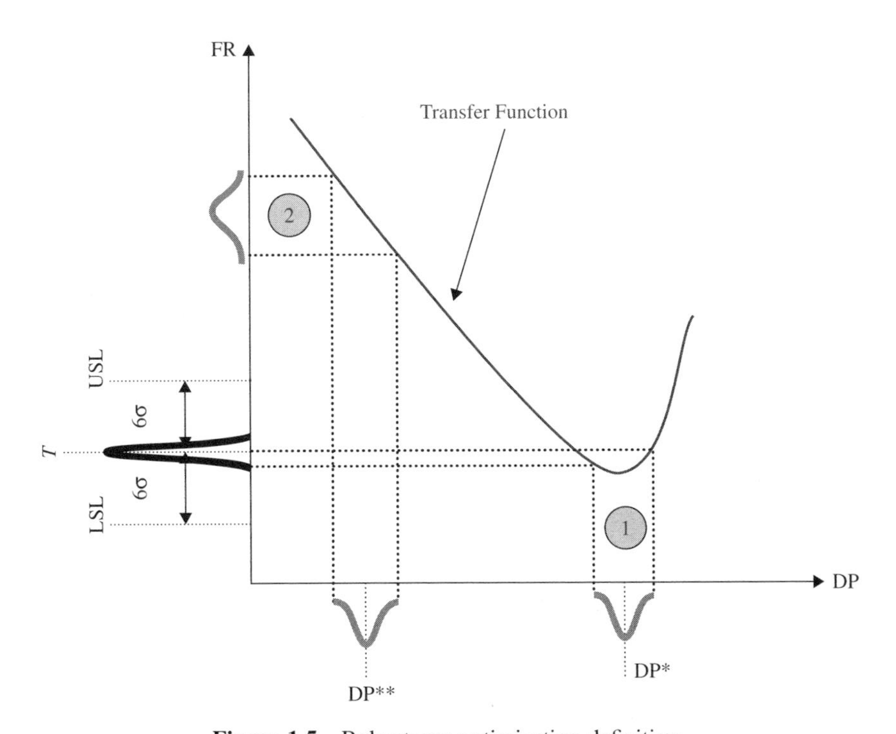

Figure 1.5 Robustness optimization definition.

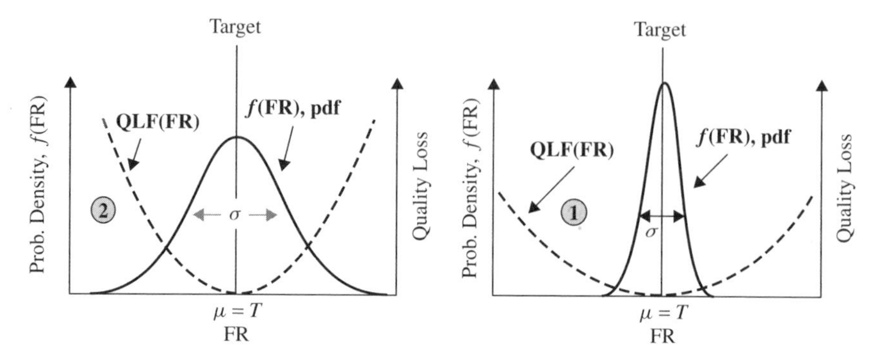

Figure 1.6 Quality loss function scenarios of Figure 1.5.

is in this case a nonlinear function in the DP. It is obvious that setting 1 produces less variation in the FR than does setting 2, by capitalizing on nonlinearity.[4] This also implies a lower degree of information content and thus a lower degree of complexity based on the information axiom. Setting 1 (DP*) will also produce a lower quality loss, similar to the scenario shown on the right in Figure 1.6. In other words, the design produced by setting 1 (DP*) is more robust than that produced by setting 2. Setting 1's robustness is evident in the amount of variation transferred through the transfer function to the FR response of Figure 1.5 and the flatter quadratic quality loss function in Figure 1.6. When the distance between the specification limits is six times the standard deviation ($6\sigma_{FR}$), a 6σ-level optimized FR is achieved. When all design FRs are released at this level, a six-sigma design is obtained.

The important contribution of robust design is the systematic inclusion into experimental design of noise variables, that is, variables over which the designer has little or no control. A distinction is also made between *internal noise*, such as component wear and material variability, and *environmental noise*, which the designer cannot control (e.g., humidity, temperature). Robust design's objective is to suppress, as far as possible, the effect of noise by exploring the levels of factors to determine their potential for making a system insensitive to these sources of variation.

The noise factors affect the FRs at different segments in the life cycle. As a result, they can cause a dramatic reduction in product reliability, as indicated by the failure rate. The bathtub curve in Figure 1.7 (see Fowlkes and Creveling, 1995) implies that robustness can be defined as reliability over time. *Reliability* is defined as the probability that a design will perform as intended [i.e., deliver the FRs to satisfy the CAs (Figure 1.1)] throughout a specified period when operated under stated conditions. One reason for early life failures is manufacturing variability. The unit-to-unit noise causes failure in the field when a product is subjected to external noise. The random failure rate of the DPs that characterizes

[4]In addition to nonlinearity, leveraging interaction between the noise factors and the design parameters is another popular empirical parameter design approach.

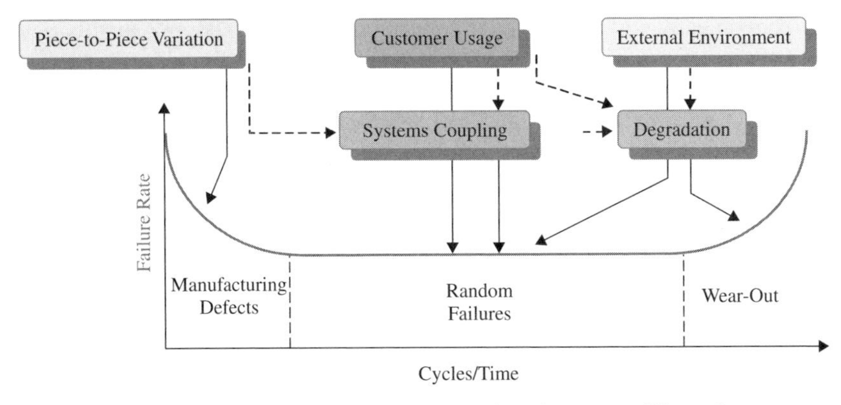

Figure 1.7 Effect of noise factors during the system life cycle.

most of a product's life is the performance of design subject to external noise. Notice that the coupling vulnerability contributes to unreliability of the design in customer hands. Deterioration noise is active at the end of product life. Therefore, a product is said to be *robust* (and therefore reliable) when it is insensitive to the effect of noise factors, even though the sources themselves have not been eliminated (Fowlkes and Creveling, 1995).

1.6 PROBLEMS ADDRESSED BY AXIOMATIC QUALITY

To stay competitive, the manufacturing industry must deliver high-quality products in a short time and at lowest cost. With increasing demands for a shorter time to market, new products that lack the support of scientific knowledge and the presence of existing experience will be encountered. In addition, it is no longer sufficient to rely solely on traditional after-the-fact quality methods and accumulated engineering knowledge alone. To design a system right during the first attempt, manufacturing companies need to base developmental activities on (1) a set of design principles, and (2) the deployment of quality methods during upstream and downstream design stages. A design entity can be tuned rapidly if the developmental activities follow a self-sustaining set of principles. The design axioms, as a special set of principles, facilitate the adaptation of new and innovative up-front solutions.

The methods, concepts, and processes presented in this book, called collectively *axiomatic quality*, are tasked with addressing both conceptual and operational design vulnerabilities. The axiomatic quality process does not substitute any other knowledge, nor does it replace the need to constantly learn, adapt, and implement new knowledge in related disciplines. Deployment of the axiomatic quality process complements the specific design knowledge needed to develop products and manufacturing systems.

The axiomatic quality process identifies premier design vulnerabilities in the concept stage and includes strategies to minimize them at the lowest possible cost. The objective is to release a robust product at six-sigma quality levels in all requirements, while emphasizing the importance of making good front-end design decisions. In doing so, the axiomatic quality supports abstraction; that is, it is able to handle high-level requirement issues with end users and address realization details with manufacturing, utilize existing information on datum design to optimize new systems, and accommodate design changes on existing baseline systems. It is recognized that most product development projects rarely start from scratch with a completely clean sheet of paper, and that reusable design elements should be leveraged.

The axiomatic quality process is well suited for the development of structure-optimized, reusable designs by providing comprehensive analysis, synthesis, and optimization tools. Experience shows that modifications of existing designs through reusable components have been completed in less the time. That is, reusability can improve the design process productivity. However, making a reusable design of a current baseline can be very different from making a reusable design for general-purpose use by others. The additional efforts come from making a reusable-component generic enough for a wide range of applications. These efforts must include making the component configurable, which can be facilitated as a result of design axiom obedience. In essence, the reusability of a design does not come from similarity of design parameters alone; it requires a design discipline and careful consideration at early stages to reach both an efficient reusable design and a design that takes advantage of reusable elements.

The presence of conceptual and operational vulnerabilities in manufacturing practices puts intense pressure on a design team to devise and implement quality-oriented design processes and methods that are compatible with developmental activities. As the adoption of six-sigma philosophy grows in industry, the implementation is challenged by situations where statistical tools are not adequate, especially during the conceptual stages. There is an immediate need to launch such a philosophy as early as possible in the development cycle, where most of the potential can be realized. Axiomatic quality satisfies this need and involves meeting aggressive six-sigma quality targets for all the FRs that are delivered by the design entity while *simultaneously* eliminating or reducing systemic vulnerabilities.

1.7 INTRODUCTION TO THE AXIOMATIC QUALITY PROCESS

The axiomatic quality process developed in this book employs tools, methods, and concepts from engineering design theory, quality engineering, and optimization. The process is a phased approach that is constructed from the voice of the customer (VOC)-to-FR mapping, CDFC, and optimization phases. The product of the VOC-to-FR mapping is a well-defined measurable array of requirements that enable both the CDFC and optimization phases. The CDFC phase is based on the integration of quality engineering and the axiomatic design method. It presents a systemic approach to address the conceptual vulnerability created between the FRs and to facilitate positioning the design entity for operational vulnerability treatment when needed.

Axiomatic quality recognizes the peculiarities of the design process, guides the creativity of the design team, and provides analytical tools and techniques that eliminate or reduce design vulnerabilities at the various developmental stages. Axiomatic design integration with robust design engineering forms the backbone of the axiomatic quality process, which, in turn, distinguishes between the following two types of system design projects:

1. *Creative design* (design at the "white paper" stage), where the CDFC phase of axiomatic quality has a higher application potential. *The CDFC phase ensures that the objective of establishing six-sigma capability (potential) in the design entity is conceptually possible and feasible.*
2. *Incremental design* (design from a datum system), where some baseline information and data can be used. The degree of deviation from a datum design by the system being utilized is the key factor when deciding on the usefulness of relative data available. Both conceptual and statistical tools can be applied.

1.8 AXIOMATIC QUALITY IN PRODUCT DEVELOPMENT

In this section we discuss the design (development) stages of the product life cycle, depicted in Figure 1.8. The major phases of the axiomatic quality approach, the scope of this book, are laid on top of the development stages. The reader may already have noticed that the axiomatic quality process has a hard stop at the start of stage 6 in its first generation. Implications of the axiomatic quality process in the last two stages are beyond the scope of the book.

Naturally, the process of product design begins when there is a need, an impetus. People create the need, whether it is a problem to be solved (incremental design) or a new invention (creative design). Design objective and scope are critical in the impetus stage. A design project objective should describe simply and clearly what is to be designed; it cannot be vague. Writing a clearly stated design objective is just one step. In stage 2 the design team must write down all the information they think they may need, in particular the voice of the customer (VOC) and the voice of the business (VOB).

With the help of the quality function deployment method (stage 2), product FRs are defined. Then they are mapped to DPs using the zigzagging process. The design parameters are later grouped into modular elements such as systems, subsystems, and components, what we will call the physical structure within the axiomatic quality process context. A functional requirement must contribute to the solution of the problem described in the design objective statement obtained from both the voice of the customer and the voice of the business. Another question that should be on the mind of the design team relates to how the end result will look. The shape, color, and texture should make the product attractive. What materials are available to the team, and at what cost? Do they have the right physical properties (DPs), such as strength, rigidity, color, and durability?

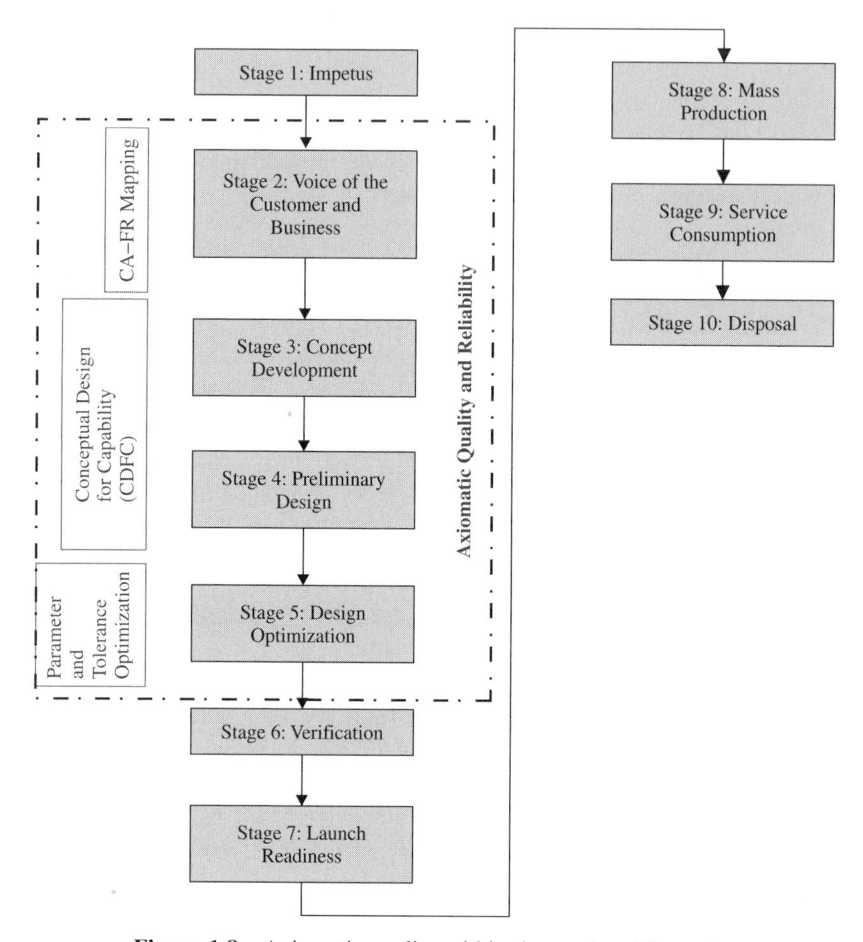

Figure 1.8 Axiomatic quality within the product life cycle.

Will the product be hard to make (PVs)? Consider what methods are needed to cut, shape, form, join, and finish the material. The design must be safe to use. It should not cause accidents.

In stage 3 the design team should produce a number of solutions. It is very important that they write or draw every idea on paper as it occurs to them. This will help them remember and describe more clearly. It is also easier to discuss these ideas with other people if drawings are available. These first drawings do not have to be very detailed or accurate. Sketches will suffice and should be made quickly. The important thing is to record all ideas and to develop solutions in the preliminary design stage (stage 4). Design mappings and hierarchy are detailed and solutions can be synthesized conceptually. The design team may find that they like several solutions. Eventually, the design team must choose one. Usually, careful comparison with the original design objective will help them to select the best subject to the constraints of cost, technology, and skills available.

Deciding among the several possible solutions is not always easy. It helps to summarize the design requirements and solutions and to put the summary in a *morphological matrix*. An overall design alternative set is synthesized from this matrix that are conceptually feasible solutions with a high potential for success (recall the information axiom). Which solution would they choose? A concept selection approach should be used. The solution selected will experience a thorough design optimization stage (stage 5). This optimization could be either deterministic or statistical in essence. On the statistical front, the design solution will be made insensitive to uncontrollable factors (i.e., noise factors) that may affect performance. Factors such as the customer use profile, environment, and production piece-to-piece variation should be considered for their capacity to fulfill the FRs. In stage 5 the team needs to make a blueprint of the optimized solution as detailed by transfer functions. This drawing must include all of the information needed to produce the product. Consideration of design documentation, dimensions, production, materials, finish, communication, marketing, and so on, should be put in place.

In stage 6, design verification, the team can make a model and later a prototype, or they can go directly to making a prototype or pilot. A *model* is a full-sized or small-scale simulation of the design entity. Models are one more step in communicating the effort that went into the design project. For most people it is easier to understand a design when it is shown in three-dimensional form. A *scale model* is used when the scope is very large. A *prototype* is the first working version of a team's solution. It is generally full size and often uses homegrown expertise. Stage 6 also includes testing and evaluation to answer certain very basic questions: Does it work? Does it meet the design objective? If failure is discovered, will modifications improve the solution? These questions have to be answered.

When satisfactory answers have been developed, the team can move to the next development and design stage. In stage 7 the team needs to prepare the production facilities where the product will be produced for launch. At this stage they should ensure that the product is still marketable. The design team, together with the project stakeholders, must finalize how many product units to make. The product may be mass-produced in low or high volume. The task of making the product is divided into jobs. Each worker trains to do his or her assigned job. As workers complete their special jobs, the product takes shape. Following stage 7, a lean production system saves time and other resources.

1.9 SUMMARY

Engineering design and manufacturing can be defined as sets of processes and activities that transform customer wants into design solutions that are useful to society. These processes are carried over several stages, beginning at the voice of the customer stage. In the concept stage, conceiving, evaluating, and selecting good design solutions are tasks that have enormous significance. It is imperative that design and manufacturing organizations conceive healthy systems with no (or minimal) vulnerabilities within a single development cycle.

Manufacturing companies usually operate in two modes: (1) *fire prevention*, conceiving feasible and healthy conceptual design entities with no or minimal conceptual vulnerabilities, and (2) *fire fighting*, solving problems that are caused by inherent conceptual or operational vulnerabilities. Unfortunately, the latter mode consumes the largest portion of an organization's human and nonhuman resources. The axiomatic quality process provides a solution approach that fits both modes of organizational operations. The full power of the axiomatic quality approach can be gained when it is implemented with a focus on the fire prevention mode. The axiomatic quality process integrates concepts and tools from the axiomatic design method, robust design approach, and six-sigma process. The objective of this book is to develop concepts, processes, and methodologies that eliminate or reduce both conceptual and operational vulnerabilities and assist design teams in producing systems that operate at high quality levels for each of their design requirements.

The axiomatic quality process is a business process that allows companies to improve their products, services, and processes in a way that minimizes waste and design resources while increasing customer satisfaction. It is a process that uses statistical techniques and design axioms to drive for results by supplementing means of decision making.

CHAPTER 2

AXIOMATIC DESIGN METHOD

2.1 INTRODUCTION

Axiomatic design is a prescriptive engineering design theory and methodology that provides a systematic and scientific basis for making design decisions. In addition to corollaries and theorems derived from them, axioms give design teams a solid basis for formalizing design problems, conceptualizing solution alternatives, eliminating bad design ideas during the conceptual stages, choosing the best design among those proposed, and improving existing designs. An *axiom* is a proposition regarded as self-evident true without proof. The term *axiom* is a slightly archaic synonym for *postulate*. Axioms can be compared to hypotheses, both of which connote apparently true but not self-evident statements (e.g., Archimedes' axiom, Newton's laws, probability axioms, field axiom).

The verb *design* refers to the process of developing an entity. *Engineering design*[1] is the process of developing a product, service, or process to meet

[1]The terms *design* and *engineering design* are defined by the National Academy of Engineering (NAE) (2002) as "the process by which human intellect, creativity, and passion are translated into useful artifacts. Engineering design is a subset of this broad design process, in which performance and quality objectives and the underlying science are particularly important. Engineering design is loosely structured, open-ended activity that includes problem definition, learning processes, representation and decision making."
See National Academy of Engineering (NAE), 2002, *Approaches to Improve Engineering Design*, available on the web at *http://www.nap.edu/books/NI000469/html/*.

Axiomatic Quality: Integrating Axiomatic Design with Six-Sigma, Reliability, and Quality Engineering, by Basem Said El-Haik
ISBN 0-471-68273-X

desired customer needs. Design entities share a common attribute, *hierarchy*, which indicates the levels of complexity in both the magnitude of development effort and the end-result ease of operation. In the context of axiomatic quality, design is an iterative decision-making process in which various physical, mathematical, and engineering sciences are applied to convert resources optimally to meet stated customer needs. Among the fundamental activities of the design process are the establishment of project objectives, synthesis, analysis, construction, testing, and launch.

Human technology evolution continually takes revolutionary steps from ancient civilizations to the twenty-first century. During the twentieth century, technology created by engineering design advanced at an exponential rate: New forms of communications, medical science, new means of travel, and the refinement and distribution of computer technology are just a few examples.[2] Today's design teams have unprecedented technology at their disposal. Modern engineers rely heavily on technology, design tools, and proven design processes. Ancient civilizations could only rely on simple tools that were truly innovative and far ahead of their time. In fact, when we compare today's design practices with ancient practices, we discover that principles of design were used to support the design process, around which they continue to evolve. What has changed is the sophistication of the means itself at both the process and technology levels.

Because the pyramids were built for a pharaoh, no expense or effort was spared. The sheer size of the pyramids and how ancient civilizations could have built them without the use of modern equipment have become the source of numerous theories. Theories of alien visitors and extinct scientifically advanced civilizations are many. However, by investigating and examining analytically how the pyramids could have been built, a much more rational answer emerges. The precision of the great pyramid of Khufu is exact; it is as precise as any modern skyscraper.[3] It can be proven that ancient Egyptians applied a sound principle-based design and production process for the pyramid's construction. They followed a design process (although not documented, it can be deduced from historical evidence) with several stages, highlighting the voice of the customer (the pharaoh) to conceive a system definition, followed by preliminary design to establish pyramid subsystem definition, and several technical reviews with the customer for next-stage commencement approval. Then they established a detailed design definition and the rest of the stages of a sound engineering design process (Lake, 1994). Studies also showed that they employed principles such as life-cycle design early on, with emphasis on developing one level of modularity

[2]See Bill Jacob's paper at *http://www.bandisoftware.com/Incose2002.pdf.*

[3]For example, the great pyramid of Khufu consists of 2,300,000 blocks each weighing, on average, 2.5 tons. The latest evidence suggests that it was built in 23 years or less, corresponding to the length of time that Khufu ruled. That translates to 340 blocks a day. With 10 hours of daylight per day, 34 blocks were laid every hour. That includes quarrying, transporting, cutting, finishing, and coating. Each base is 230.33 m and the height is 146.59 m. The pyramid's orientation is 3 ft 6 in. off true north. The base is level within 1 in. The greatest distance between the length of the sides is 1.75 in., truly astonishing (Lehner, 1997).

at a time (i.e., system to subsystem to component). In addition, the Egyptians considered the design of production tools, facilities, layouts, and procedures. Tools were designed for the tasks at hand. For example, copper chisels, horizontal and vertical levels, and wooden T-squares have all been discovered around the area of the great pyramid. For validation, each successive step was tested and compared to the entire effort to ensure uniformity. Because rise and run were controlled one level at a time through stone markings, errors in previous levels could be corrected in subsequent levels. Evidence of verification can be seen in other, imperfect pyramids (prototypes), whose rise and run were not properly controlled; the tops of these pyramids are twisted to align the size of the pyramid at its top. Indeed, many researchers concluded that the pyramid design team followed a group of design principles that the team set for themselves based around their design.

A *design principle* is a fundamental idea on which a design process could be based. For example, the principle of *concurrent engineering* calls for the continuous participation of design and manufacturing teams in almost all aspects of design activities and as early as possible. A design principle helps drive efficient and effective design project management, faster development cycle time, lower cost of development, and improved customer satisfaction. A common set of design principles provides manufacturing companies and design houses with a common foundation, enabling them to interpret and apply their design and development process from a common point of understanding and to make the right development and business decisions. Design principles need not be mixed with design axioms. In the context of this book, a principle can be promoted to an axiom when it possesses universal applicability (i.e., acceptance in all design domains). A principle is primarily domain specific (e.g., software vs. hardware).

Design decision making has a significant impact on lead time, function and form, quality, and cost of the end result. Studies suggest that decisions made during the early stages of the design phase commit 80% of the total costs associated with developing and manufacturing a product (Fredrikson, 1994). Furthermore, when manufacturing systems and processes are designed poorly, the productivity—throughout—decreases substantially.[4] Despite the fundamental importance of proper design, most manufacturing firms and design houses do not have a rational design practice and thus produce poor-quality products and prolong the development cycle. This problem can be solved only if design teams understand what constitutes good design and how to produce such designs.

In the following sections we present the concepts of axiomatic design to enable the reader to follow the remaining chapters. In these sections we focus on the process of performing axiomatic design and on exploring the axiomatic attributes of a design entity. At its most basic level, axiomatic design consists of five concepts: domains, design hierarchy, zigzagging, and the two design axioms. All are explained below.

[4]Several case studies show that two serially clustered machine systems can have a much lower throughput rate than that of the slowest machine by itself.

2.2 AXIOMATIC DESIGN METHOD

The axiomatic design method establishes a scientific theoretical basis that gives structure to the design process. Axiomatic design offers perspectives that most conventional algorithmic design approaches fail to achieve. Algorithmic methods such as design for assembly and design for manufacturability are goal oriented in that the design activities are devised around existing best practices and their integration in an algorithmic process setup. New design problems dictate the creation of new algorithms, or in the best case, modified algorithms. Algorithmic methods are most successful in conventional and simple design situations, mostly incremental design type. The practice of engineering using design algorithms is both time consuming and problem dependent. When the design problem is complex (i.e., a large number of FRs with numerous hierarchical levels), it might be difficult to fit the problem into an algorithmic format. In algorithmic design methods, the selection of a solution entity for a function from its pool of possible physical embodiment alternatives is usually motivated by economic considerations.

Axiomatic design introduces a different perspective to design theory. The new view offered by axiomatic design and axiomatic quality is not limited to the product-conceptualizing stage but is extended to include the detailed design and manufacturing process domain. Axiomatic design delivers these premises through the concepts of *generalization* and *abstraction* (Section 2.2.2).

Through an evolutionary process, Suh has simplified his original 12 axioms to two basic axioms, several corollaries, and theorems that designs need to satisfy (Suh, 1990):

The Independence Axiom Maintain the independence of the FRs.
The Information Axiom Minimize the information content in a design.

By definition, axioms are truths with no exceptions and can be hypothesized from a large pool of observations. For example, Newton's laws of motion and the first law of thermodynamics are axioms for the concepts of force and energy, respectively. The independence axiom states that the DPs and FRs are related such that a specific DP can be adjusted to satisfy its corresponding FR without affecting other FRs. The information axiom states that the independent (uncoupled) design alternatives that minimize the information content are the *best*. Axiomatic design method best practice is always to satisfy the axioms in sequence (i.e., the independence axiom, then the information axiom). The sequence followed in satisfying design axioms implies some dependency and the possibility of combining the two axioms. However, the risk of such a simplification could be choosing a coupled design with less information rather than a completely uncoupled design (Suh, 1990). It is always wise to seek uncoupled designs that have less information content. However, if more than one design solution is conceived, the solutions, which completely satisfy the independence axiom, are ranked according to the information content they possess. The solution with the lowest information content is the *best* according to the information axiom. When complete independence cannot be achieved within the binding technology and cost constraints, the design team is left with one option, coupling vulnerability minimization.

Upon satisfying the independence axiom, we are after simplicity by minimizing the information content of a design (the information axiom). Design information content can be defined as a measure of *complexity*. In this case it is related to the probability of certain events occurring in the process domain (in this case, manufacturing or producing the DPs successfully) when information is supplied. Violation of the information axiom generates complexity, another design vulnerability.

2.2.1 Design Domains

In addition to the *customer attributes domain* (the CA domain), three fundamental design domains are recognized by the axiomatic design method (Suh, 1990, 1995): the *functional domain* (the FR domain), the *physical domain* (the DP domain), and the *process domain* (the PV domain). These domains are linked through several mappings, as shown in Figure 2.1. The domain on the left relative to the domain on the right represents "*what* we want to achieve"; the domain on the right represents the design solution of "*how* we propose to satisfy the requirements specified in the left domain" (Suh, 1990). The first mapping, a relationship, is from the customer domain to the functional domain. The *physical mapping* is a mapping from the functional domain to the physical domain, where possible design parameters are determined for each function. The *process mapping* is a mapping from the physical domain to the process domain. Axiomatic design process analysis (mapping) is bidimensional (domain and codomain; Figure 2.1), and the selection process of a solution entity is based on axiom satisfaction and is subject to design constraints.

Suh (1990) defines the design as "the creation of a synthesized solution in the form of products, processes or systems that satisfy perceived needs through the mapping between FRs in the functional domain and DPs of the physical domain, through the proper selection of DPs that satisfy FRs." The set of FRs is defined as

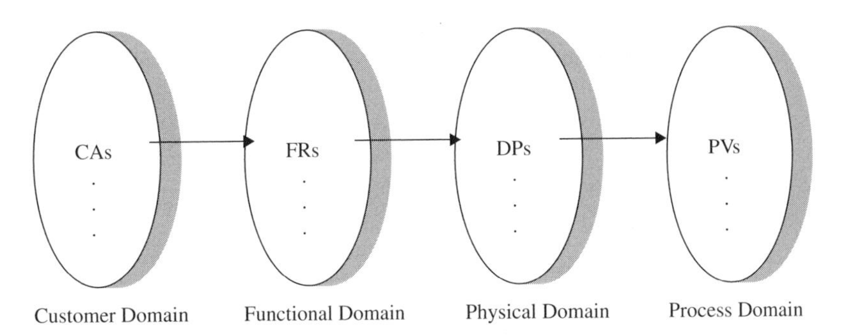

Figure 2.1 Design mappings and domains.

the "minimum set of independent requirements for a product or process that completely characterizes the design objective" (Suh, 1990). The FRs must be defined in a solution-neutral environment, without considering the physical domain DPs at the time of their definition. Defining FRs are always, by definition, mixed with *constraints*. FRs are not physical constraints, and they do not bound the design solution entities. The axiomatic design method differentiates between two types of constraints: *input constraints* (specifications on size, weight, cost, etc.) and *system interfacial constraints* (shape, laws of nature, machine capability, etc.). One of the major characteristic that may be used to distinguish FRs from constraints are the fact that constraints are functions of DPs, FRs, or other constraints with tolerance around them, whereas FRs are completely independent by definition. The domain of solutions for typical engineering design optimization problems is limited due to restrictions placed on the values that variables (DPs *or* PVs) can assume. An example of such a constraint might be a size (packaging) constraint or a budget (monetary) constraint. The solution to the problem must be within the feasible region as defined by the constraints. For example, a linear programming problem involves optimization of a function y, which is linearly dependent on the independent variables $(x_1, x_2, \ldots, x_n)$ as

$$y = a_1x_1 + a_2x_2 + \cdots + a_nx_n$$

Further, the function is subject to several constraints that are linear:

$$\begin{aligned} a_{11}x_1 + a_{12}x_2 + \cdots + a_{1n}x_n &\le C_1 \\ a_{21}x_1 + a_{22}x_2 + \cdots + a_{2n}x_n &\le C_2 \\ &\vdots \end{aligned}$$

where the a's and C's are constants. The function y needs to be maximized or minimized subject to the constraint set defined. This formulation is called a *linear programming* (LP)[5] *problem*. In axiomatic quality, however, some optimization formulation will be experienced, however, nonlinear in essence.

In summary, the design problem can be prescribed in four domains, which may be generalized as the customer domain, functional domain, physical domain, and process domain. Associated with each domain are design elements: CAs, FRs, DPs, and PVs. The number of domains is always four, but the nature of the design elements in each domain change depending on the problem field. Gebala and Suh (1992) list examples of the breakdown of problems.

[5]The LP problem can be solved by a graphical method—by plotting the constraints to define the solution domain and then plotting the objective function to obtain the solution. Typically, the solution is found at one of the extreme points (where two constraints intersect). The graphical method loses its utility as the number of variables increases (>3). The most efficient method for solving LP problems is the simplex method.

2.2.2 Design Hierarchy and Zigzagging Process

The axiomatic design method proceeds from a high level of abstraction to detailed modularity elements (i.e., from systems to subsystems to component to feature). This activity of definition and detailing produces a prescriptive model of design hierarchy for the design entity in each of three domains: functional, physical, and process. The decisions that are made at higher levels affect the statement of the design definition at lower levels. That is, the design team goes through a process whereby they zigzag between domains to decompose the design problem (Figure 2.2). For example, in designing an automobile speed control system, a top-level DP is the transmission system: automatic or manual. Further decomposition of the problem will produce the following lower-level DPs: gearing subsystem, transmission fluid subsystem, clutch subsystem, pump subsystem, and so on.

At a given level of design hierarchy, there exists a set of FRs defined as the minimum set of requirements needed at that level. Defining acceptable FRs may involve several iterations. Axiomatic design employs a limited number of logical questions in the mapping process. The use of a complete set of logical questions leads much more quickly to a correct and creative set of FRs. In addition to the employment of logical questions, Suh's definition requires the maintenance of functional requirement independence.

Before these FRs can be zagged, the corresponding hierarchical level DPs and PVs must be selected. Once a corresponding DP can satisfy an FR and a PV can satisfy a DP, that FR can be decomposed into a set of lower-level

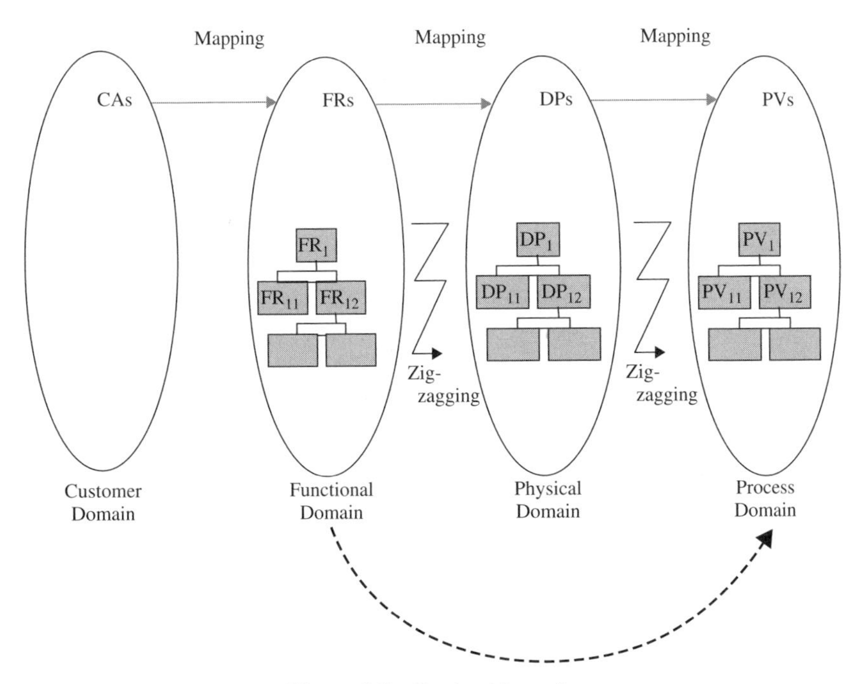

Figure 2.2 Design hierarchy.

requirements, and the process becomes iterative. The design team will develop different solutions for each DP to satisfy an FR and select the best alternative (option) at each hierarchical level. Led by design axioms combined with experience and knowledge, the design team should be able to conceive, select, and optimize the "best," even at the conceptual stages. By following the zigzagging approach while checking the soundness of the design at each level, the design team should continue until no more decisions regarding the design need to be made. Once a set of FRs has been formulated and feasible sets of DPs have been synthesized at a given hierarchical level, the two design axioms are applied to evaluate the design concepts proposed. Application of the independence axiom may be described in terms of the design matrixes (Suh, 1990). A design matrix **[DM]** prescribes the relationships between the FR array and the corresponding DP array at the same hierarchical level.

The axiomatic quality process addresses the two main ingredients of a design process, analysis and synthesis. We analyze the design conceptually by simplifying a design problem, i.e., breaking it down into functional elements prescribed in the design mappings, generating design concepts, and applying creative thinking techniques [e.g., TRIZ theory of inventive problem solving] to develop innovative solutions. We synthesize conceptually, designing by combining design parameters systematically into modular groups that form an overall physical structure. The term *physical structure* is a graphical description that takes the form of a block diagram depicting energy, material, and information functional requirements traded between the modular elements. It represents the embodiment of the design mappings and uses the synthesis activity of the DPs to separate into structural modules (components, subsystems, and systems). A physical structure helps the design team achieve a technical consensus as to how the design entity delivers the functional requirements of the mappings developed.

With each copy of this book, purchasers can access a copy of Acclaro DFSS Light®[6] by downloading from the Wiley ftp site

ftp://ftp.wiley.com/public/sci_tech_med/axiomatic_quality/

This is a training version of Acclaro DFSS software tool kit from Axiomatic Design Solutions, Inc. (ADSI), of Brighton, MA. Under license from MIT, ADSI is the only company dedicated to supporting axiomatic design methods with services and software solutions. Acclaro software, a Microsoft Windows™ based solution implementing DFSS quality frameworks around axiomatic design processes, won Industry Week's Technology of the Year.

The Acclaro DFSS Light® enables the design team to meet the following needs in managing the design process:

- Create a process for requirements cascading (hierarchy) and analysis by implement zigzagging and decomposition trees to capture and track design logic
- Collaborate at the earliest possible time frame

[6]Acclaro DFSS Light® is one of the software products of Axiomatic Design Solutions, Inc. It is protected under both copyright and pending patents. Acclaro is a registered trademark of Axiomatic Design Solutions, Inc. (website: *http://www.axiomaticdesign.com/default.asp*).

- Introduce a systematic process for design synthesis before document-build-test cycles
- Analyze design quality *before* committing to any cost by applying axioms
- Track the complete traceability from the customer attributes to design mappings to design release
- Visualize and quantify impact assessments when requirements change

2.3 INTRODUCTION TO THE INDEPENDENCE AXIOM

The Independence Axiom Maintain the independence of the FRs.

In the context of axiomatic design, the array of FRs is the minimum set of independent requirements that completely characterizes the design objective, the CAs. *Design* is defined as the creation of a synthesized solution to satisfy perceived needs through mapping between the FRs in the functional domain and the DPs in the physical domain and through mapping between the DPs and the PVs in the process domain. A violation of the independence axiom occurs when an FR is mapped to a DP that is coupled with another FR. Such a practice creates a design vulnerability called *coupling*, which implies a lack of controllability and adjustability by both the design team and the customer.

The mapping process can be written mathematically as the following matrix equations:

$$\{\mathbf{FR}\}_{m\times 1} = [\mathbf{A}]_{m\times p}\{\mathbf{DP}\}_{p\times 1} \tag{2.1}$$

$$\{\mathbf{DP}\}_{p\times 1} = [\mathbf{B}]_{p\times n}\{\mathbf{PV}\}_{n\times 1} \tag{2.2}$$

or, equivalently,

$$\{\mathbf{FR}\}_{m\times 1} = [\mathbf{C}]_{m\times n}\{\mathbf{PV}\}_{n\times 1} \tag{2.3}$$

where $\{\mathbf{FR}\}_{m\times 1}$ is a vector of independent functional requirements with m elements, $\{\mathbf{DP}\}_{p\times 1}$ a vector of design parameters with p elements, $\{\mathbf{PV}\}_{n\times 1}$ a vector of process variables with n elements, $\mathbf{A}_{m\times p}$ a physical design matrix, $\mathbf{B}_{p\times n}$ a process design matrix, and $[\mathbf{C}]_{m\times n} = [\mathbf{A}][\mathbf{B}]$ an overall design matrix. In general and throughout the book, we use physical mapping for illustration and derivation purposes. Nevertheless, the formulation, derivations, and conclusions are equally applicable to process mapping.

Before proceeding, we define the following terminology relative to the independence axiom, to ground readers regarding terminology and concepts introduced briefly in earlier sections.

- *Functional requirements* (FRs) are a minimum set of independent requirements that completely characterize the functional needs of a design solution in the functional domain within the constraints of safety, economy, reliability, and quality.

How are FRs defined? In the context of Figure 2.1, first mapping, customers define a product using features or attributes that are saturated by some or a lot of linguistic uncertainty. For example, in automotive product design, customers use the terms *quiet, stylish, comfortable*, and *easy to drive* in describing the features of their dream car. The challenge is how to translate these features into functional requirements and then into solution entities. Quality function deployment is the tool adopted here to accomplish an actionable set of FRs.

In defining their wants and needs, customers use some vague and fuzzy terms that are difficult to interpret or attribute to specific engineering terminology, in particular the FRs. In general, FRs are technical terms extracted from the voice of the customer. Customer expressions are not dichotomous or crisp in nature but something in between. As a result, uncertainty may lead to inaccurate interpretation and therefore to vulnerable or unwanted design. There are many classifications for customer linguistic inexactness. In general, two major sources of imprecision in human knowledge—linguistic inexactness and stochastic uncertainty (Zimmerman, 1985)—are usually encountered. Stochastic uncertainty is well handled by probability theory. Imprecision can arise from a variety of sources: incomplete knowledge, ambiguous definitions, inherent stochastic characteristics, measurement problems, and so on.

This brief introduction to linguistic inexactness is warranted to enable design teams to appreciate the task at hand, to assess their understanding of the voice of the customer, and to seek clarification where needed. Ignorance of such facts may cause several failures in the design project, as well as a team's efforts altogether. The worst failure is the possibility of introducing inexactness into design activities, including analysis and synthesis of the wrong requirements.

- *Design parameters* (DPs) are the elements of the design solution in the physical domain that are chosen to satisfy the FRs specified. In general terms, standard and reusable DPs (grouped into design modules within the physical structure) are often used and usually have a higher probability of success, thus improving the quality and reliability of the design.
- *Constraints* (Cs) are bounds on acceptable solutions.
- *Process variables* (PVs) are the elements of the process domain that characterize the process that satisfies the DPs specified.

The design team will conceive a detailed description of what functional requirements the design entity needs to perform to satisfy customer needs, a description of the physical entity that will realize those functions (the DPs), and a description of how this object will be produced (the PVs).

2.4 INTRODUCTION TO THE INFORMATION AXIOM

The Information Axiom Minimize the information content in a design.

The second axiom of axiomatic design provides a selection metric based on design information content. The selection problem between alternative design solution entities (concepts) of the same design variable (project) will occur in many situations. Even in the ideal case, a pool of uncoupled design alternatives; the design team needs to select the best solution. The selection process is criteria based, hence the information axiom, which states that the design that results in the highest probability of FR success [$\text{Prob}(\text{FR}_1), \text{Prob}(\text{FR}_2), \ldots, \text{Prob}(\text{FR}_m)$] is the best design. Information and probability are tied together via entropy, H (Chapter 4), which may be defined as

$$H = -\log_{v}(\text{Prob}) \tag{2.4}$$

Note that the probability [Prob in (2.4)] takes the Shannon entropy (1948) form of a discrete random variable supplying the information, the source. Note also that the logarithm is to the base v, a real nonnegative number. If $v = 2e$,[7] H is measured in *bits* (*nats*). The expression of information and hence design complexity in terms of probability hints as to the fact that FRs are random variables themselves and have to be met within a tolerance range acceptable to the customer. The array **{FR}** elements are also functions (the physical mapping) of random variables; the array **{DP}**, in turn, comprises functions (the process mapping) of another vector of random variables, the array **{PV}**—hence the transferred variation phenomenon discussed in Section 1.5 and depicted in Figure 1.5.

The PV downstream variation can be induced by several sources, such as manufacturing process variation, including tool degradation and environmental factors, the noise factors. This fact facilitates mathematical formulation of the axiomatic quality process and enables several venues of design vulnerability treatment, but with an axiomatic flavor. For example, assuming statistical independence, the overall (total) design information content of a given design hierarchical level is additive since its probability of success is a multiple of the probability of success of the individual FRs belonging to that level; that is, to reduce complexity, we need to address the largest contributors to the total (the sum). When the statistical independence assumption is not valid, the system probability of success is not multiplicative; rather, it is conditional.

A solution entity is characterized as complex when the probability of success of the total design (all hierarchical levels) is low. The manufacture of complex design solution entities requires more information. That is, complexity is a design vulnerability that is created in the design entity due to violation of the information axiom. Note that complexity here has two arguments: the number of FRs as well as their probability of success.

Information content is related to tolerances and process capabilities since probabilities are. The probability of success is the probability of meeting design specifications, the area of intersection between the *design range* (voice of the customer) and the *system range* (voice of the process). The system range is

[7] e is the base of the natural logarithm.

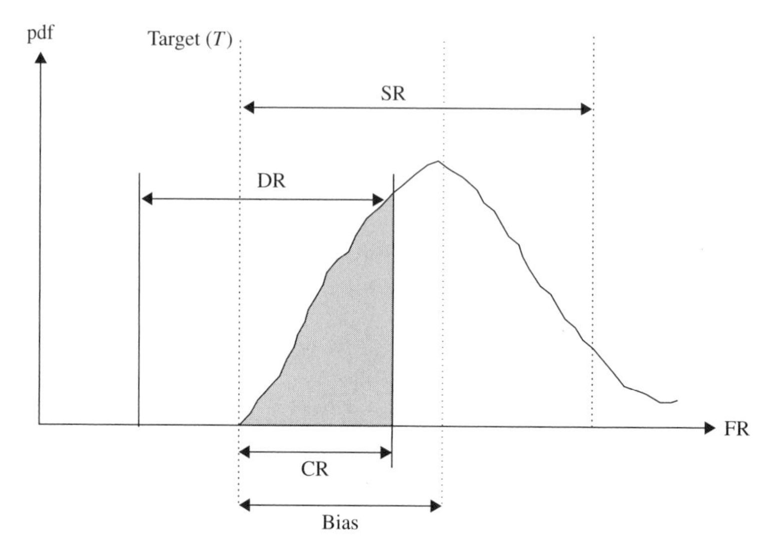

Figure 2.3 Probability of success definition.

denoted "SR" and the design range is denoted "DR" (see Figure 2.3). The overlap between the design range and the system range is called the *common range*, "CR." The *probability of success* is defined as the area ratio of the common range to the system range, CR/SR. Substituting this definition in (2.4), we have

$$H = \log_{v} \frac{\text{SR}}{\text{CR}} \tag{2.5}$$

2.5 AXIOMATIC DESIGN THEOREMS AND COROLLARIES

Study of the most famous examples (such as Euclidean geometry, Newton's laws, thermodynamics, and the axiomatic branch of modern mathematics[8]) of the axiomatic disciplines reveals several common threads. For example, Euclid's axiomatic geometry opens with a list of definitions, postulates, then axioms, before proving propositions. The aim is to present geometrical knowledge as an ordered list of proven facts, a historical paradigm of disciplines of axiomatic origin. Newton's laws were set up deliberately to emulate the Euclidean style. The laws open with a list of definitions and axioms, before proving propositions. Whereas the axioms are justified empirically, consequences of the axioms are meant to be drawn deductively. Modern mathematics and empirical knowledge are two streams that can be observed in disciplines that emerge from an axiomatic origin.

[8]Axiomatic theories in modern mathematics include modern axiomatic geometry (Euclidean and non-Euclidean geometries), Peano's axioms for natural numbers, axioms for set theory, axioms for group theory, and order axioms: linear ordering, partial ordering, axioms for equivalence relations—not the sort of axiomatic theory we consider in this book.

In axiomatic design, the goal is to systematize our design knowledge regarding a particular subject matter by showing how particular propositions (derived theories and corollaries) follow the axioms, the basic propositions. To prove a particular proposition, we need to appeal to other propositions that justify it. But our proof is not done if those other propositions themselves need justification. Ultimately, to avoid infinite regress, we will have to start our proofs with propositions that do not themselves need justification. What sorts of propositions are not in need of justification? Answer: the axioms. Therefore, differentiation of axioms from other postulates is needed. The label *axiom* is used to name propositions that are not in need of justification. Nevertheless, historically, various distinctions have been made between axioms and postulates. We will encounter two ways of drawing the distinction, one based on logical status, the other based on status relative to the subject matter of the theory. Axioms are self-evident truths. For example, the independence and information axioms are both entertained in axiomatic design. They are self-evident and have been learned from a large pool of observations. Whereas a postulate is a synthetic proposition, the contradiction of which, although difficult to imagine, nevertheless remains conceivable, the axiom is an analytic proposition whose denial is not accepted. As such, a science must start from indemonstrable principles; otherwise, the steps of demonstration would be endless.

"One of the major causes for the dismal state of design is simply mental block: the notion that design, unlike the natural sciences, cannot stand on a scientific basis. This hypothesis is both unnecessary and incorrect" (Suh, 1990). The use of design principles is a vehicle capable of gearing the design activities to fruitful systematic results while providing the desired design scientific basis. The two design axioms suggested by (Suh, 1990) are prominent examples of empirical design principles proven to be beneficial, as evidenced by application growth and industrial coverage. The employment of axioms in design seems to be promising because history tells us that knowledge based on axioms will continue to evolve through theorems and corollaries as long as the axioms are maintained. A subset of axiomatic design corollaries and theories, developed primarily by Suh, that we utilize in this book are listed below. The rest can be found in Suh (2001). Additional theorems are also developed in this book by the author.

2.5.1 Axiomatic Design Corollaries[9]

Corollary 2.1: Decoupling of Coupled Designs Decouple or separate parts or aspects of a solution if FRs are coupled or become interdependent in the designs proposed.

Corollary 2.2: Minimization of FRs Minimize the number of FRs and constraints.

[9]A *corollary* is an immediate consequence of a result already proved. Corollaries usually state more complicated theorems in language simpler to use and apply.

Corollary 2.3: Integration of Physical Parts Integrate design features in a single physical part if FRs can be independently satisfied in the solution proposed.
Corollary 2.4: Use of Standardization Use standardized or interchangeable parts if the use of the parts is consistent with FRs and constraints.
Corollary 2.5: Use of Symmetry Use symmetrical shapes and/or components if they are consistent with the FRs and constraints.
Corollary 2.6: Largest Design Tolerance Specify the largest allowable tolerance in stating the FRs.
Corollary 2.7: Uncoupled Design with Less Information Seek an uncoupled design that requires less information than a coupled design in satisfying a set of FRs.
Corollary 2.8: Effective Reangularity of a Scalar The effective reangularity,[10] R, for a scalar coupling matrix element is unity.

2.5.2 Axiomatic Design Theorems[11] of General Design

Theorem 2.1: Coupling Due to Insufficient Number of DPs When the number of DPs is less than the number of FRs, either a coupled design results or the FRs cannot be satisfied.
Theorem 2.2: Decoupling of Coupled Design When a design is coupled due to the greater number of FRs than DPs (i.e., $m > p$), it may be decoupled by the addition of new DPs so as to make the number of FRs and DPs equal each other if a subset of the design matrix containing $p \times p$ elements constitutes a triangular matrix.
Theorem 2.3: Redundant Design When there are more DPs than FRs, the design is either a redundant design or a coupled design.
Theorem 2.4: Ideal Design In an ideal design, the number of DPs is equal to the number of FRs, and the FRs are always maintained independent of each other.
Theorem 2.5: Need for New Design When a given set of FRs is changed by the addition of a new FR, by substitution of one of the FRs with a new one or by selection of a completely different set of FRs, the design solution given by the original DPs cannot satisfy the new set of FRs. Consequently, a new design solution must be sought.
Theorem 2.6: Path Independence of Uncoupled Design The information content of an uncoupled design is independent of the sequence by which the DPs are changed to satisfy the given set of FRs. (See Section 1.3 for more details.)
Theorem 2.7: Path Dependency of Coupled and Decoupled Designs The information contents of coupled and decoupled designs depend on the sequence by which the DPs are changed to satisfy the given set of FRs. (See Section 1.3 for more details.)

[10] *Reangularity* (R) is a measure-coupling vulnerability and is defined as the orthogonality between DPs in terms of the absolute value of the product of the geometric sines of all the angles between the various DP pair combinations of the design matrix. See Chapter 3 for more details.

[11] A *theorem* can be defined as a statement that can be demonstrated to be true by accepted mathematical operations and arguments. In general, a theorem is an embodiment of some general principle that makes it part of a larger theory. The process of showing a theorem to be correct is called a *proof.*

Theorem 2.8: Independence and Tolerance A design is an uncoupled design when the designer-specified tolerance is greater than

$$\sum_{\substack{i \neq j \\ i=1}}^{p} \frac{\partial \mathrm{FR}_i}{\partial \mathrm{DP}_j} \Delta \mathrm{DP}_j \qquad i = 1, \ldots, m; \quad j = 1, \ldots, p$$

in which case the nondiagonal elements of the design matrix can be omitted from design consideration.

Theorem 2.9: Design for Manufacturability For a product to be manufacturable, the design matrix for the product, **[A]** (which relates the FR vector for the product to the DP vector of the product), times the design matrix for the manufacturing process, **[B]** (which relates the DP vector to the PV vector of the manufacturing process), must yield either a diagonal or a triangular matrix. Consequently, when either **[A]** or **[B]** represents a coupled design, the independence of the FRs and robust design cannot be achieved. When they are full triangular matrixes, they must both be upper triangular or both be lower triangular for the manufacturing process to satisfy the independence of functional requirements.

Theorem 2.10: Modularity of Independence Measures Suppose that a design matrix **[DM]** can be partitioned into square submatrixes that are nonzero only along the main diagonal. Then the reangularity, R, and semangularity,[12] S, for **[DM]** are equal to the product of their corresponding measures for each nonzero submatrix.

Theorem 2.11: Invariance Reangularity, R, and semangularity, S, for a design matrix **[DM]** are invariant under alternative orderings of the FR and DP variables, as long as orderings preserve the association of each FR with its corresponding DP.

Theorem 2.12: Sum of Information The sum of information for a set of events is also information, provided that proper conditional probabilities are used when the events are not statistically independent.

Theorem 2.13: Information Content of the Total System If each DP is probabilistically independent of other DPs, the information content of the total system is the sum of the information of all individual events associated with the set of FRs that must be satisfied.

Theorem 2.14: Information Content of Coupled Versus Uncoupled Designs When the state of FRs is changed from one state to another in the functional domain, the information required for the change is greater for a coupled process than for an uncoupled process.

Theorem 2.15: Design–Manufacturing Interface When the manufacturing system compromises the independence of the FRs of the product, either the design of the product must be modified or a new manufacturing process must be designed and/or used to maintain the independence of the FRs of the products.

[12]*Semangularity*, S, on the other hand, is an angular measure of pair axes between DPs and FRs (see Chapter 3 for more details).

Theorem 2.16: Equality of Information Content All information content that is relevant to the design task is equally important regardless of its physical origin, and no weighing factor should be applied.

2.5.3 Theorems for Design of Large Systems

Theorem 2.17: Importance of High-Level Decisions The quality of design depends on the selection of FRs and the mapping from domains to domain. Wrong decisions made at the highest levels of the design hierarchy cannot be rectified through the lower-level design decisions.
Theorem 2.18: Best Design for Large Systems The best design among the designs proposed for a large system that satisfy m FRs and the independence axiom can be chosen if the complete set of the subsets of the **{FR}** vector that the large system must satisfy over its life is known a priori.
Theorem 2.19: Need for Better Design for Large Systems When the complete set of the subsets of FRs that a given large flexible system must satisfy over its life is not known a priori, there is no guarantee that a specific design will always have the minimum information content for all possible subsets, and thus there is no guarantee that the same design is the best at all times.
Theorem 2.20: Improving the Probability of Success The probability of choosing the best design for a large flexible system increases as the known subsets of FRs that the system must satisfy approach the complete set that the system is likely to encounter during its lifetime.
Theorem 2.21: Infinite Adaptability Versus Completeness A large flexible system with an infinite adaptability (or flexibility) may not represent the best design when the large system is used in a situation where the complete set of the subsets of FRs that the system must satisfy is known a priori.
Theorem 2.22: Complexity of Large Systems A large system is not necessarily complex if it has a high probability of satisfying the FRs specified for the system.
Theorem 2.23: Quality of Design The quality of design of a large flexible system is determined by the quality of the database, the proper selection of FRs, and the mapping process.

2.6 CASE STUDY: DEPTH CHARGE INITIATOR (NORDLUND, 1996)

A depth charge initiator is a mechanical clock combined with a pressure sensor used for depth charging and consisting of more than 350 parts. The design project objective is to design an initiator that sends a signal to the detonator only when the depth charge hits a target and is intended to explode. The depth charge is launched from a ship using a special launcher (Figure 2.4). The baseline design of a depth charge initiator does not satisfy the customer for cost and reliability, an incremental design assignment (see Section 1.6). This case study shows an industrial application of axiomatic design by demonstrating the use of design

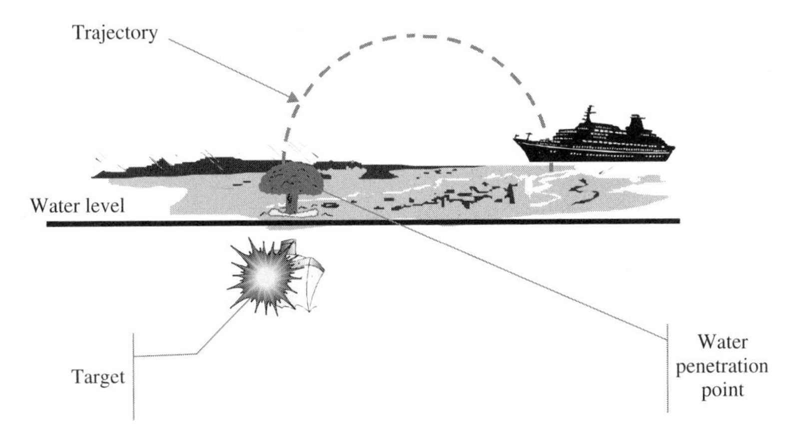

Figure 2.4 Depth charge initiator operation.

mappings and design axioms to effectively search for nonvulnerable solutions. It also explains how design constraints, when cascaded, can force a designer to compromise on coupling situations.

Due to a cost constraint, baseline design standard logic needs to be used. A schematic of an initiator is shown in Figure 2.5. The system requires an electrical power source, three independent arming conditions (ACs), and an ignition signal. When all these requirements exist, the initiator signals the detonator to detonate the warhead. Safety regulations state that at least one of the arming conditions should be detected under water from its first occurrence until detonation. This means that the system will prevent detonation when the arming condition state is out of the water. Furthermore, the system's sensors must be robust: must not react erroneously to humidity, electromagnetic radiation, darkness, vibrations, accelerations, temperature, or other noise factors.

Specifically, the design task is to design a sensor system which ensures that the initiator sends a signal to the detonator when the depth charge hits a target and is

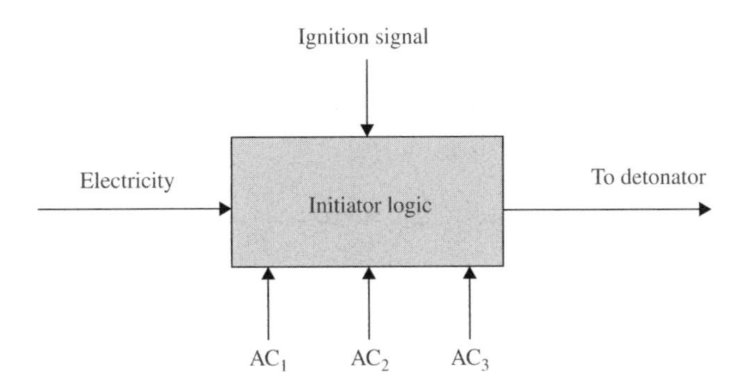

Figure 2.5 Initiator system logic diagram.

intended to explode, but not under any other circumstances. The aim is to come up with an alternative design to an existing arming, initiating, and detonating system for a depth charge.

Hierarchical Level 1 Analysis: Design Mapping Based on the general problem description, a more concrete description of the customer attributes (CAs) was captured.

CA_1 = lower cost/part count [fewer parts means less information content (the information axiom)]
CA_2 = simpler working concept
CA_3 = more reliable working principle

The highest-level FRs in the hierarchy's highest level are (1) to initiate detonation of the warhead, and (2) to convey the driving gas pressure in the barrel to the entire depth charge, making it accelerate and begin its ballistic trajectory. We have the following design matrix:

$$FR_1 = \text{initiate detonator}$$

$$FR_2 = \text{launch depth charge}$$

$$DP_1 = \text{electrical system}$$

$$DP_2 = \text{mechanical launcher system}$$

$$\begin{Bmatrix} FR_1 \\ FR_2 \end{Bmatrix} = \begin{bmatrix} X & 0 \\ 0 & X \end{bmatrix} \begin{Bmatrix} DP_1 \\ DP_2 \end{Bmatrix} \tag{2.6}$$

Hierarchical Level 2 Analysis: Design Mapping

Decomposing FR_2: Launch Depth Charger

$$FR_{21} = \text{provide force to launch device}$$

$$FR_{22} = \text{send device in desired direction}$$

$$FR_{23} = \text{convey force to entire device}$$

and the second-hierarchical-level DPs are

$$DP_{21} = \text{explosive}$$

$$DP_{22} = \text{barrel}$$

$$DP_{23} = \text{chassis}$$

The design equation is

$$\begin{Bmatrix} FR_{21} \\ FR_{22} \\ FR_{23} \end{Bmatrix} = \begin{bmatrix} X & 0 & 0 \\ 0 & X & 0 \\ 0 & 0 & X \end{bmatrix} \begin{Bmatrix} DP_{21} \\ DP_{22} \\ DP_{23} \end{Bmatrix} \tag{2.7}$$

These design matrixes indicate that the independence axiom is satisfied. In addition, the design for the launcher (including the explosive, barrel, and chases) is already done and will not be decomposed further. At this second level, there are a number of constraints that apply to all DPs in the zigzagging process. They are:

C_1 = safety

C_2 = weight

C_3 = position of the center of gravity

C_4 = outside measures (geometry) have to fit within the chassis

C_5 = environmental endurance

Decomposing FR_1*: Initiate Detonator* With reference to Figure 2.5, the functional requirement FR_1, defined above as "initiate detonator," may be decomposed with DP_1, defined as "electrical system," in mind in a zagging step as

FR_{11} = provide electricity

FR_{12} = activate AC_1

FR_{13} = activate AC_2

FR_{14} = activate AC_3

FR_{15} = send ignition signal

Based on the analysis above, the following environmental DPs were chosen because they satisfy the independence axiom in order to satisfy the design constraints. They could be used to activate FR_{12}, FR_{13}, and FR_{14} and set the ignition signal to the detonator (FR_{15}).

1. Gas pressure (in the launcher barrel), which exists only upon launching, making it a unique and independent event
2. Passage of the launcher muzzle
3. Presence of water: clearly distinguishes between air and water phases, making it suitable as a state to detect
4. Water pressure, which distinguishes between the air and water phases
5. Hitting the target

DP_{11} = battery with electrolyte in ampoule

DP_{12} = leave barrel (event 1)

DP_{13} = entering water (event 2)

DP_{14} = water pressure (state)

DP_{15} = impact on target (event 3)

The design equation is given as

$$\begin{Bmatrix} FR_{11} \\ FR_{12} \\ FR_{13} \\ FR_{14} \\ FR_{15} \end{Bmatrix} = \begin{bmatrix} X & 0 & 0 & 0 & 0 \\ X & X & 0 & 0 & 0 \\ X & 0 & X & 0 & 0 \\ X & 0 & 0 & X & 0 \\ X & X & X & X & X \end{bmatrix} \begin{Bmatrix} DP_{11} \\ DP_{12} \\ DP_{13} \\ DP_{14} \\ DP_{15} \end{Bmatrix} \tag{2.8}$$

The constraints at this level are:

C_{11} = safety, with the implication of creating a partially decoupled design to ensure that the events have happened in a desired sequence (it must not be possible to detect the ignition signal before all other FRs are satisfied)
C_{12} = probability of function, with the implication that simple solid-state mechanisms will be used

This case study investigated the different environments that this product would encounter while at the customer/users, to take advantage of the free environmental factors and conditions as design parameters. Figure 2.4 outlines all the environments the depth charge will experience before it is expected to detonate, excluding storage, transportation, and loading activities. It is in these environments that a search is made for environmental factors that can be used as arming conditions (DPs). That is, the study demonstrates that the environment also provides a rich inventory of physical design parameters.

Hierarchical Level 3 Analysis: Design Mapping The five FRs are then decomposed to arrive at a detailed design of each DP that will produce arming conditions (AC_1, AC_2, and AC_3). Only the zigzagging of FR_1, FR_{12}, and FR_{13} is discussed by Nordlund (1996).

Decomposing FR_{11}*: Provide Electricity* The decomposition of FR_{11} gives the following FRs:

$$FR_{111} = \text{sense activation time}$$
$$FR_{112} = \text{supply electrolyte}$$

and the following DPs:

$$DP_{111} = \text{gas pressure}$$
$$DP_{112} = \text{impact piston}$$

The design equation is

$$\begin{Bmatrix} FR_{111} \\ FR_{112} \end{Bmatrix} = \begin{bmatrix} X & 0 \\ X & X \end{bmatrix} \begin{Bmatrix} DP_{111} \\ DP_{112} \end{Bmatrix} \tag{2.9}$$

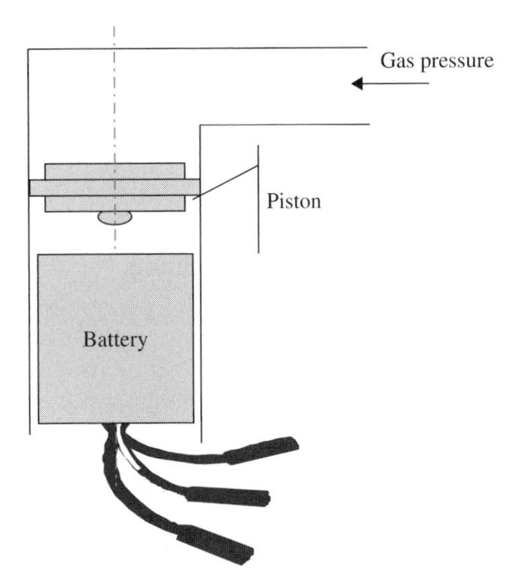

Figure 2.6 FR_{11} system schematic.

A concept realizing this solution is shown in Figure 2.6. The gas pressure enters the rear of the depth charge and is led to chamber when an impact piston is forced to hit one end of a battery. The impact should suffice to break a glass ampoule containing an electrolyte. When the electrolyte diffuses, the battery becomes active.

Decomposing FR_{15}*: Provide an Initiation Signal*

$$FR_{151} = \text{sense target impact}$$

$$FR_{152} = \text{send signal to detonator}$$

$$DP_{151} = \text{accelerometer}$$

$$DP_{152} = \text{switch activated by accelerometer}$$

with the following design matrix:

$$\begin{Bmatrix} FR_{151} \\ FR_{152} \end{Bmatrix} = \begin{bmatrix} X & 0 \\ X & X \end{bmatrix} \begin{Bmatrix} DP_{151} \\ DP_{152} \end{Bmatrix} \tag{2.10}$$

This design can be realized using off-the-shelf components (DPs).

Decomposing FR_{12}*: Generate Arming Condition 1 (*AC_1*)*

$$FR_{121} = \text{sense launch}$$

$$FR_{122} = \text{activate circuit after leaving barrel}$$

$$\mathrm{DP}_{121} = \text{rod sensing barrel}$$

$$\mathrm{DP}_{122} = \text{switch activated by rod}$$

$$\begin{Bmatrix} \mathrm{FR}_{121} \\ \mathrm{FR}_{122} \end{Bmatrix} = \begin{bmatrix} X & 0 \\ X & X \end{bmatrix} \begin{Bmatrix} \mathrm{DP}_{121} \\ \mathrm{DP}_{122} \end{Bmatrix} \tag{2.11}$$

Hierarchical Level 4 Analysis: Design Mapping

Decomposing FR_{121}*: Sense Launch*

$$\mathrm{FR}_{1211} = \text{push rod toward barrel}$$

$$\mathrm{FR}_{1212} = \text{extend rod when leaving barrel}$$

$$\mathrm{FR}_{1213} = \text{prevent rod from moving back after launch}$$

with the following DPs and design matrix:

$$\mathrm{DP}_{1211} = \text{gas pressure}$$

$$\mathrm{DP}_{1212} = \text{piston}$$

$$\mathrm{DP}_{1213} = \text{latch mechanism}$$

$$\begin{Bmatrix} \mathrm{FR}_{1211} \\ \mathrm{FR}_{1212} \\ \mathrm{FR}_{1213} \end{Bmatrix} = \begin{bmatrix} X & 0 & 0 \\ X & X & 0 \\ 0 & 0 & X \end{bmatrix} \begin{Bmatrix} \mathrm{DP}_{1211} \\ \mathrm{DP}_{1212} \\ \mathrm{DP}_{1213} \end{Bmatrix} \tag{2.12}$$

A mechanism to integrate this design is given in Figure 2.7. The gas pressure is channeled into a chamber that that has a piston. On the low-pressure side of

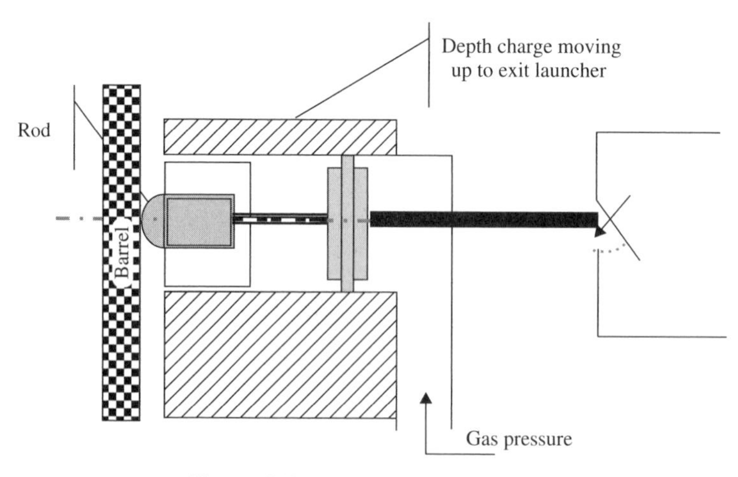

Figure 2.7 AC1 generating system.

the piston, there is a rod sliding along the barrel. When launching, gas pressure builds up in the chamber, forcing the rod toward the barrel. When the depth charge leaves the launcher, the rod moves farther out. This causes an electric switch to close. For the switch not to reopen later, a latch mechanism (electrical or mechanical) to hold it in place needs to be introduced.

2.7 SUMMARY

Axiomatic quality is a new quality and reliability discipline that hinges on the axiomatic design method developed by Suh. There are five principal concepts in axiomatic design:

1. *Design domains.* These are the customer attributes domain, the functional requirements domain, the design parameters domain, and the process variables domain. The requirements specified in one domain are mapped in the design phases to a set of parameters (variables) in an adjacent domain (see Figures 2.1 and 2.2).
2. *Hierarchies.* The output of each domain evolves from abstract concepts to detailed information in a top-down or hierarchical design manner.
3. *Zigzagging.* The design team members go through a process whereby they zigzag between domains in decomposing the design problem. The result is that the hierarchical development process in each domain is conducted in conjunction with those in the other domains.
4. *Design axioms.* There are two design axioms about the relations that should exist between FRs and DPs, which provide a rational basis for evaluation of proposed design solution alternatives and subsequent selection of the best alternative.
   a. *Independence axiom.* Maximize the independence of the FRs.

      Use of the independence axiom is described in terms of the design matrixes within a hierarchical level. Violation of the independence axiom produces coupling, a design vulnerability. A design matrix can be categorized as a diagonal matrix (uncoupled design), or a triangular matrix (decoupled design), or a full matrix with nonzero elements above and below the diagonal (coupled design). An uncoupled design is best according to the independence axiom.

   b. *Information axiom.* Minimize the information content of the design (maximize the probability of success).

      A violation of this axiom produces complexity, another design vulnerability.

CHAPTER 3

INDEPENDENCE AXIOM

3.1 INTRODUCTION

In today's highly competitive world, the voice of the customer is more important than ever. This new paradigm stresses the importance of design theory and the need to bring consumer wants to the forefront in product development, as depicted in Figure 1.8. In addition to the voice of the customer, the axiomatic quality process integrates several key disciplines into the product development process, the most important of which are axiomatic design, quality engineering, and design for six-sigma.

The ultimate goal of the axiomatic design method, the design theory adopted here, is to establish a scientific design basis and to improve the design process by providing design teams with theoretical and logical foundations for both analysis and synthesis development activities. In accomplishing this goal, axiomatic design provides a systematic approach to minimize designer psychological inertia and to find the best solution that accomplishes the design intent, the voice of the customer.

An axiom in mathematics and logic is a general statement accepted without proof as the basis for logical deduction of other statements (e.g., corollaries and theorems) which later forms a logical system of its own. Examples of axioms used widely in mathematics are those related to operations (e.g., the associative and commutative laws of set theory). It is sometimes said that an axiom or postulate is a self-evident statement but that the truth of the statement need not be readily

Axiomatic Quality: Integrating Axiomatic Design with Six-Sigma, Reliability, and Quality Engineering, by Basem Said El-Haik
ISBN 0-471-68273-X

evident. An axiomatic system is valid and sustained if the set of axioms satisfies the following attributes:

- *The Axioms Are Independent.* No one axiom statement may be deduced from any combination of the others.
- *The Axioms Are Consistent.* It is not possible to deduce from them contradictory theories and corollaries.
- *The Axiomatic Set Is Complete.* Any true statement within the system may be deduced from the axioms.

Axiomatic design is a methodology based on a set of two axioms, the independence axiom, the subject of this chapter, and the information axiom, the subject of Chapter 4, which constitute an axiomatic system. This theoretical validity is supported by the visible growth and wide acceptance of the methodology. The implementation validity of axiomatic design methodology is evident by the many applications in hardware, service organizations, systems, and software in numerous industries that have appeared in the literature in the last two decades. Axiomatic design principles have been used in the design of quality systems (Suh, 1995) and general system design (Suh, 1995, 1997). Nakazawa and Suh (1984) presented a method for process planning based on the information measures of the information axiom. Kim and Suh (1987) applied axiomatic design synthesis to injection molding. Gebala and Suh (1992) presented an axiomatic design medical application for wounded skin treatment. Kim et al. (1991) applied design axioms in software design. Albano et al. (1993) integrated the axioms and other concepts with the notion of an interface index to create a framework for facility design planning. Hillstrom (1994) showed how axiomatic design links to modular design practices. Killander (1995) proved that concurrent product development could be achieved only when uncoupled independent concepts are conceived. Igata (1996) applied axiomatic design to rapid-prototyping support on real-time control software. Hintersteiner and Tate (1998) clarified the role of axiomatic design system architecture in control theory. Suh et al. (1998) provided a manufacturing system design using the axiomatic design methodology. Babic (1999) applied the method to design flexible manufacturing systems. Suh (2001) showed how industrial applications contributed to advancement of the methodology. Arcidiacono et al. (2002) demonstrated that design for six-sigma techniques used for calculating a given process capability can interface with the axiomatic design schematization of the product. Yang and El-Haik (2003) presented a design algorithm for six-sigma based on axiomatic design methodology. Lentz et al. (2002) addressed a process used commercially to validate the architecture of a new flagship product using concepts of axiomatic design.

In this chapter we provide a formulation for the independence axiom, including a description of axiomatic design concepts and coupling design vulnerability, highlighting possible implications for design theory and practice. For illustrative purposes, several examples and case studies are presented.

3.2 INDEPENDENCE AXIOM AND THE ZIGZAGGING APPROACH

The independence axiom is stated as follows:

The Independence Axiom Maintain the independence of the FRs.

The independence of FRs may be checked while conducting physical and process mappings using the zigzagging process, an axiomatic design element introduced in Chapter 2. The zigzagging method uses design matrixes to identify design hierarchy as well as a firsthand assessment of conceptual vulnerability at each hierarchal level. Employment of the zigzagging process should also witness extensive activities of the independence axiom concepts utilizing associated design theories and corollaries.

Upon the completion of customer attributes-to-FR mapping (Figure 2.1), a vector of FRs is obtained. It is a good practice then to check the FR definition for independence and accuracy. Another good practice is to set FR definition guidelines that provide a template of each requirement for team use that is consistent with the operational definition. This practice will also minimize the voice of the customer linguistics inexactness and eliminate any confusion the team already has.

In the axiomatic design context, the design process is a continuous mapping activity. Physical mapping maps from the functional domain to the physical domain. Process mapping maps from the physical domain to the process domain, documenting the manufacturing activities required to produce the solution entity. The mapping process may be described mathematically using the concept of vector space.

Like peeling an onion, the mapping process for many design projects is carried out over many stages of zigzagging or decomposition. High-level FRs and DPs are abstractions of lower-level detailed FRs and DPs. That is, the high-level FRs and their mapped-to DPs, conceived initially from the CA (customer attributes) domain, need to be cascaded to lower levels for further detailing and therefore actionable clarity. The mapping process from the FR domain to the physical domain is carried out in the following zigzagging steps. First we *zig* from the functional domain to the physical domain to determine the mapped-to higher-level DPs and then *zag* back to the functional domain to determine the FRs of those DPs in another level but with lower hierarchical ranking. After completion of this first iteration, we *zig* to the physical domain to identify the mapped-to DPs of the second-level FRs. This mapping process between domains can be expressed mathematically as discussed in Chapter 2. For example, the physical mapping between the functional domain and the physical domain may be written as $\{\mathbf{FR}\}_{m\times1} = [\mathbf{A}]_{m\times p}\{\mathbf{DP}\}_{p\times1}$, where $\{\mathbf{FR}\}_{m\times1}$ is the vector of independent FRs with m elements, $\{\mathbf{DP}\}_{p\times1}$ is the vector of design parameters with p elements, and $[\mathbf{A}]$ is the design matrix $[\mathbf{DM}]$. Note that the bold symbols are vector quantities unless otherwise specified. A design matrix entry is denoted by X where a nonzero mapping relationship exists and by

0 where it does not. Upon differentiation, an element $A'_{ij} \in [\mathbf{A}']$ is a sensitivity coefficient of the functional requirement FR_i with respect to the design parameter DP_j (i.e., $A'_{ij} = \partial \mathrm{FR}_i / \partial \mathrm{DP}_j$). In this case, $\mathbf{A}'$ is called the *sensitivity matrix*.

The shape and dimension of matrix $\mathbf{A}$ is used to classify the design into one of the following categories: *uncoupled, decoupled, coupled*, and *redundant*. For the first two categories, the number of functional requirements, m, equals the number of design parameters, p. In a *redundant design* we have $m < p$. A design that complies completely with the independence axiom is called an *uncoupled* (*independent*) *design*. The resulting design matrix in this case, $\mathbf{A}$, is a square-diagonal matrix with $m = p$ and $A_{ij} \neq 0$ when $i = j$ and 0 elsewhere, as in (3.1). An uncoupled design is an ideal[1] (square matrix) design with many attractive attributes. First, it enjoys the path independence property, which enables the traditional quality methods the objectives of reducing functional variability and mean adjustment to target, through only one parameter per FR, its respective DP. Second, the complexity of the design is additive (assuming statistical independence) and can be reduced through axiomatic treatment of the individual DPs, which should be conducted separately. This additivity property is assured because complexity may be measured by design information content, which in turn is a probabilistic function. Third, cost and other constraints are more manageable (i.e., less binding) and are met with significant ease, including high degrees of freedom for controllability and adjustability.

$$\begin{Bmatrix} \mathrm{FR}_1 \\ \vdots \\ \mathrm{FR}_m \end{Bmatrix} = \begin{bmatrix} A_{11} & 0 & \cdots & 0 \\ 0 & A_{22} & \cdots & \cdot \\ \cdot & \cdot & \cdots & 0 \\ 0 & \cdot & 0 & A_{mm} \end{bmatrix} \begin{Bmatrix} \mathrm{DP}_1 \\ \vdots \\ \mathrm{DP}_m \end{Bmatrix} \quad \text{(uncoupled design)} \quad (3.1)$$

$$\begin{Bmatrix} \mathrm{FR}_1 \\ \vdots \\ \mathrm{FR}_m \end{Bmatrix} = \begin{bmatrix} A_{11} & 0 & \cdots & 0 \\ A_{21} & A_{22} & 0 & \cdot \\ \cdot & \cdot & \cdots & 0 \\ A_{m1} & A_{m2} & \cdots & A_{mm} \end{bmatrix} \begin{Bmatrix} \mathrm{DP}_1 \\ \vdots \\ \mathrm{DP}_m \end{Bmatrix} \quad \text{(decoupled design)} \quad (3.2)$$

$$\begin{Bmatrix} \mathrm{FR}_1 \\ \vdots \\ \mathrm{FR}_m \end{Bmatrix} = \begin{bmatrix} A_{11} & A_{12} & \cdots & A_{1p} \\ A_{21} & A_{22} & \cdots & \cdot \\ \cdot & \cdot & \cdots & A_{(m-1)p} \\ A_{m1} & \cdot & A_{m(p-1)} & A_{mp} \end{bmatrix} \begin{Bmatrix} \mathrm{DP}_1 \\ \vdots \\ \mathrm{DP}_p \end{Bmatrix} \quad \text{(coupled design)} \quad (3.3)$$

A violation of the independence axiom occurs when an FR is mapped to a DP that is coupled with another FR. A design that satisfies the independence axiom, however, with path dependence[2] (or sequence) is called a *decoupled design* as in (3.2). In a decoupled design, matrix $\mathbf{A}$ is square triangular (lower or upper;

[1]Recall Theorem 2.4 in Section 2.5.2, which states that "in an ideal design, the number of DPs is equal to the number of FRs, and the FRs are always maintained independent of each other."
[2]See Theorem 2.7 in Section 2.5.2, as well as Section 1.3.

sparse or otherwise). In an extreme situation, **A** could be a complete (i.e., nonsparse full lower or upper) triangular matrix. For example, in a *full lower triangular matrix*, the maximum number of nonzero entries is, $p\,(p-1)/2$, where $A_{ij} \neq 0$ for $j = 1, i$ and $i = 1, \ldots, p$. A lower (upper) triangular decoupled design matrix is characterized by $A_{ij} = 0$ for $i < j$ (for $i > j$). A rectangular design matrix with $(m > p)$ is classified as a *coupled design*, as in (3.3).

In the uncoupled linear design case or constant-sensitivity design matrix, the independence requirement implies that FRs are a basis for the subspace of the design in the functional domain. This means that each functional requirement could be adjusted and controlled by only one DP without affecting any other FR (see Figure 1.3*a*). This view is subject to the homogeneity and compatibility levels among the FRs. A homogeneous (compatible) set of FRs occurs when the FRs form a field in which all the FRs are tied together in a circuit of physical mapping relationships.

A simple analysis can be used to give more clarity in nonlinear (i.e., sensitivity matrix entries are a function of the DPs) coupled or decoupled design situations by substituting an approximation for the function A'_{ij} around a point (i.e., design settings) of interest within the design space.

3.2.1 Coupling Measures

Since coupling is defined on a continuous scale, it is fundamental to derive measures of coupling in order to evaluate its degree in a given design mapping. Rinderle (1982) and Suh and Rinderle (1982) proposed the use of reangularity, R, and semangularity, S, as coupling measures; R and S are defined in (3.4) and (3.5), respectively. R is a measure of the orthogonality between the DPs in terms of the absolute value of the product of the geometric sines of all the angles between the various DP pair combinations of the design matrix. As the degree of coupling increases, R decreases. Semangularity, S, on the other hand, is an angular measure of the parallelism of the DP and FR pair (see Figure 1.2). When $R = S = 1$, the design is uncoupled completely. The design is decoupled when $R = S$ (Suh, 1990):

$$R = \prod_{\substack{j=1,p-1 \\ k=1+i,p}} \sqrt{\frac{1 - \left(\sum_{k=1}^{p} A_{kj} A_{kj}\right)^2}{\left(\sum_{k=1}^{p} A_{kj}^2\right)\left(\sum_{k=1}^{p} A_{kj}^2\right)}} \tag{3.4}$$

$$S = \frac{\prod_{j=1}^{p} |A_{jj}|}{\sqrt{\sum_{k=1}^{p} A_{kj}^2}} \tag{3.5}$$

The independence axiom is best satisfied if **A** is a diagonal matrix depicting an uncoupled design. For a decoupled design, the independence axiom can be satisfied if the DPs can be set (adjusted) in a specific order conveyed by the matrix to maintain independence. A design that violates the independence axiom as it distances itself from uncoupled and decoupled categories is by definition a coupled design. The vulnerability of coupling is assured whenever the number of DPs, p, is less than the number of FRs, m (see Theorems 2.1 and 2.2 in Section 2.5.2). In other words, the desired bijection one-to-one mapping property between two design domains cannot be achieved without an axiomatic treatment. An axiomatic treatment can be produced by the application of design theories and corollaries deduced from the axioms. The conceptual design for capability is a phase of the axiomatic quality process dedicated to this task.

For a unifunctional design entity ($m = 1$), the independence axiom is always satisfied. The design sequence in this case proceeds to optimization and verification and can be characterized as trivial compared with a higher level of modularity (e.g., a subsystem or a system). Regardless of whether deterministic or probabilistic, optimization of a multifunctional module is complicated by the presence of coupling (lack of independence). Uncoupled design matrices may be treated as independent modules for optimization (where DPs are the variables), and extreme local or global DP settings in the direction of goodness can be found. In a decoupled design, optimization of a modular element cannot be carried out in a single routine. Many optimization algorithms (in fact, m routines) need to be invoked sequentially, starting from the DP at the head of the triangular matrix and proceeding to the base.

3.3 DESIGN MAPPINGS AND DESIGN STRUCTURES

Hierarchy is built by the decomposing design into a number of simpler functional design matrices that, collectively, meet the high-level functional requirements conceived from the voice of the customer. A design structure can be defined as an input–output or cause-and-effect relationship of functional modular elements. The zigzagging method of axiomatic design captures the design mappings in a mathematical format. Graphically, it may be depicted in a block diagram that consists of nodes connected by arrows depicting the mapping relationships (Chapter 6). A structure should capture all DPs within the scope and ensure correct flow down to critical elements. The DPs are usually grouped together to synthesize modular structure elements such as components, subsystems, and systems. Corollary 2.3 has ample opportunities for deployment as the structure is formed. There are two recognized structures in axiomatic quality:

1. The physical structure, synthesized from the FRs and DPs. The physical structure is a focus in the context of this book, and the Greek letter ψ (lowercase psi) is used to denote this structure mathematically going forward.
2. The process structure, pieced together from the DPs and the PVs, which can be satisfied by a six-sigma process mapping diagram, value stream mapping, IDEF-x techniques, and others.

The physical structure is usually developed first to detail the design concept. The preliminary work to verify structural choices should help the design team to get started on concept generation. The team needs to select the best solution entity element in terms of DPs to meet or exceed requirements. Technology and structure choices are sometimes closely interlinked via the physical and process mappings when conducted following design axioms. New technologies (DPs) can enable new structures, and different technology availability may suggest different mappings of FRs. The pursuit of linked technology and structure options may reveal new opportunities for customer *delighters* (see Section 6.3.1). Conversely, because axiomatic-driven structures often have very long life spans, they need to be relatively insensitive to technology choices. An axiomatic-driven structure should enable reimplementation of new technologies without undue impact on either the structure or the design mappings. Therefore, to assure the insensitivity of the structure to future unknown technology, they need to be derived using design axioms. It is wise to examine the robustness of a structure against current, known technology and design alternatives. Structures need to be robust against customer use, misuse, and abuse; errors in requirements and specifications; unanticipated interactions with other portions of the solution entity; or process variations. The FRs should be verified over a range of parameters which exceed known requirements and specifications. Determining the sensitivity of FRs due to changes in operating conditions (including local environment and solution entity interactions) over the expected operating range is an essential task for transfer function optimization within the axiomatic quality process (Chapters 6 and 9).

3.4 CASE STUDY 1: AXIOMATIC DESIGN OF A WATER FAUCET (SWENSON AND NORDLUND, 1995)

The water faucet case study introduced in Chapter 1 has been selected as the first case study to present in this chapter because (1), it is a very familiar design to everyone and (2), it illustrates nicely several axiomatic design concepts. The study objectives are many, among which are an explanation of how to use axiomatic design effectively during product development while demonstrating the use of its associated theorems. The authors also hint at the implication of the information axiom in this case study.

For the water faucet, the customer cares about two elements: the water temperature and the flow. According to the independence axiom, we should be able to do this with two design parameters, one for each requirement. However, there is a difficulty. Since the water comes in two pipes (hot and cold), some may think it is easy to control the volume of hot and cold water, but these are not the things we want to control. This is a case study where a coupled baseline design exists (Figures 3.1 and 3.2).

The two functional requirements of the water faucet are:

$$\mathrm{FR}_1 = \text{control the flow } (Q) \text{ of water}$$

$$\mathrm{FR}_2 = \text{control the temperature } (T) \text{ of the water}$$

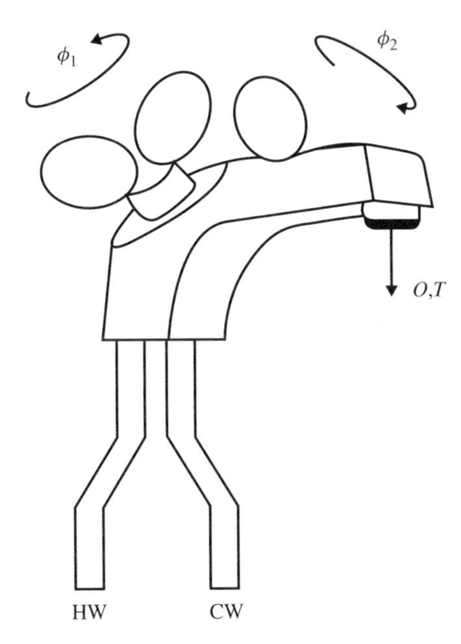

Figure 3.1 Water faucet baseline design.

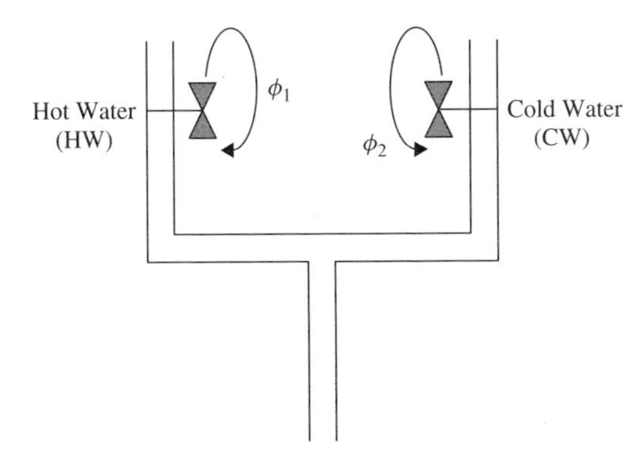

Figure 3.2 Water faucet baseline schematic.

In the baseline design, there are two valves (two design parameters) that need to be turned to deliver the two requirements noted above. To satisfy Theorem 2.4 (Section 2.5.2), we know that the number of DPs have to equal or exceed the number of FRs as a prerequisite for the independence axiom. In this case there are two DPs: DP_1 = valve ϕ_1 and DP_2 = valve ϕ_2.

The functional requirements are always independent by definition. In this case, flow is definitely a physical quantity that is different from (independent of) temperature. Therefore, by the independence axiom, DPs must be chosen to maintain independence between the FRs. The baseline faucet schematic in Figure 3.2 does not satisfy the independence axiom, as either DP_1 or DP_2 affect both FR_1 and FR_2. The design mapping for this system is given by

$$\begin{Bmatrix} FR_1 \\ FR_2 \end{Bmatrix} = \begin{bmatrix} X & X \\ X & X \end{bmatrix} \begin{Bmatrix} DP_1 \\ DP_2 \end{Bmatrix} \tag{3.6}$$

Equation (3.6) is revealing in that the baseline faucet is a coupled design. The design sought is uncoupled; that is, according to the independence axiom, an independent design with a design matrix where all the diagonal elements are X's and all the off-diagonal elements are 0's. A decoupled design is usually represented by a triangular design matrix. Uncoupled and decoupled designs are acceptable according to the independence axiom. Coupled designs violate the independence axiom.

Hierarchical Level 1 Analysis A valve can be introduced that controls the flow (Q). Additional DPs, the hot and cold water valves, have been connected such that a turn causes one valve to close as the other opens, therefore controlling the temperature (T). The design equation for this proposed design is given as

$$\begin{Bmatrix} FR_1 \\ FR_2 \end{Bmatrix} = \begin{bmatrix} X & 0 \\ 0 & X \end{bmatrix} \begin{Bmatrix} DP_1 \\ DP_2 \end{Bmatrix} \tag{3.7}$$

The design matrix in (3.7) is better than the baseline design since it is functionally uncoupled. However, according to Corollary 2.3 (Section 2.5.1), it is desirable to integrate the design in a single physical structure if the FRs can be satisfied independently. The aim is to identify an integration of the design parameters that would require the two valves.

The desired customer balance between hot and cold water can be achieved by moving a connecting rod that connects the two valves in the system where the design parameter for the temperature is the displacement D (Figure 3.3). The connecting rod is made with an adjustable length to control the flow by turning the two threaded ends of the connecting rod in opposite directions, ϕ; hence

$$\begin{Bmatrix} Q \\ T \end{Bmatrix} = \begin{bmatrix} X & 0 \\ 0 & X \end{bmatrix} \begin{Bmatrix} \phi \\ D \end{Bmatrix} \tag{3.8}$$

According to (3.8), design mapping, the flow, Q, is controlled by turning the rod ends an angle ϕ relative to each other, and the temperature, T, is controlled by the position of the rod, D. However, in this design it is important to ensure that the mechanism controlling ϕ moves the valve mechanism equally to avoid affecting the temperature. The hot and cold flow areas need to be detailed further in the connecting rod design.

Hierarchical Level 2 Analysis Let A_w and A_c be the flow areas of hot and cold areas, respectively, in the design shown in Figure 3.4. The figure indicates

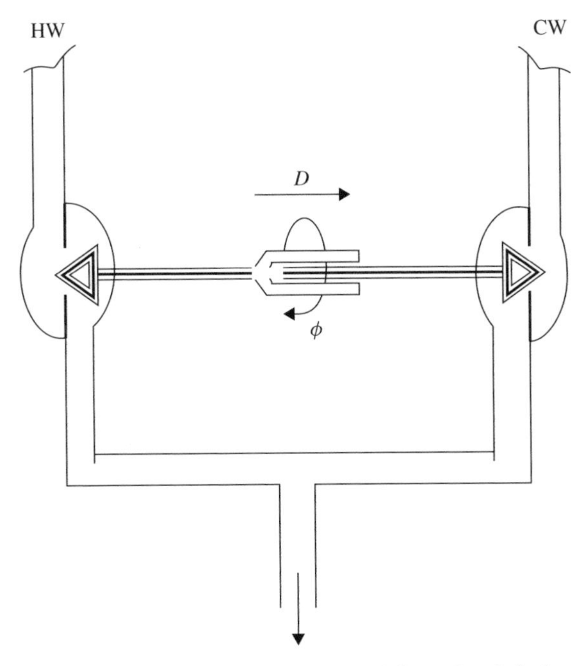

Figure 3.3 Proposed connecting rod for valve 2 design.

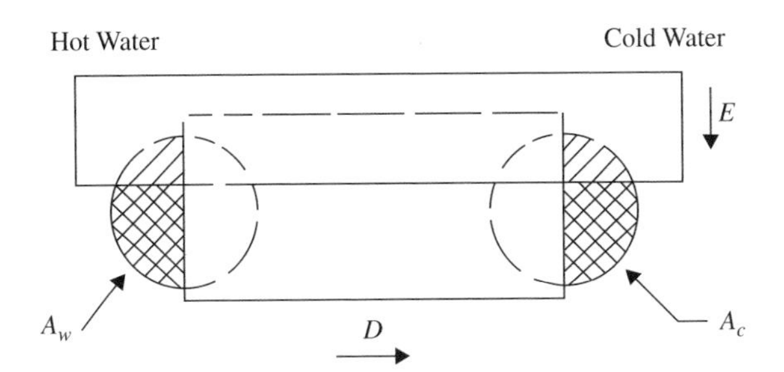

Figure 3.4 Proposed connecting of faucet design.

that flow and temperature control can be achieved when two plates are used that can be moved in a plane: hence, the equations

$$Q = f(A_w + A_c)$$
$$T = f\left(\frac{A_w}{A_c}\right) \tag{3.9}$$

That is, the flow is a function of the total area, and the temperature is a function of the ratio of the areas for hot and cold water. In moving the top plate along E we control the flow, and in moving the other plate along D we control the temperature. The resulting design matrix is a very good solution for a bathroom sink faucet:

$$\begin{Bmatrix} Q \\ T \end{Bmatrix} = \begin{bmatrix} X & 0 \\ 0 & X \end{bmatrix} \begin{Bmatrix} E \\ D \end{Bmatrix} \tag{3.10}$$

As applied to the water faucet, a modular subsystem with two moving parts has a lower probability of success than a system with one moving part. Assume that the probability of manufacturing success for a moving part is 0.99; then a system consisting of one moving part would have the information content $H_1 = -\ln(0.99) = 0.01$ nat, while a system of two parts would have an information content $H_2 = 0.02$ nat, assuming statistical independence by Theorem 2.12 (Section 2.5.2). It is obvious that $H_2 > H_1$ (i.e., is more complex).

Based on these assumptions about the probabilities of success, the authors set out to try to realize a design with one movable part that has two degrees of freedom of movement, which could be used to satisfy the two FRs. Axiomatic design does not provide any methodological support to integrate DPs in a way that maintains functional independence while minimizing information content. The design team has to rely on experience and analogies to do this part of the work. The design team has found that the theory of inventive problem solving (TRIZ) can be useful in coming up with uncoupled design solutions integrating DPs in a way that leads to lower information content. TRIZ is currently implemented in a number of different computer tools, such as TechOptimizer and Goldfire Innovator (both trademarks of Invention Machine Corporation). Using the *Effects* database of TRIZ, the authors generated the solution in Figure 3.5, which shows the principle for a design integrating design parameters in one physical part called the *Reuleaux triangle*. This triangle is formed from curves which, when rotated in a square, make contact with all four sides. A curve of constant width is constructed by drawing arcs from each polygon vertex of an equilateral triangle between the other two vertexes. The Reuleaux triangle has the smallest area for

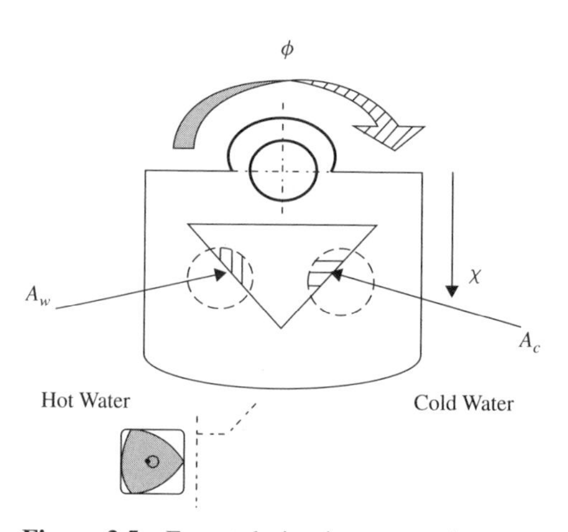

Figure 3.5 Faucet design in one moving part.

a given width of any curve of constant width. The area of each meniscus-shaped portion of the Reuleaux triangle is a circular segment with opening angle $\pi/3$.[3] The main useful feature in our case is that we have the same width for any angular position if the triangle. This feature determines the range of applications for objects having a Figure 3.5 profile.

Both functional requirements can be satisfied with the movable plate. The plate has one triangular hole that affects A_w and A_c. Turning the plate ϕ controls the temperature, while moving the entire plate along X controls the flow. The resulting design mapping is still uncoupled. This solution represents a designed valve where flow and temperature can be controlled independent of each other in one moving part.

This case study has demonstrated a general approach to improving a baseline design, where the design team always starts by asking the question "what do you want this system to do?" rather than "what can this system do?" The case also demonstrates use of the design axioms to identify the need for a new solution and to a certain extent, how the axiom and some of its associated theorems force the design team to look for certain types of solutions.

Finally, the information axiom advises the design team to look for simpler solutions with a higher probability of success while satisfying the independence axiom. In mechanical design the information content can be minimized through intelligent integration of the design parameters into a simple physical embodiment that is easy to manufacture and maintains functional independence (see Corollary 2.3 in Section 2.5.1).

3.5 CASE STUDY 2: IMPLEMENTATION METHODOLOGY FOR TRANSITION FROM TRADITIONAL TO CELLULAR MANUFACTURING USING AXIOMATIC DESIGN (DURMUSOGLU et al., 2002)

In this case study, a framework to transform a traditional production system from process orientation to cellular orientation based on axiomatic design principles is introduced. A feedback mechanism for continuous improvement is also suggested for evaluating and improving the cellular design against preselected performance criteria (Figure 3.6). The criteria, which are developed based on the independence axiom, provide necessary steps in transforming an existing process-oriented system into a cellular manufacturing system.

Transition to cellular manufacturing follows after all cellular manufacturing steps are completed successfully. At this stage, the production is achieved through a cellular manufacturing system. Databases and information for comparing system performance need to be generated with set target goals on some business metrics. Based on target values and achievements, new target values are established and appropriate system modifications and changes are affected through cellular manufacturing system improvement principles provided in the procedure

[3]See more details about the Reuleaux triangle at *http://www.mathworld.wolfram.com/Reuleaux-Triangle.html*.

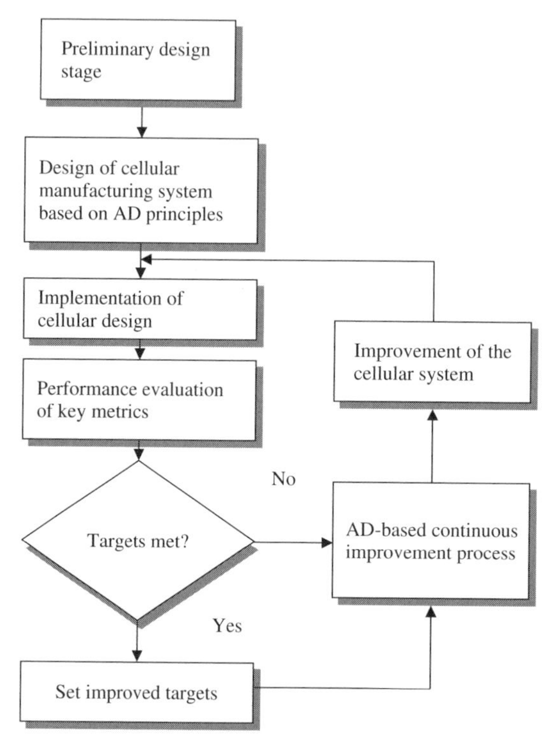

Figure 3.6 Cellular manufacturing design process with feedback mechanism.

proposed. These principles are also based on axiomatic design concepts. These continuous feedback and improvement principles are also in agreement with the spirit of lean manufacturing and *kaizen* activities. A complete functional requirement to design parameter mappings and hierarchy for the design of a cellular manufacturing system through axiomatic design principles is provided.

3.5.1 Axiomatically Driven Cellular Manufacturing System

The first step is to define the FRs of the system at the highest level of its hierarchy in the functional domain. At this stage, many FRs may be established. Depending on the functional definition, each FR established at this stage may lead to a completely different cellular manufacturing design. The authors have selected the following as the highest FR.

Hierarchical Level 1 Analysis

FR = provide a flexible production in line with customer needs

Customer needs are summarized in more product variety, smaller batch sizes with highest quality, and more frequent deliveries at lower costs. These requirements are forcing companies to reevaluate their classical manufacturing systems for more flexibility in response to these customer needs. The flexibility of a manufacturing system is measured by its speed and agility to respond to rapidly changing customer needs.

In step 2, mapping to the DP domain is next using the zigzagging approach. At this step, DPs that satisfy the FRs established in the preceding step are defined by zigging. To make the correct DP selection, the DPs corresponding to the FRs established before must be generated exhaustively. The following DP has been selected to satisfy the FR provided above: DP = cellular manufacturing system design. A production system that can answer customer needs in an efficient way through elimination of waste, reduction of lead time, and improved quality is a cellular manufacturing system designed with lean principles in mind.

Hierarchical Level 2 Analysis In step 3, the zigzagging process is continued for further hierarchical level prescription of the design and to demand further clarification. If the DPs proposed for satisfying the FRs defined earlier cannot be implemented without further clarification, the axiomatic design principles recommend returning to the functional domain (in a zagging step) for decomposing the FRs into their lower functional requirement set as follows:

FR_1 = classify and group products/components for simple material flow

FR_2 = define production strategy based on product specifications

FR_3 = rearrange resources to minimize waste

FR_4 = provide means to control production based on customer demand

In step 4 we need to find the corresponding DPs by mapping FRs in the physical domain (a zigging step). In satisfying the four FRs defined above, the authors move to the physical domain from the functional domain (a zigging step) and obtain

DP_1 = procedure for defining product families

DP_2 = procedure for selecting production strategy

DP_3 = product-oriented layout

DP_4 = pull production control system

The next logical step (step 5) is to determine a design matrix [**DM**] that provides relationships between the FR and DP mapping elements. It is critical to ensure that the [**DM**], as established, satisfies the independence axiom. The design equation is given as

$$\begin{Bmatrix} FR_1 \\ FR_2 \\ FR_3 \\ FR_4 \end{Bmatrix} = \begin{bmatrix} X & 0 & 0 & 0 \\ X & X & 0 & 0 \\ X & X & X & 0 \\ X & X & X & X \end{bmatrix} \begin{Bmatrix} DP_1 \\ DP_2 \\ DP_3 \\ DP_4 \end{Bmatrix} \tag{3.11}$$

A quick look reveals that the design is decoupled and thus satisfies the independence axiom. In the **[DM]** above, an X represents a strong relationship between the corresponding FR–DP pair; a 0 indicates the absence of such a relationship.

Hierarchical Level 3 Analysis In step 6, the zigzagging process continues with FR_1, FR_2, FR_3, and FR_4 by going from the physical to the functional domain again and determining the corresponding DPs.

Step 6a. FR_1: Products/Components Branch FR_1 as defined above (classify and group products/components for simple material flow) may be decomposed with DP_1 (procedure for defining product families) in mind as:

FR_{11} = determine high-volume products/components to group

FR_{12} = determine operations and machine types for producing each product family

FR_{13} = form product families

FR_{14} = determine the final number of machine groups

The corresponding DPs may be stated as

DP_{11} = product-quantity Pareto analysis

DP_{12} = machine-component incidence matrix

DP_{13} = product grouping techniques

DP_{14} = cost analysis and economic justification techniques

In this step, product families that will be manufactured economically through cellular manufacturing and their corresponding machine groups are determined by using Pareto analysis, followed by product family assignment techniques. The design matrix **[DM]** for the vectors of FRs and DPs above is

$$\begin{Bmatrix} FR_{11} \\ FR_{12} \\ FR_{13} \\ FR_{14} \end{Bmatrix} = \begin{bmatrix} X & 0 & 0 & 0 \\ X & X & 0 & 0 \\ 0 & X & X & 0 \\ 0 & X & X & X \end{bmatrix} \begin{Bmatrix} DP_{11} \\ DP_{12} \\ DP_{13} \\ DP_{14} \end{Bmatrix} \tag{3.12}$$

Once again, this is a decoupled design satisfying the independence axiom.

Step 6b. FR_2: Production Strategy Branch FR_2 as defined above (define production strategy based on product specifications) may be decomposed with DP_2 (procedure for selecting production strategy) in mind as

FR_{21} = determine the master process

FR_{22} = select the most appropriate process elements

FR_{23} = determine the training/education needs required

FR_{24} = motivate labor participation

The corresponding DPs may be stated as

DP_{21} = master process selection

DP_{22} = production resources selection procedure

DP_{23} = multipurpose labor training programs

DP_{24} = gain-sharing program

At this stage, production resources are determined following establishment of the master process based on product specifications. Once the resource selection is complete, the education and training requirements of the workers can be established. For ensuring the full participation of workers, appropriate gain-sharing programs must be established and announced to workers to seek their dedication and involvement. The decoupled design equation for this requirement is given as

$$\begin{Bmatrix} FR_{21} \\ FR_{22} \\ FR_{23} \\ FR_{24} \end{Bmatrix} = \begin{bmatrix} X & 0 & 0 & 0 \\ X & X & 0 & 0 \\ 0 & X & X & 0 \\ 0 & X & X & X \end{bmatrix} \begin{Bmatrix} DP_{21} \\ DP_{22} \\ DP_{23} \\ DP_{24} \end{Bmatrix} \tag{3.13}$$

Step 6c. FR_3: Resource Rearrangement Branch FR_3, defined above (rearrange resources to minimize waste) may be decomposed with DP_3 (product-oriented layout) in mind as

FR_{31} = minimize material handling

FR_{32} = eliminate wasted motion of operators

FR_{33} = minimize waste due to imbalance in the system

The corresponding DPs may be stated as

DP_{31} = material flow-oriented layout

DP_{32} = arrangement of stations to facilitate operator tasks

DP_{33} = balanced resources in response to Takt time

(Takt time = available time/demand)

Lean manufacturing principles are the guiding principles of this design step. In this step the focus is on waste elimination. Therefore, in rearranging the resource waste due to motion, material handling and imbalances between resources are minimized. Without this step, the cell designed will not provide the performance expected. Once again, the decoupled design equation is given as

$$\begin{Bmatrix} FR_{31} \\ FR_{32} \\ FR_{33} \end{Bmatrix} = \begin{bmatrix} X & 0 & 0 \\ X & X & 0 \\ X & X & X \end{bmatrix} \begin{Bmatrix} DP_{31} \\ DP_{32} \\ DP_{33} \end{Bmatrix} \tag{3.14}$$

Step 6d. FR_4: Production Control Branch FR_4 defined above (provide means to control production based on customer demand) may be decomposed with DP_4 (pull production control system) in mind as

FR_{41} = ensure smooth and steady production in assembly line

FR_{42} = provide material/information flow

FR_{43} = provide continuous feedback information flow

The corresponding DPs may be stated as

DP_{41} = leveled/mixed production

DP_{42} = card system (*kanban*)

DP_{43} = information/report system and visual management tools

Satisfying customers by providing the right amount, just in time, can only be accomplished through the pull system. However, just-in-time systems require a steady pull on all products in a family. To ensure a steady pull, a leveled/mixed production schedule must be established. This leads us into developing the appropriate Heijunka schedule and the necessary visual management tools, including the kanban system, for successful implementation. The design equation and matrices are as follows:

$$\begin{Bmatrix} FR_{41} \\ FR_{42} \\ FR_{43} \end{Bmatrix} = \begin{bmatrix} X & 0 & 0 \\ X & X & 0 \\ X & X & X \end{bmatrix} \begin{Bmatrix} DP_{41} \\ DP_{42} \\ DP_{43} \end{Bmatrix} \tag{3.15}$$

3.6 SUMMARY

Axiomatic design is a design methodology that takes the independence axiom, the subject of this chapter, and the information axiom (Chapter 4) as the basis for its axiomatic system. An axiomatic system dictates that the axioms be independent of each other, consistent, and complete (Section 3.1). Such properties are all satisfied in the axiomatic design method founded by Suh (1984).

In this chapter we stressed the mathematical formulation of the independence axiom to lay down the background deemed necessary for axiomatic quality process concepts and derivations. We explored the coupling vulnerability, a conceptual vulnerability that is created in the design entity when the independence axiom is violated. Coupling implies limited design controllability and governance for both the design teams and the end users, the customers. It will result in lower reliability and quality levels and makes the operational robustness hard to achieve. In addition, we presented the zigzagging method, the axiomatic design method used for design mapping, detailing, and requirements cascading. Several case studies were provided.

CHAPTER 4

INFORMATION AXIOM AND DESIGN COMPLEXITY

4.1 INTRODUCTION

The introduction of the axiomatic design method contributed to the advancement of design practices by directing design research toward more useful arenas. One of these arenas is concerned with elimination and minimization of vulnerabilities that are established in a design entity as a result of violation of design axioms. A major vulnerability is complexity, which can be addressed by the information axiom. In this chapter we identify three components of design complexity: sensitivity, variability, and correlation. We use information measures to quantify complexity and derive mathematical relationships that quantify these components within the context of the axiomatic quality process.

Historically, there have been many attempts in design research that span different arenas as part of the effort to understand natural phenomena. Design was considered as an art conducted by a person with some unexplained talents. Ironically, the history of design is very closely related to the history of science. The effort in design research had shifted to include, in addition to technical knowledge, streams of development for the purpose of enhancing the design process. A large body of research in the design arena has been published in German. Unfortunately, only a limited portion has been translated

Axiomatic Quality: Integrating Axiomatic Design with Six-Sigma, Reliability, and Quality Engineering, by Basem Said El-Haik
ISBN 0-471-68273-X

into English. Most of these efforts are listed in Hubka (1980), the German Guidelines VDI (Verein Deutscher Ingenieure, 1986), and Pahl and Beitz (1988). One major developmental stream is related to the attributes of the design entity itself as prescribed by design principles. The latest contribution in this stream is the axiomatic design method proposed by Suh (1984, 1990, 1995, 1996, 1997, 2001). A major concern of the principles arena is design vulnerability, which is induced in the solution entity when a principle-based criterion is violated. For example, the vulnerability of functional coupling will weaken design controllability and adjustability when the independence axiom cannot be satisfied (Chapter 3). The shortcomings of the current design entities can be overcome by the efficient deployment of basic design principles: in particular, those identified as axioms.

Information content is defined as a measure of complexity and is related to the probability of certain events occurring when information is supplied. According to the information axiom, the independent design that minimizes the information content is the *best* (see also Corollary 2.7, Section 2.5.1). However, the exact deployment of design axioms might not be feasible, due to technological and/or cost limitations. Under these circumstances, different degrees of conceptual vulnerability are established in the measures (criteria) related to the unsatisfied axioms. For example, a degree of design complexity may exist as a result of an information axiom violation. Such a vulnerable design entity may have questionable quality and reliability performance even after thorough operational optimization. Quality and reliability improvements in weak conceptual entities usually produce marginal results. Before such efforts are made, conceptual vulnerability should be reduced, if not eliminated. Indeed, the presence of functional coupling and complexity vulnerabilities aggravates the symptomatic behavior of design solution entities. Coupling measures are functions of the sensitivities, the partial derivatives of the FRs with respect to the mapped-to DPs (see Chapter 3). Our consideration of Theorem 4.1 led to the quantification of two components of complexity: vulnerability and variability. *Complexity due to vulnerability* is related to coupling, since both take sensitivity coefficients as arguments, whereas *complexity due to variability* deals with the inherent variability of the DPs or PVs and their correlation. The presence of correlation increases the degree of total design complexity and must be minimized. Correlation can be classified as the third component of complexity, in addition to the components identified by Theorem 4.1, and is quantified for normal sources of complexity in Theorem 4.2. To reduce complexity, we need to attack the three components altogether.

The objectives of this chapter are (1) to present axiomatic design traditional information concepts and measures as defined by Suh (2001) and (2) to explore a new theory of design complexity developed by the author. The new theory distinguishes between several components of complexity in engineering design of a statistical nature and derives mathematical relationships to quantify them. The reader is encouraged to read Chapter 5 to explore additional relationships between complexity and robustness measures.

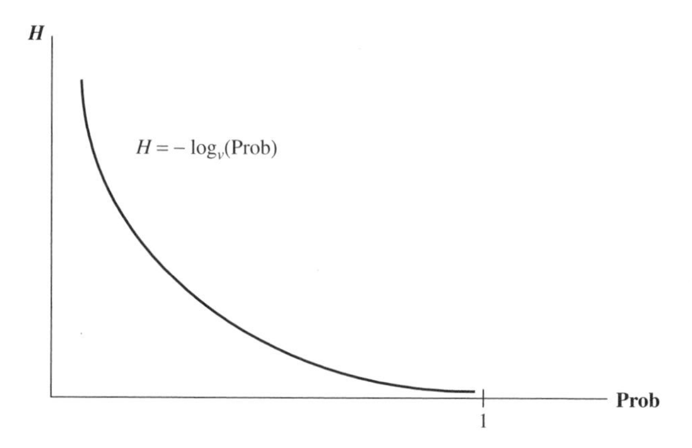

Figure 4.1 Entropy (information content) as a function of probability.

4.2 TRADITIONAL FORMULATION OF THE INFORMATION AXIOM: SUH'S DEFINITION

The information axiom is stated:

The Information Axiom Minimize the information content in a design.

As discussed in Chapter 2, this axiom provides a selection role based on design information content. That is, among the design solutions that are equally acceptable in the light of the independence axiom, the one with the highest probability of manufacturing (production) success is the *best*. The highest probability of success indicates the lowest amount of information needed to manufacture (produce) the design. Information and probability are tied together via entropy, H. Probability here refers to the probability of satisfying a given FR in the process domain. Entropy is defined as $H = -\log_{\nu}(\text{Prob})$[1] in (2.4), taking the form of Shannon entropy (1948) of a discrete random variable supplying the information (Figure 4.1). Ideally, the information content should be zero; that is, the probability is unity. A design solution entity is characterized as complex when the probability of success is low due to the lower probability levels associated with satisfying the design FRs. Complex design solutions are much more difficult to manufacture and involve complicated manufacturing systems, processes, and procedures. Complexity also results in additional operational costs and worsened quality levels.

Since the PVs are random variables, the DPs are random variables as dictated by the process mapping **[B]**. The same arguments can be extended to the FRs. Therefore, the FR vectors at any hierarchical level are random variables and need to be delivered within some acceptable tolerance as determined by customers. We use the notation $T \pm \Delta\text{FR}$ to indicate the target value T and two-sided

[1] $\nu = 2(e)^1$; H is measured in bits (nats).

half-tolerance width ΔFR around the target, both specified by customers. The FRs variation can be induced by several noise sources (such as manufacturing process variation, customer usage, and degradation over time) through design mappings. Some noise factors can be identified, isolated, and even eliminated, but others cannot. The ability of a low-cost design to work as intended regardless of uncontrollable outside influences is called *robust design* (Taguchi, 1986).

The information axiom deals with design information content, which is a function of the number of FRs and DPs (solution size) and their inherent and correlated variation. Shannon entropy can be used to quantify information content. Shannon (1948) proved that as long as the communication rate was below channel capacity, the probability of error should not increase. Through his study on random processes, Shannon defined a level of complexity, called the *entropy*, below which the signal cannot be compressed. The principle of entropy was generalized and extended to many disciplines and used as a measure of uncertainty. Entropy takes the probability as an argument in its logarithmic form. Suh (1990) proposed the mathematical form of Shannon entropy as a measure of complexity in the context of the information axiom and defined the probability of success as the probability of meeting the specifications (Section 2.4).

Assuming statistical independence, the overall (total) design information of a given hierarchical level, L, with FRs being a vector of size m, is additive since the probability of design success is the multiplication of individual FRs' probability of success belonging to that level. That is,

$$H_L = -\log_v\left(\prod_{i=1}^{m} \text{Prob}_i\right) = -\sum_{i=1}^{m} \log_v(\text{Prob}_i)$$

In the absence of statistical independence, the probability of success is conditional, not multiplicative.

In Chapter 2 we learned that information is related to tolerances and process capabilities. Suh (1990, 2001) defined the probability of success as the probability of meeting design specifications, the area of intersection between the *design range* (DR; the voice of the customer) and the *system range* (SR; the voice of the process) (see Figure 4.2). The overlap between design range and system range is called the *common range*, CR. The probability of success is defined as the area under the pdf curve of the ratio common range to system range, CR/SR, yielding $H = \log_v$ (SR/CR) in (2.5).

4.2.1 Complexity Reduction Techniques

Several techniques were suggested by Suh (2001) to reduce information content and therefore design complexity. These are summarized below.

Bias Elimination The term *bias* is defined as the difference between the mean of an FR, μ_{FR}, in the system range distribution and the target value T defined by the customer, as depicted in Figure 4.2. For a uni-FR system or uncoupled

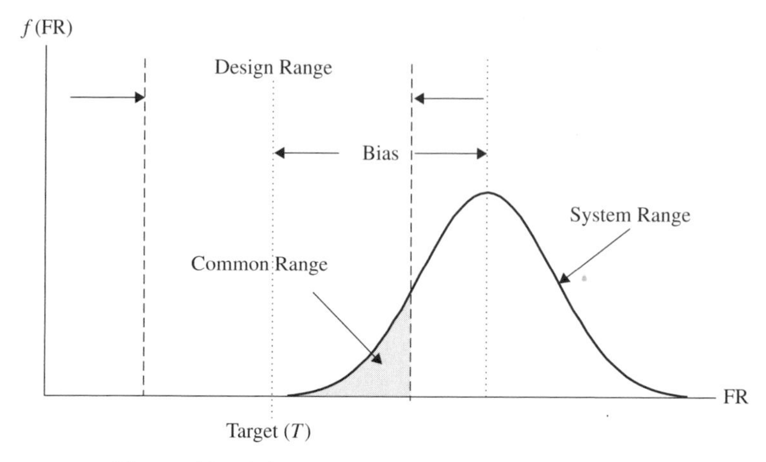

Figure 4.2 Suh's definition of probability of success.

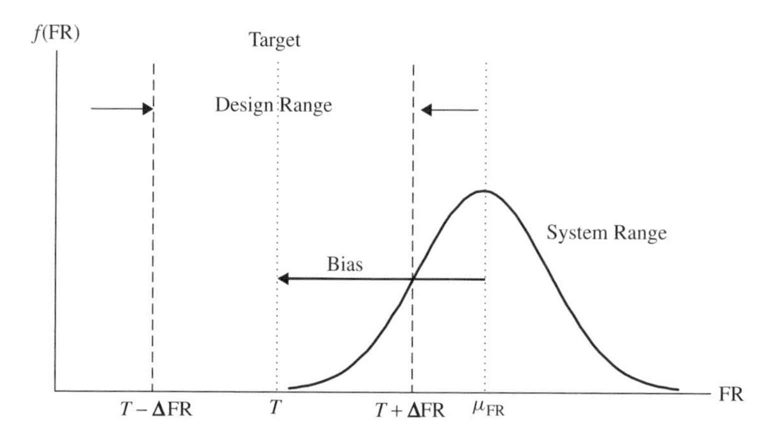

Figure 4.3 Bias elimination technique.

design, the bias elimination can be done by changing the appropriate DP mean value to set the FR to target value (Figure 4.3) in the spirit of the mapping **[A]** expressed by a mapping (transfer function) (Figure 1.5). For multiple FRs, the bias may not be eliminated unless the design satisfies the independence axiom. That is, in coupled design, each time a DP is changed to eliminate the bias for a certain FR, the bias for other FRs cannot be avoided. In a decoupled design case, the bias elimination must be done in the right sequence, as revealed by the design matrixes in the concerned design hierarchical level.

This technique has a very fundamental link to the robust design method through the quality loss function concept introduced in Chapter 1 and is explored further in Chapter 5. The quality loss, denoted as $L(\cdot)$, of an FR has two components—the mean (μ_{FR}) deviation from the targeted performance value (T), the bias, and the variability (σ_{FR}^2)—and can be approximated by a quadratic polynomial of the functional requirement system and design range parameters in the form

$$L = K[\sigma_{FR}^2 + (\mu_{FR} - T_{FR})^2] = K[\sigma_{FR}^2 + (\text{bias})^2]$$

as in (5.2), where K is a constant determined by loss associated with functional specification limits. This is a symmetrical quality loss function (QLF) because it is assumed that there is a constant K for the entire loss function. The value of K determines the slope of the QLF: The larger the value of K, the steeper the parabola. In addition, the presence of a target value T implies a nominal-the-best functional requirement classification (Chapter 5). Other quality loss forms are cited in Chapter 5. For example, given the following parameters of a current-regulating design (FR = current in amperes): $K = \$500$, $T \pm \Delta\text{FR} = 10.00 \pm 0.04$ A. Assuming a product with a system range average value μ_{FR} of 10.2 mm and a variance $\sigma_{FR}^2 = 0.1\ \text{A}^2$, we get $L(\text{FR}, T) = \$500[(10.2 - 10)^2 + (0.1)] = \$70$.

Variance Reduction This is accomplished by making the design more robust (i.e., immune to variation) through DP settings. The concepts introduced in Section 1.5 and several techniques in Chapter 9 are aggressive methods used to accomplish this task. Assuming a single-requirement system with the linear mapping (transfer function) $\text{FR}_1 = A_{11} \times \text{DP}_1$, the smaller the "stiffness" (the magnitude of A_{11}), the larger the allowable tolerance on DP_1. The design in FR_1 is more robust against DP variability, with less stiffness resulting in less complexity, as depicted in Figure 4.4.

In contrast to the stiffness (sensitivity) technique, variance reduction, robustness, and complexity minimization can be achieved by leveraging transfer function (mapping) nonlinearity, if any. Assuming a nonlinear function $\text{FR}_1 = f(\text{DP}_1, \text{DP}_2, \ldots)$ in DP_1, for example, the task is to find a "design window" with a large allowable tolerance on DP_1, for example. The design is more robust against the DP_1 variability in this case, as presented in Figure 4.5. It is

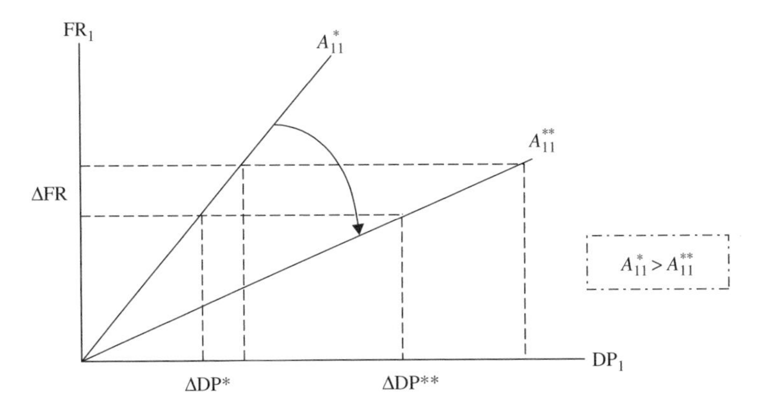

Figure 4.4 Variance reduction technique of linear FR.

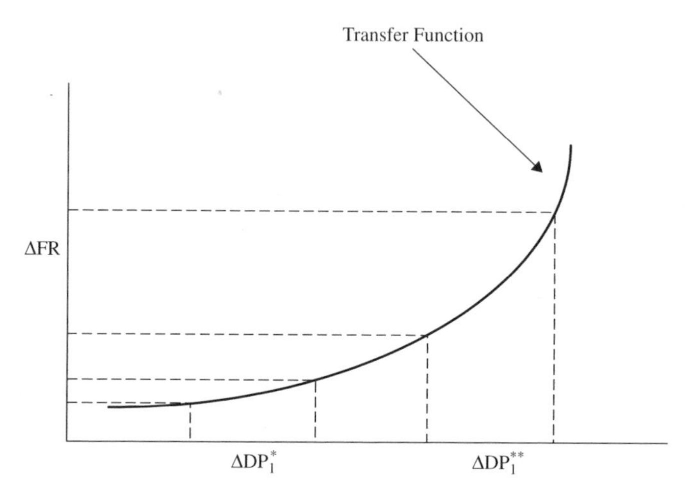

Figure 4.5 Variance reduction technique of nonlinear FR.

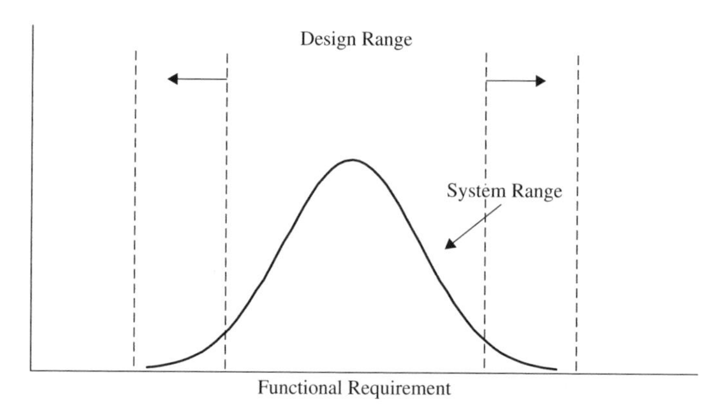

Figure 4.6 Technique for increasing design range.

obvious that the setting DP_1^* is more robust than DP_1^{**} by the amount of variation transferred through the mapping, expressed here as a nonlinear transfer function in DP_1. This discussion emphasizes the concepts expressed in Section 1.3.

One can leverage interaction of the DPs (also known as control factors at lower design hierarchical levels) and noise factors to reduce the variance of an FR. The use of heteroscedacticity is another option. In this case the standard deviation of the distribution is a function of the mean, so can move the mean and lower the standard deviation.

Design Range Widening (Figure 4.6) This technique is a direct application of Corollary 2.6 (Section 2.5). In some special cases, the design range can be widened without jeopardizing the overall design goal. A robust design technique that can accomplish this task is Taguchi's tolerance design, where a balance between the design range and the system range is achieved through the overall cost, to enhance the ability to limit the variability from the target (T) values obtained by transfer function optimization. Some sensitivity analysis needs to be conducted with the customer to check the validity of this technique prior to adoption.

Fixing Extra DPs In a redundant design (more DPs than FRs, or $p > m$), the variance can be reduced by reducing the variation associated with the extra DPs, that is, fixing the level of the extra DPs by a noise factor treatment such as a control mechanism (feedback loops, open control loops) or robust parameter and tolerance design. Choose to fix the DPs such that the remaining design matrix provides an ideal uncoupled design that will minimize variance. For example, in Figure 4.7, the variance of FR_1 can be written as

$$\sigma_{\text{FR}_1}^2 = \sum_{j=1}^{6} A_{1j}^2 \sigma_{\text{DP}_j}^2 + 2 \sum_{j=1}^{6} \sum_{l=1}^{j-1} A_{1j} A_{1l} \operatorname{cov}(\text{DP}_j, \text{DP}_l)$$

When DP_4, DP_5, and DP_6 are fixed, the variance term drops to

$$\sigma_{\text{FR}_1}^2 = \sum_{j=1}^{3} A_{1j}^2 \sigma_{\text{DP}_j}^2 + 2 \sum_{j=1}^{3} \sum_{l=1}^{j-1} A_{1j} A_{1l} \operatorname{cov}(\text{DP}_j, \text{DP}_l)$$

which is definitely less than or equal to the starting variance prior to the "fixing" task. The term "cov" denotes the statistical covariance operator.

Use of Axiomatic Design Theorems and Corollaries The following relevant theorems and corollaries are selected from Section 2.5.1.

Corollary 2.2: Minimization of FRs Minimize the number of FRs and constraints.

- *Advantage*: Reducing the number of FRs results in design simplification and hence complexity reduction.

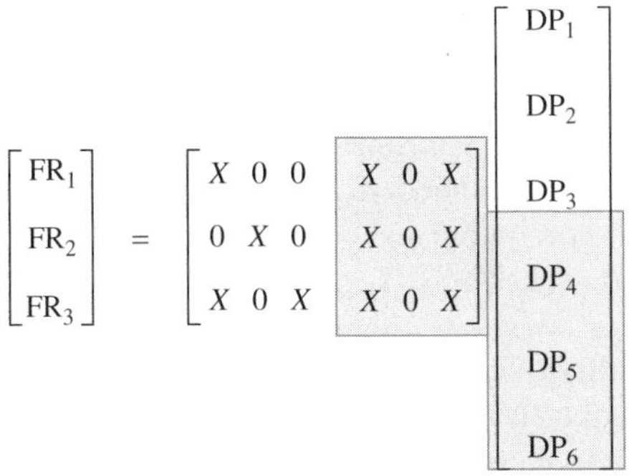

$$\begin{bmatrix} \text{FR}_1 \\ \text{FR}_2 \\ \text{FR}_3 \end{bmatrix} = \left[\begin{array}{ccc|ccc} X & 0 & 0 & X & 0 & X \\ 0 & X & 0 & X & 0 & X \\ X & 0 & X & X & 0 & X \end{array}\right] \begin{bmatrix} \text{DP}_1 \\ \text{DP}_2 \\ \text{DP}_3 \\ \text{DP}_4 \\ \text{DP}_5 \\ \text{DP}_6 \end{bmatrix}$$

Figure 4.7 Technique for fixing extra DPs.

Corollary 2.5: Use of Symmetry Use symmetrical shapes and/or arrangements if they are consistent with the FRs and constraints.

- *Advantage*: Symmetry is created by replicating a design parameter (features) on the same part. Standardized features within a part have positive effects on the design similar to those created by reusing standardized parts in assembled systems. They reduce the information content of the product. Using standard reusable parts boosts these advantages further.

Corollary 2.6: Largest Design Tolerance Specify the largest allowable tolerance in stating the FRs.

- *Advantage*: Reduce the information content and therefore design complexity by using the largest tolerances possible.

Corollary 2.7: Uncoupled Design with Less Information Seek an uncoupled design that requires less information than a coupled design in satisfying a set of FRs.

- *Advantage*: When possible, the designer should strive to minimize information and interdependence between design components.

Theorem 2.12: Sum of Information The sum of information for a set of events is also information, provided that proper conditional probabilities are used when the events are not statistically independent.

- Always seek to reduce the sum of information.

Theorem 2.13: Information Content of the Total System If each DP is probabilistically independent of other DPs, the information content of the total system is the sum of the information of all individual events associated with the set of FRs that must be satisfied.

- Reduce information content (a special case of Theorem 2.12).

Other helpful theorems:

Theorem 2.14: Information Content of Coupled Versus Uncoupled Designs When the state of FRs is changed from one state to another in the functional domain, the information required for the change is greater for a coupled process than for an uncoupled process.

Theorem 2.16: Equality of Information Content All information content that is relevant to the design task is equally important regardless of its physical origin, and no weighting factor should be applied.

4.3 COMPLEXITY VULNERABILITY

Complexity in design has many facets, including the lack of transparency of the transfer functions (mappings) between inputs and outputs in the design structures, the difficulty of the physical attributes employed, and the relatively large number of assemblies and components involved (Pahl and Beitz, 1988). The term *complexity* is used in most literature in a pragmatic sense. It is easier to have an idea about complexity by shaping where it does exist and how it affects us rather than what it really means. Linguistically, *complexity* is defined as a quality of an object with many interwoven elements, aspects, details, or attributes that makes the entire object difficult to understand in a collective sense. Complexity is a universal attribute that exists, to some degree in all objects. The severity of complexity varies according to the number of phenomena in the object that are explored. Ashby (1973) defines *complexity* as "the quantity of information required to describe the vital system." Simon (1981) defines a *complex system* as an object that "is made up of a large number of parts that interact in a nonsimple way." These definitions imply different levels of communication between the interrelated elements of an interactive *hierarchical system*. Simon (1981) illustrated that *hierarchy* reflects some level of communication or interaction between related entities. In a physical system, the higher the level of interaction, the shorter is the relative spatial propinquity and the higher the complexity.

In a seminal paper, Weaver (1948) distinguished between two types of complexity: disorganized and organized. Systems of *disorganized complexity* are characterized by a huge number of variables. The effects of these variables and their interaction can only be explained by random and stochastic processes using empirical statistical methods rather than by any deterministic approach. The objective is to describe the system in an aggregate and average sense. Statistical mechanics is a good example of a discipline that addresses this type of complexity. Analytical approaches work well in the case of *organized simplicity*, where systems are characterized by a small number of significant variables that are tied together in deterministic relationships (transfer functions). Weak variables may exist but have little bearing in explaining the phenomena. Organized simplicity is the extreme of the complexity spectrum at the lower end. This discussion is beneficial because it allows the classification of problems in which axiomatic quality is effective (see Chapter 8 for more details).

It is safe to say that most design problems often belong to a separate standalone category placed in between the two extremes, called *organized complexity*. This category of problem may have solutions that utilize statistical and deterministic–analytical methods at different development stages. Design problems are more susceptible to analytical, nonempirical approaches in the early development stages and to statistical methods in the optimization phases, due to the unanticipated effect of the noise factors. Organized complexity suggests the utilization of a new paradigm for simplification that makes use of information and complexity measures.

The amount of manufacturing information involved is also a measure of the level of design complexity. In the context of design, there is a component of complexity that is judged by the level of obedience of the manufacturing operations to the specifications required (i.e., complexity due to capability). Another component of complexity can be attributed to the correct selection of machining processes that fit the design entity (i.e., complexity due to compatibility). Compatibility is concerned with engineering and scientific knowledge. The selection of incapable process and/or equipment to attain a certain DP will increase the complexity encountered in delivering the design entity. Compatibility is the essence of product engineering, along with manufacturing system engineering and material science—both of which are beyond the scope of this chapter.

In the context of this book, complexity in physical design entities is related to the information required to meet each FR, which in turn is a function of the information content of the DPs and PVs, through a system of design mappings. The ability to satisfy an FR is a function of associated DP tolerances and PV machine capabilities, since designs do not always achieve the performance targets. Thus, a FR should be met with tolerance [i.e., FR $\in (T \pm \Delta\text{FR})$]. The amount of complexity encountered in an FR is related to the probability of successful manufacture of its mapped-to DPs. Since probability is related to complexity, the use of an entropy information measure as a measure of complexity is more than justified.

4.4 THEORETICAL FOUNDATION OF THE NEW COMPLEXITY THEORY

The introduction of the entropy principle was the origin of information theory (Shannon, 1948). The original concept of entropy was introduced in the nineteenth century in the context of heat theory (Carnap, 1977). Clausius used entropy as a measure of the disorganization of a system. The first fundamental form was developed by Boltzmann in 1896 in his work in the theory of ideal gases, when he developed a connection between the macroscopic property of entropy and the microscopic state of a system. The concept of entropy is used in thermodynamics to supplement the second law of thermodynamics.

Hartley (1928) introduced a logarithmic measure of information in the context of communication theory. Hartley, and later Shannon (1948), introduced their measure for the purpose of measuring information in terms of uncertainty. Hartley's information measure is essentially the logarithm of the cardinality or source alphabet size (Definition 4.1); Shannon formulated his measure in terms of probability theory. Both measures are information content measures, and hence are measures of complexity.

Hartley's information measure (Hartley, 1928) can be used to explore the concepts of information and uncertainty in a mathematical framework. Let X be a finite set with a cardinality $|X|$, where the cardinality of a set is the number of elements contained in the set. A sequence can be generated from set X by successive selection of its elements. Once a selection is made, all but one of the possible elements that might have been chosen are eliminated. Before a selection is made, ambiguity is experienced. The level of ambiguity is proportional to the number of alternatives available. Once a selection is made, no ambiguity is sustained. Thus, the amount of information obtained can be equated to the amount of ambiguity eliminated. Hartley's information measure I is given by $I = \log_2 |X|_s$ (bits), where s is the sequence of selection. The conclusion is that the amount of uncertainty needed to resolve a situation or the amount of complexity to be reduced in a design problem are equivalent to the potential information involved. A reduction of information of I bits represent a reduction in complexity or uncertainty of I bits.

Definition 4.1 A source of information is an ordered pair $= (X, P)$, where $X = \{x_i\}$ is a finite set, known as a *source alphabet*, and P is a probability distribution on X. We denote the probability of x_i by Prob_i.

The elements of set X provide specific representations in a given context. For example, X may represent the set of all possible tolerance subintervals of a certain DP. The association of set X with probabilities suggests consideration of a discrete random variable as a source of information. It conveys information about the variability of its behavior around some central tendency. Suppose that we select at random an arbitrary element of X, say x_i, with probability Prob_i. Before the sampling occurs, there is a certain amount of uncertainty associated with the outcome. However, an equivalent amount of information is gained about the source after sampling, and therefore, uncertainty and information are related. If $X = \{x_1\}$, there is no uncertainty and thus no information gained. At the other extreme, maximum uncertainty occurs when the alphabets carry equal probabilities of being chosen. In this situation, maximum information is gained by sampling. This amount of information reveals the maximum uncertainty that preceded the sampling process.

According to Definition 4.1, DPs, PVs, and FRs with discrete support are sources of information, and Shannon entropy for the FRs can be written as

$$H_b(\text{Prob}_1, \text{Prob}_2, \ldots, \text{Prob}_m) = -\sum_{i=1}^{m} \text{Prob}_i \log_v(\text{Prob}_i) \tag{4.1}$$

where $v > 1$. The function H is called v-ary *entropy*. If $v = 2e$, H has the units of bits (nats), respectively (a nat $= 1.44$ bits). Also, when $\text{Prob} = 0$, the product $\text{Prob} \log_v(\text{Prob}) = 0$.

Shannon entropy is a measure for a discrete source of information, a discrete random variable, and can be used as a complexity measure when the argument Prob_i is defined as the probability of success. In the case of a continuous information source (a continuous random variable), X, with $f(x)$ as a probability density function (pdf), the Boltzmann information measure, $h(f)$, can be used and is defined as

$$h_v(f) = -\int_S f(x) \log_v f(x)\, dx \qquad \text{if the integral exist} \tag{4.2}$$

where S is the support set [i.e., $S = \{x/f(x) \geq 0\}$] of the random variable.

The information content is a measure of the amount of complexity encountered in achieving an FR and takes the probability of successful manufacture of its mapped-to DPs as arguments. The probability distribution implies manufacturing variation and machine capability assessment and indicates that an FR is always associated with a tolerance.

4.5 NEW COMPLEXITY THEORY

The interpretation of *Shannon entropy* is as follows: When the probabilities are small, we are surprised by an event happening; we are uncertain if rare events will happen, and thus their occurrences carry considerable amounts of information. Therefore, we should expect entropy to decrease with an increase in probabilities. Shannon entropy is the expected value of the function $\log_{\upsilon}$ (1/Prob) of a discrete information source. Boltzmann entropy, on the other hand, may be used for continuous information sources which are the cases encountered most frequently. It has an appealing mathematical form that may be considered the continuous analogy to Shannon's entropy when Prob_i is replaced with the pdf $f(\cdot)$. However, there are two major issues in adopting Boltzmann entropy as a complexity measure: Shannon entropy does not converge to a Boltzmann measure, and for some pdf's there is no closed-form integral. The first issue can be reconciled by employing Boltzmann entropy in a differential sense coupled with discretization schemes (Chapter 9), and the second may be solved by approximation.

By employing the concept of Boltzmann entropy, components of complexity can be identified. For example, let FR be a normal source of information, $\text{FR} \sim \text{Normal}(\mu_{\text{FR}}, \sigma_{\text{FR}}^2)$ and $f(\text{FR}) = (1/\sqrt{2\pi\sigma_{\text{FR}}^2})e^{-(\text{FR}-\mu_{\text{FR}})^2/2\sigma_{\text{FR}}^2}$. The complexity in the interval $[\mu_{\text{FR}} - \Delta\text{FR}, \mu_{\text{FR}} + \Delta\text{FR}]$ is given by

$$\begin{aligned} h(f) &= -\int_{\mu-\Delta\text{FR}}^{\mu+\Delta\text{FR}} f(\text{FR}) \ln f(\text{FR})\, d\text{FR} \\ &= -\int_{\mu-\Delta\text{FR}}^{\mu+\Delta\text{FR}} f(\text{FR}) \left[\frac{-(\text{FR}-\mu_{\text{FR}})^2}{2\sigma_{\text{FR}}^2} - \ln\sqrt{2\pi\sigma_{\text{FR}}^2} \right] d\text{FR} \\ &= \ln\sqrt{2\pi e \sigma_{\text{FR}}^2} \qquad \text{nats} \end{aligned} \tag{4.3}$$

Equation (4.3) indicates that in the case of a normal source of information, complexity is a function of variability. Therefore, variability is a component of complexity. A reduction in the variance will reduce not only the probability of manufacturing nonconfirming parts, but also the information required to manufacture the part. The reader may recall that variation reduction is a technique suggested to reduce information (Section 4.2.1). However, variability is not the only component of complexity. In fact, sensitivity adds to complexity according to the following theorem:

Theorem 4.1 Complexity of a design has two premier components: variability and coupling. The total design complexity of a linear design is given by

$$h(\{\mathbf{FR}\}) = h(\{\mathbf{DP}\}) + \ln|[\mathbf{A}']| \tag{4.4}$$

where $|[\mathbf{A}]|$ is the determinant of the nonsingular sensitivity matrix $\mathbf{A}'$, the Jacobian.

Proof Assume the case of an unifunctional requirement in which $\{\mathbf{FR}\} = [\mathbf{A}]\{\mathbf{DP}\}$, with $f(\text{DP})$ as the pdf. Then $f_{\text{FR}}(\text{FR}) = (1/|A|) f_{\text{DP}}(\text{FR}/A)$. Using a Boltzmann measure over any given interval of interest,

$$\begin{aligned} h(\mathbf{FR}) &= -\int f_{\text{FR}}(\mathbf{FR}) \ln f_{\text{FR}}(\mathbf{FR})\, d\mathbf{FR} \\ &= -\int \left[\frac{1}{|A'|} f_{\text{DP}}(\mathbf{A}^{-1}\mathbf{FR}) \ln \frac{1}{|A'|} f_{\text{DP}}(\mathbf{A}^{-1}\mathbf{FR})\right] d\text{DP} \\ &= -\int \left[f_{\text{DP}}(\mathbf{DP}) \ln f_{\text{DP}}(\mathbf{DP})\, d\mathbf{DP}\right] + \ln|\mathbf{A}'| \\ &= h(\mathbf{DP}) + \ln|[\mathbf{A}']| \end{aligned}$$

There are two components of FR complexity that can be identified in (4.4). The first component is due to variability induced by the DPs$[= h(\text{DP})]$, and the second is due to sensitivity $(= \ln|[\mathbf{A}']|)$. The sensitivity complexity component of Theorem 4.1 has a broader meaning than the numerical values of the sensitivity coefficients, the arguments of the sensitivity matrix determinant. Three ingredients collectively make the sensitivity–complexity component: mapping, additivity, and dimension. The mapping ingredient refers to the binary variable Y_{ij}, denoting the mapping process between the functional domain and the physical domain and is defined as $Y_{ij} = 1$ if $\text{FR}_i \rightarrow \text{DP}_j$ and 0 otherwise. In other words, the mapping variable represents the position of the nonzero sensitivity coefficients in the design matrix $\mathbf{A}$. The additivity ingredient refers to the sign of nonzero $A'_{ij} = \partial\text{FR}_i/\partial\text{DP}_j$ elements. The dimension ingredient refers to the size of the design problem (i.e., the number of the FRs, m). We view our interpretation of the complexity component due to sensitivity as the mathematical translation of Simon's (1981) complexity definition.

The theme of Theorem 4.1 is that the design team experiences two complexity components in attaining an FR (in the physical mapping) or a DP (in the process mapping) if they do not know: how its mapped-to variables vary (the variability component) and at what scale (the sensitivity component). For an uncoupled design, the value of $|[\mathbf{A}]|$ is a product of the diagonal elements, $|[\mathbf{A}']| = \prod_{i=1}^{p} A'_{ii}$, and the complexity component due to sensitivity is $\sum_{i=1}^{p} \ln|A'_{ii}|$. The total independent design complexity (assuming that all DPs are normal information sources) equals $\sum_{i=1}^{p} \ln(\sqrt{2\pi e}\sigma_i^* A'_{ii})$ nats.

The procedure used to prove Theorem 4.1 can be extended to the case of an uncoupled nonlinear design. Assume that $\text{FR} = q(\text{DP})$ is a physical mapping (transfer function) with the DP having the pdf $f(\mathrm{DP})$. The first step is to derive the pdf of the random variable FR. We may use $f(\mathrm{FR}) = \sum_{c=1}^{C} \left| \frac{dq^{-1}}{d\mathrm{FR}} \right|_c f(q^{-1}(\mathrm{FR}))$ to derive the pdf of the FR (Bowker and Lieberman 1959), where q^{-1} is the inverse function of q. The absolute value of the inverse function derivatives is taken with respect to FR and has C terms, the roots of $q^{-1}(\mathrm{FR}) = \mathrm{DP}$: for example, the power (P) in a 1-Ωresistor $= I^2$, where I is the current with pdf $f(I)$ [i.e., $P = q(I) = I^2$]. Using Bowker's formula, we have

$$f(P^*) = \begin{cases} \frac{1}{2\sqrt{P^*}} \left[f\left(\sqrt{P^*}\right) + f\left(-\sqrt{P^*}\right) \right] & P^* > 0 \\ 0 & P^* < 0 \end{cases}$$

where P^* is the generic variable of P. Once $f(P)$ is obtained, we can substitute for it in the Boltzmann entropy equation to obtain a closed form for complexity, if an integral exists.

Corollary 4.1 For independent process mapping $h(\{\mathbf{DP}\}) = h(\{\mathbf{PV}\}) + \ln|[\mathbf{B}']|$. Then, by substitution in (4.4), the total design complexity is given by

$$\begin{aligned} h(\{\mathbf{FR}\}) &= h(\{\mathbf{DP}\}) + \ln|[\mathbf{A}']| \\ &= h(\{\mathbf{PV}\}) + \ln|[\mathbf{B}']| + \ln|[\mathbf{A}']| \\ &= h(\{\mathbf{PV}\}) + \ln|[\mathbf{B}'][\mathbf{A}']| \\ &= h(\{\mathbf{PV}\}) + \ln|[\mathbf{C}']| \end{aligned} \tag{4.5}$$

where $[\mathbf{C}] = [\mathbf{A}][\mathbf{B}]$, the overall design matrix.

4.5.1 Coupled Design Complexity

Uncoupled solution entities are rarely found in the current development processes yield, due primarily to the late conception of design axioms or their inefficient deployment. We would expect coupled entities to be the majority, with occasional decoupled incidents. Consequently, it would be logical to extend Theorem 4.1 to the case of coupled design. The challenge in this case would be in obtaining the joint probability of functions of random variables. Basically, let $m = p$ and $\{\mathbf{FR}\}_{p \times 1}$ be the vector of FRs with p components, which might also considered as functions of p jointly distributed continuous random variables $\mathrm{DP}_1, \mathrm{DP}_2, \ldots, \mathrm{DP}_p$:[2]

$$\begin{aligned} \mathrm{FR}_1 &= q_1(\mathrm{DP}_1, \mathrm{DP}_2, \ldots, \mathrm{DP}_p) \\ \mathrm{FR}_2 &= q_2(\mathrm{DP}_1, \mathrm{DP}_2, \ldots, \mathrm{DP}_p) \\ &\vdots \\ \mathrm{FR}_p &= q_p(\mathrm{DP}_1, \mathrm{DP}_2, \ldots, \mathrm{DP}_p) \end{aligned} \tag{4.6}$$

[2]This is the transfer function format of the mapping $\{\mathbf{FR}\} = [\mathbf{A}]\{\mathbf{DP}\}$.

The requirements placed on the transfer function q (or mapping **[A]**) is that the Jacobian is nonzero, which dictates the existence of continuous first-order derivatives in the design space defined by the tolerance ranges of the DPs. If the DPs are jointly distributed continuous random variables with continuous pdf, $f_{\mathrm{DP}_1,\mathrm{DP}_2,\ldots,\mathrm{DP}_p}(\mathrm{dp}_1, \mathrm{dp}_2, \ldots, \mathrm{dp}_p)$, the random variables FR_1, FR_2, ..., FR_p in (4.6) are jointly distributed with the following continuous pdf (see Theorem 11.1 for a proof):

$$f_{\mathrm{FR}_1,\mathrm{FR}_2,\ldots,\mathrm{FR}_p}(\mathrm{fr}_1, \mathrm{fr}_2, \ldots, \mathrm{fr}_p) = f_{\mathrm{DP}_1,\mathrm{DP}_2,\ldots,\mathrm{DP}_p}(\mathrm{dp}_1, \mathrm{dp}_2, \ldots, \mathrm{dp}_p) \left|[A']\right|^{-1} \tag{4.7}$$

where fr_1, fr_2, ..., fr_p, dp_1, dp_2, ..., dp_p are the generic continuous variables of the respective design requirements and parameters, respectively. When the design is linear, the Jacobian is constant and (4.4) is still valid. Otherwise, the Jacobian cannot be factored out from the integration and a closed form will be case dependent. The complexity components of sensitivity and variability may not be separated and will be lumped in one term.

The separation of complexity components in the case of linear design facilitates the development of schemes to reduce complexity. Options are crisper than in these of a nonlinear design. Approximation to the linear design might be sought at the expense of numerical complications. In many situations, the design team may need to explore the nonlinear design space by searching for subspaces where design is either uncoupled or decoupled. In other situations, the design nonlinearity may inhibit design uncoupling and decoupling such that only coupled design can be obtained. In either case, the design team should make every effort to reduce both complements of complexity. The effort to reduce design complexity can be complicated by the presence of an elevated degree of coupling and by the form of the joint pdf. The jointly distributed forms may be further complicated by the presence of correlation between the DPs. Correlation among the DPs introduces another component of complexity in the design entity. In other words, design complexity is established in an entity when the DPs not only vary but vary (due to correlation) with other DPs. The presence of correlation adds another component to the total design complexity that needs to be quantified, in addition to the components of variability and sensitivity discussed so far. These components are hard to capture by discrete-source information measures.

4.6 COMPLEXITY DUE TO STATISTICAL CORRELATION

The fact that complexity is a function of variability opens the door for additional derivation and development of comprehensive complexity models. As the degrees of variability and correlation increase, the degree of design complexity increases. In this case, the achievement of the two major operational optimization (Chapter 9) tasks, mean adjustability to targeted performance and variability optimization, will not be trivial. Correlation is a component of variability that affects the component $h(\{\mathbf{DP}\})$ in Theorem 4.1. Our approach in this section will be aligned along the same thought processes, including the employment of a normal distribution, for facilitation purposes. The main objective is to show that design complexity has a correlation component among the DPs (in physical mapping) and between the PVs (in process mapping).

From a statistical perspective, DPs and FRs are functions of random variables or, more precisely, functions of complexity sources. The bivariate normal distribution is used in the derivation presented here. Selection of the normal distribution is based on many properties. In addition to its simplistic logarithmic relationship, there are many attractive features of a multivariate normal distribution. Chief among them is the fact that a random variant that can be explained as a linear combination of independent and identical distribution random variables, or large-sample distributions of test statistics, will be distributed as a multivariate normal according to the central limit theorem. In the bivariate case, the joint density is given by

$$\begin{aligned}\phi(\mathrm{DP}) &= \frac{1}{\sqrt{(2\pi)^p|\Sigma|}} e^{-1/2(\mathrm{DP}-M)'|\Sigma|^{-1}(\mathrm{DP}-M)} \\ &= \frac{1}{\sqrt{(2\pi)^p|\Sigma|}} e^{-1/2\sum_{i=1}^{p}[(\mathrm{DP}_i-\mu_i)/\sigma_i]^2}\end{aligned} \quad (4.8)$$

where $\mathrm{DP}' = \{\mathrm{DP}_1, \ldots, \mathrm{DP}_p\}$, $M' = \{\mu_1, \ldots, \mu_p\}$, and

$$\Sigma = \begin{bmatrix} \sigma_1^2 & 0 & \cdots & 0 \\ 0 & \sigma_2^2 & \cdots & \\ \cdot & \cdots & \cdots & \\ 0 & \cdots & \cdot & \sigma_p^2 \end{bmatrix} \quad (4.9)$$

Corollary 4.2 The complexity of a multivariate normal distribution is given by

$$h(\mathrm{DP}_1, \ldots, \mathrm{DP}_p) = \ln\sqrt{(2\pi e)^p|\Sigma|} \qquad \text{nats} \quad (4.10)$$

where $|\Sigma|$ is the determinant of the variance matrix.

Proof

$$\begin{aligned}h(\phi) &= -\int \phi \ln\phi \, dx \\ &= -E(\ln\phi)\end{aligned}$$

$$\ln\phi = \left(-\frac{1}{2}\right)(\mathbf{DP}-\mathbf{M})'\sum^{-1}(\mathbf{DP}-\mathbf{M}) - \ln\sqrt{(2\pi)^p|\sum|},$$

Then

$$h(\phi) = -\int f\left[\left(-\frac{1}{2}\right)(\mathrm{DP}-M)'|\Sigma|^{-1}(\mathrm{DP}-M) - \ln\sqrt{(2\pi)^p|\Sigma|}\right] d\mathrm{DP}$$

$$= \frac{1}{2} E \left[\sum_{i=1}^{p} \sum_{j=1}^{p} (\mathrm{DP}_i - \mu_i) \sigma_{ij}^{-2} (\mathrm{DP}_j - \mu_j) + \ln \sqrt{(2\pi)^p |\Sigma|} \right]$$

$$= \frac{1}{2} \left\{ \sum_{i=1}^{p} \sum_{j=1}^{p} E[(\mathrm{DP}_i - \mu_i)(\mathrm{DP}_j - \mu_j)] \sigma_{ij}^{-2} + \ln \sqrt{(2\pi)^p |\Sigma|} \right\} \quad (4.11)$$

Of a total of $p(p-1)/2$ terms, only the p diagonal terms are nonzero; hence,

$$\begin{aligned} h(\phi) &= \frac{1}{2} \sum_{i=1}^{p} \sigma_i^2 \sigma_i^{-2} + \ln \sqrt{(2\pi)^p |\Sigma|} \\ &= \frac{1}{2} \sum_{i=1}^{p} 1 + \ln \sqrt{(2\pi)^p |\Sigma|} \\ &= \frac{p}{2} + \ln \sqrt{(2\pi)^p |\Sigma|} \\ &= \ln \sqrt{(2\pi e)^p |\Sigma|} \qquad \text{nats} \end{aligned}$$

For $p = 2$ we have

$$h(\mathrm{DP}_l, \mathrm{DP}_j) = \ln(2\pi e \sigma_l \sigma_j) \qquad \text{nats} \quad (4.12)$$

Consider a correlated bivariate normal distribution. Let ρ_{jl} be the correlation coefficient between two arbitrary DPs, say DP_j and DP_l, defined as the ratio $\mathrm{cov}(\mathrm{DP}_j, \mathrm{DP}_l)/\sigma_j \sigma_l$. In this case we have

$$\mu = \begin{Bmatrix} \mu_j \\ \mu_l \end{Bmatrix} \qquad \Sigma = \begin{bmatrix} \sigma_j^2 & \rho_{jl}\sigma_j\sigma_l \\ \rho_{jl}\sigma_j\sigma_l & \sigma_l^2 \end{bmatrix} \quad (4.13)$$

and the joint distribution is given by

$$f(\mathrm{DP}_j, \mathrm{DP}_l) = \frac{1}{2\pi\sigma_j\sigma_l\sqrt{1-\rho_{jl}^2}} e^{\{(1-\rho_{jl}^2)[(\mathrm{DP}_j-\mu_j)/\sigma_j^2)-2\rho_{jl}(\mathrm{DP}_j-\mu_j)(\mathrm{DP}_l-\mu_l)/\sigma_j\sigma_l+(\mathrm{DP}_l-\mu_l)/\sigma_l^2)]\}/2} \quad (4.14)$$

The complexity correlation component is given in Theorem 4.2.

Theorem 4.2 The design complexity component due to correlation between two DPs that are bivariate normal jointly distributed is given by

$$\ln(2\pi e\sqrt{1-\rho^2}) \quad \text{and} \quad \rho \neq 1, -1 \quad (4.15)$$

Proof Let ρ be the correlation coefficient between DP_1 and DP_2, $\rho = \text{cov}(DP_1, DP_2)/\sigma_1\sigma_2$. In this case we have

$$\mu = \begin{Bmatrix} \mu_1 \\ \mu_2 \end{Bmatrix} \qquad \Sigma = \begin{bmatrix} \sigma_1^2 & \rho\sigma_1\sigma_2 \\ \rho\sigma_1\sigma_2 & \sigma_2^2 \end{bmatrix}$$

and the generic joint distribution is given by

$$\begin{aligned} &\phi(DP_1, DP_2) \\ &\quad = (1/2\pi\sigma_1\sigma_2\sqrt{1-\rho^2})e^{-1/2(1-\rho^2)\sum_{k=1}^{2}[(DP_1-\mu_1)/\sigma_1^2)-2\rho(DP_1-\mu_1)(DP_2-\mu_2)/\sigma_1\sigma_2} \end{aligned} \tag{4.16}$$

Let $z_j = \dfrac{(DP_j - \mu_j)}{\sigma_j}$ where $j = 1, 2$. Then the standard normal joint distribution is given by

$$\phi(z_1, z_2) = \frac{1}{2\pi\sqrt{1-\rho^2}} e^{\{1/[2(1-\rho^2)]\}(z_1^2 - 2\rho z_1 z_2 + z_2^2)} \tag{4.17}$$

Let

$$A = \frac{1}{2\pi\sqrt{1-\rho^2}} \quad \text{and} \quad B = \frac{-1}{2(1-\rho^2)}$$

Then

$$\phi(z_1, z_2) = Ae^{B(z_1^2 - 2\rho z_1 z_2 + z_2^2)} \quad \text{and} \quad \ln(\phi(z_1, z_2)) = \ln A + B(z_1^2 - 2\rho z_1 z_2 + z_2^2) \tag{4.18}$$

By definition we have

$$\begin{aligned} h(\phi) &= -\int_{-\infty}^{\infty}\int_{-\infty}^{\infty} \phi(z_1, z_2) \ln \phi(z_1, z_2)\, dz_1\, dz_2 \\ &= -\int_{-\infty}^{\infty}\int_{-\infty}^{\infty} Ae^{B(z_1^2 - 2\rho z_1 z_2 + z_2^2)}[\ln A + B(z_1^2 - 2\rho z_1 z_2 + z_2^2)]\, dz_1\, dz_2 \\ &= -\ln A - AB \int_{-\infty}^{\infty}\int_{-\infty}^{\infty} e^{B(z_1^2 - 2\rho z_1 z_2 + z_2^2)}(z_1^2 - 2\rho z_1 z_2 + z_2^2)\, dz_1\, dz_2 \end{aligned} \tag{4.19}$$

Let

$$I_1 = \int_{-\infty}^{\infty}\int_{-\infty}^{\infty} e^{B(z_1^2 - 2\rho z_1 z_2 + z_2^2)} z_1^2\, dz_1\, dz_2 \tag{4.20}$$

$$I_2 = -\int_{-\infty}^{\infty}\int_{-\infty}^{\infty} e^{B(z_1^2 - 2\rho z_1 z_2 + z_2^2)} 2\rho z_1 z_2\, dz_1\, dz_2 \tag{4.21}$$

$$I_3 = \int_{-\infty}^{\infty}\int_{-\infty}^{\infty} e^{B(z_1^2 - 2\rho z_1 z_2 + z_2^2)} z_2^2\, dz_1\, dz_2 \tag{4.22}$$

However, $I_1 = I_3$:

$$\begin{aligned} I1 &= \int_{-\infty}^{\infty}\int_{-\infty}^{\infty} e^{B[(z_1-\rho z_2)^2+(1-\rho^2)z_2^2]} z_1^2 \, dz_1 \, dz_2 \\ &= \int_{-\infty}^{\infty} e^{B(1-\rho^2)z_2^2} \int_{-\infty}^{\infty} e^{B(z_1-\rho z_2)^2} z_1^2 \, dz_1 \, dz_2 \\ &= \int_{-\infty}^{\infty} e^{-z_2^2/2} \int_{-\infty}^{\infty} e^{B(z_1-\rho z_2)^2} z_1^2 \, dz_1 \, dz_2 \end{aligned} \tag{4.23}$$

Let $II_1 = \int_{-\infty}^{\infty} e^{B(z_1-\rho z_2)^2} z_1^2 \, dz_1$ and $x = z_1 - \rho z_2 => dx = dz_1$; then

$$\begin{aligned} II_1 &= \int_{-\infty}^{\infty} x^2 e^{Bx^2} \, dx + 2\rho z_2 \int_{-\infty}^{\infty} x e^{Bx^2} \, dx + \rho^2 z_2^2 \int_{-\infty}^{\infty} e^{Bx^2} \, dx \\ &= 2\int_{0}^{\infty} x^2 e^{Bx^2} \, dx + 2\rho z_2 \int_{-\infty}^{\infty} x e^{Bx^2} \, dx + 2\rho^2 z_2^2 \int_{0}^{\infty} e^{Bx^2} \, dx \end{aligned} \tag{4.24}$$

but

$$\begin{aligned} \int_0^{\infty} x^{2n} e^{-ax^2} \, dx &= \frac{(1)(3)(5)\cdots(2n-1)}{2^{n+1}a^n}\sqrt{\frac{\pi}{a}} \quad \text{and} \quad \int_0^{\infty} e^{-a^2x^2} \, dx \\ &= \frac{1}{2a}\sqrt{\pi} \Rightarrow \end{aligned}$$

Let $B = -C$; then

$$I_{11} = \frac{1}{2C}\sqrt{\frac{\pi}{C}} + 2\rho z_2 \int_{-\infty}^{\infty} x e^{-Cx^2} \, dx + \rho^2 z_2^2 \sqrt{\frac{\pi}{C}} \tag{4.25}$$

But $\int_{-\infty}^{\infty} x e^{-Cx^2} \, dx = 0 \Rightarrow$

$$I_{11} = \frac{1}{2C}\sqrt{\frac{\pi}{C}} + \rho^2 z_2^2 \sqrt{\frac{\pi}{C}}$$

Thus,

$$\begin{aligned} I_1 &= \int_{-\infty}^{\infty} e^{-z_2^2/2}\left(\frac{1}{2C}\sqrt{\frac{\pi}{C}} + \rho^2 z_2^2 \sqrt{\frac{\pi}{C}}\right) dz_2 \Rightarrow \\ &= \frac{\pi}{\sqrt{2}C^{3/2}} + \rho^2 \pi \sqrt{\frac{2}{C}} = I_3 \end{aligned} \tag{4.26}$$

$$\begin{aligned} I_2 &= -\int_{-\infty}^{\infty}\int_{-\infty}^{\infty} e^{B(z_1^2-2\rho z_1 z_2+z_2^2)} 2\rho z_1 z_2 \, dz_1 \, dz_2 \\ &= -2\rho \int_{-\infty}^{\infty} z_2 e^{B(1-\rho^2)z_2^2} \int_{-\infty}^{\infty} e^{B(z_1-\rho z_2)^2} z_1 \, dz_1 \, dz_2 \end{aligned} \tag{4.27}$$

But $\int_{-\infty}^{\infty} e^{B(z_1-\rho z_2)^2} z_1\, dz_1 = \rho z_2 \sqrt{\dfrac{\pi}{C}} \Rightarrow$

$$\mathrm{I}_2 = -2\rho \int_{-\infty}^{\infty} z_2 e^{-z_2^2/2} \left(\rho z_2 \sqrt{\frac{\pi}{C}}\right) dz_1\, dz_2 = -2\sqrt{2}\rho^2 \frac{\pi}{\sqrt{C}} \Rightarrow \tag{4.28}$$

$$\begin{aligned}
h(\phi) &= -\ln A - AB\{\mathrm{I}_1 - \mathrm{I}_2 + \mathrm{I}_3\} \\
&= -\ln A - AB\{2\mathrm{I}_1 - \mathrm{I}_2\} \\
&= -\ln A - AB\left[2\left(\frac{\pi}{\sqrt{2}C^{3/2}} + \rho^2\pi\sqrt{\frac{2}{C}}\right) - 2\sqrt{2}\rho^2\frac{1}{\sqrt{C}}\right] \\
&= -\ln A - AB\frac{2\pi}{\sqrt{2}C^{3/2}} + \frac{2\sqrt{2}\rho^2\pi}{\sqrt{C}} - \frac{2\sqrt{2}\rho^2\pi}{\sqrt{C}} \\
&= -\ln A - AB\frac{2\pi}{\sqrt{2}C^{3/2}} \\
&= -\ln\frac{1}{2\pi\sqrt{1-\rho^2}} + \frac{1}{4\pi(1-\rho^2)^{3/2}}\frac{2\pi}{\sqrt{2}\{1/[2(1-\rho^2)]\}^{3/2}} \\
&= \ln(2\pi e\sqrt{1-\rho^2}) \quad \text{and} \quad \rho \neq 1, -1
\end{aligned} \tag{4.29}$$

By substituting $\rho_{jl} = 0$ (the uncorrelated bivariate case), we obtain the normalized version of (4.12). Equation (4.15) shows the effect of DP correlation on complexity with complexity differential $\Delta h_{\text{uncorr}} = -\frac{1}{2}\ln(1-\rho^2)$, $\Delta h_{\text{uncorr}} = h_{\text{uncorr}}(\phi)\Big|_{\rho\neq 0} - h_{\text{uncorr}}(\phi)\Big|_{\rho=0}$, between the uncorrelated case ($\rho_{jl} = 0$) and the correlated case ($\rho_{jl} \neq 0$, $\rho_{jl} \in [-1, 1]$).

Figure 4.8 exhibits some characteristics that require further comment. First, it is apparent that in the case of independent DPs ($\rho_{jl} = 0$) there is no complexity differential [i.e., $\Delta h_{\text{uncorr}}(\rho = 0) = 0$]. The peak occurs at $\rho = 0$, which represents the complexity level of the uncorrelated case at 2.838 nats. Second, the complexity is symmetrical about $\rho_{jl} = 0$.

The unnormalized complexity h_{corr} is given by

$$\ln(2\pi e\sqrt{1-\rho^2}\sigma_l\sigma_j) \quad \text{and} \quad \rho \neq 1, -1 \tag{4.30}$$

The total design complexity due to correlation of normal sources is given by

$$h_{\text{total corr}} = \sum_{l=1}^{p-1}\sum_{j=1+l}^{p} \ln(2\pi e\sqrt{1-\rho_{jl}^2}\sigma_l\sigma_j) \tag{4.31}$$

The total design complexity in the case of linear design is given by

$$h_{\text{total}} = \sum_{l=1}^{p} \ln(\sqrt{2\pi e}\sigma_l) + \sum_{l=1}^{p} \ln|A_{ll}| + \sum_{l=1}^{p-1}\sum_{j=1+l}^{p} \ln(2\pi e\sqrt{1-\rho_{jl}^2}\sigma_l\sigma_j) \tag{4.32}$$

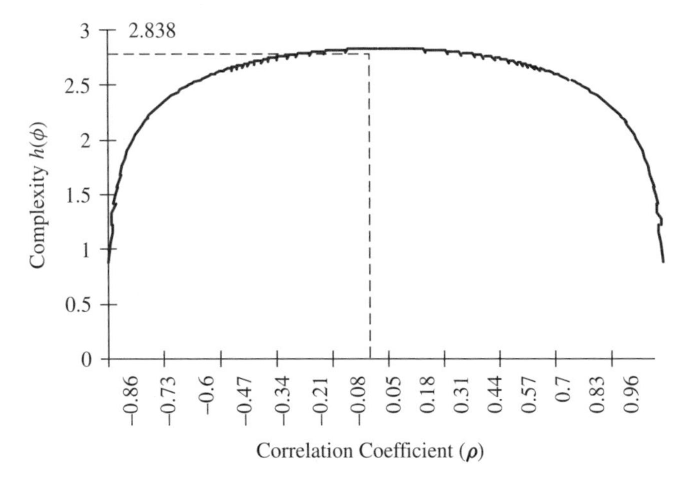

Figure 4.8 Complexity due to correlation $h(\phi)$ versus correlation coefficient (ρ).

The total design complexity is the sum of all three components of uncoupled linear design. However, the sensitivity and variability components may be inseparable and the additivity is lost in coupled nonlinear design mappings.

It is often useful to test the hypothesis $H_0 : \rho = 0$, $H_1 : \overline{H}_0$. The test statistic, t_0, for this hypothesis is given as $t_0 = r\sqrt{N-2}/\sqrt{1-r^2}$, where r is the sample (of size N) correlation coefficient and is given by

$$r = \frac{\sum_{n=1}^{N} \mathrm{DP}_k(\mathrm{DP}_l - \overline{DP}_l)}{\left[\sum_{n=1}^{N}(\mathrm{DP}_l - \overline{DP}_l)^2 \sum_{n=1}^{N}(\mathrm{DP}_k - \overline{DP}_k)^2\right]^{1/2}} \tag{4.33}$$

In this case, t_0 is distributed as Student's t with $n-2$ degrees of freedom [i.e., $\sim t(n-2)$] if H_0 is true. Other hypotheses of the form $H_0 : \rho = \rho_0$, $H_1 : \overline{H}_0$ can be tested as well (Hines and Montgomery, 1980). The correlation coefficient can be estimated and substituted for ρ in (4.31) to quantify the correlation component.

4.7 SUMMARY

In this chapter the information axiom was presented together with its formulation and implication. The traditional axiomatic design measures were reviewed in Section 4.2. A new theory of information content measures, and therefore complexity measures, was presented starting in Section 4.5. The new theories are reproduced below together with the corollaries derived. Proofs are embedded under each.

Theorem 4.1 The complexity of a design has two premier components: variability and coupling. The total design complexity in the case of linear design is given by

$$h(\{\mathbf{FR}\}) = h(\{\mathbf{DP}\}) + \ln |[\mathbf{A}]| \tag{4.4}$$

where $|[A]|$ is the determinant of the nonsingular design matrix A.

Corollary 4.1 For independent process mapping, $h(\{\mathbf{DP}\}) = h(\{\mathbf{PV}\}) + \ln |[\mathbf{B}]|$; then by substitution in (4.4), the total design complexity is given by

$$\begin{aligned} h(\{\mathbf{FR}\}) &= h(\{\mathbf{DP}\}) + \ln |[\mathbf{A}]| \\ &= h(\{\mathbf{PV}\}) + \ln |[\mathbf{B}]| + \ln |[\mathbf{A}]| \\ &= h(\{\mathbf{PV}\}) + \ln |[\mathbf{B}][\mathbf{A}]| \\ &= h(\{\mathbf{PV}\}) + \ln |[\mathbf{C}]| \end{aligned} \tag{4.5}$$

Corollary 4.2 The complexity of a multivariate normal distribution is given by

$$h(\mathrm{DP}_1, \ldots, \mathrm{DP}_p) = \ln \sqrt{(2\pi e)^p |\Sigma|} \qquad \text{nats} \tag{4.10}$$

Theorem 4.2 The complexity due to correlation between two DPs that are distributed as a normalized bivariate normal is given by

$$\ln(2\pi e \sqrt{1 - \rho^2}) \quad \text{and} \quad \rho \neq 1, -1 \tag{4.15}$$

CHAPTER 5

QUALITY ENGINEERING: AXIOMATIC PERSPECTIVE

5.1 INTRODUCTION

Design theory has been a critical and pivotal function of engineering design, as it is experiencing a paradigm shift from the "test–fix" approach to "do it right the first time" practice. At present, validation testing and finished product acceptance testing procedures have been put in the "back seat" of the entire product development process.

One of the major challenges for design teams is to cope with randomness and uncertainties. Since the 1940s, the major methods that have been used by engineers are the methods of reliability testing and online quality control. In the last decade, a new method of dealing with randomness and uncertainties was brought forward and captured attention. This new method, known as *robust design*, was developed by Genichi Taguchi. Taguchi's implementation of robust design is referred to as *Taguchi methods* or *quality engineering*. Taguchi methods were introduced to the West in the 1980s. The last 20 years have witnessed much discussion of the Taguchi methods and many industrial applications.

The basic idea of robust design is different from that of reliability and online quality control. Robust design is a method to enhance quality through improving product development and manufacturing process design. On the one hand, reliability and online quality control share some basic premises; for instance, both of them use the probability of conformance as their measures. On the other hand,

Axiomatic Quality: Integrating Axiomatic Design with Six-Sigma, Reliability, and Quality Engineering, by Basem Said El-Haik
ISBN 0-471-68273-X

reliability engineering has made great strides in the postlaunch development of modern products and technologies. Without the support of reliability engineering, it is impossible to build such sophisticated engineering wonders as space shuttles and nuclear power stations. However, it is easy to recognize that the current reliability theory and practice have some flaws: for example, the apparent gap between reliability engineering and design theory, including difficulties in connecting reliability models to failure mechanism models. See the axiomatic reliability formulation in Chapter 11.

In today's marketplace, there is fierce competition for products and services, which puts considerable pressure on shortening the product development cycle. The desire and expectation for high-quality, reliable goods grows daily. The acceptance of products by customers depends on the ability of the design to provide a superior quality level that is cost-effective in terms of its competitors. Customers pay close attention to the performance level of their product purchases and will notice any degradation in product performance over time. Therefore, the ability to develop products of high quality and reliability speedily becomes a challenge for manufacturers. The development of a strategy that integrates design theory, reliability engineering, and robust design is a necessity to attain this ability: hence the axiomatic quality process.

The objective of this chapter is to provide the mathematical foundation of axiomatic quality by deriving relationships and highlighting interfaces between robustness measures (e.g., the quality loss function and signal-to-noise ratio) and axiomatic measures such as complexity. This will enable axiomatic quality process development and demystify the connectivity between the two. A formulation that explores the relationship between axiomatic design and reliability engineering is the subject of axiomatic reliability in Chapter 11.

In this chapter we review quality engineering methods to prepare the background for the axiomatic quality formulation presented in Chapter 6. The fundamental relationship between robustness and axiomatic measures (in particular, design complexity) is discussed. We present the basic principles of robust design and develop explicit linkages between robust design and axiomatic design theory and outline the mathematical relationships that integrate these disciplines to enhance product development process in terms of quality, cost, and cycle time.

5.2 ROBUST DESIGN (QUALITY ENGINEERING): OVERVIEW

Traditional quality practice is devoted primarily to the downstream portion of the design process, with emphasis placed on inspection schemes. The concentration is now moving upstream to the concept design stage. This shift was initiated, partially, by the robust design philosophy. The concept of robust design as proposed by Taguchi is based on desensitizing the solution entity to environmental, manufacturing, deterioration, and use conditions (Taguchi and Wu, 1980; Kacker, 1985; Taguchi, 1986). In the absence of certain nuisance factors, a solution entity should perform ideally as intended. However, there are impacts from PV manufacturing variation, customer usage, wear, and time that result in performance deviation in the manufacturing and use environments. Such sources of variation are called *noise factors* (NFs). Noise factors may have an unpredictable effect on the FRs. The effect of noise factors on the FRs was addressed, traditionally, by applying control countermeasures, by tightening tolerances, by foolproofing, or by a combination of these measures. Compensation for NF effects using feedback, feedforward, and adaptive controls are typical engineering countermeasures for the effect of noise factors. For example, the NF effects on FRs can be compensated by a closed-loop control system, an additional DP, with the equivalent function "compensates noise" being added to complement the physical design structure, ψ. Noise compensation is an uneconomical noise treatment method. The countermeasure suggested by robust design is desensitization.

Robust design represents a central theme in current design practice. Robustness means that a design solution entity delivers its intended functional requirements under all operating conditions (different causes of variation) throughout its intended life. The implications of robustness for manufactured systems are significant. If artificial systems and products can be made more robust, costly redesign and readjusting (tuning) can be reduced or eliminated. If a higher order of adaptation, such as robust design, can be achieved, existing systems can perform their function longer and better.

In Taguchi's philosophy, robust design consists of three phases (Figure 5.1). It begins with the *concept design phase* followed by the *parameter design* and *tolerance design phases*. Unfortunately, the concept phase did not receive the attention it deserves in the quality engineering community: hence development of the axiomatic quality method.

The goal of parameter design is to minimize the expected quality loss by selecting design parameter settings. The tools used are quality loss function, design of experiment, statistics, and optimization. Parameter design optimization is carried out in two sequential steps: variability minimization of σ_{FR}^2 and mean (μ_{FR}) adjustment to target T_{FR}. The first step is conducted using the DPs that affect variability; the second step is accomplished via DPs that affect the mean but do not influence variability adversely. The objective is to carry both steps at low cost by exploring the opportunities in the design space.

Parameter design is the most used phase of the robust design method. The objective is to design a solution entity by making the FRs insensitive to the variation. This is accomplished by selecting optimal levels of DPs based on testing and using an optimization criterion. Parameter design optimization criteria include both the quality loss function and the signal-to-noise (SN) ratio. The optimum levels of the DPs are the levels that maximize the SNs and are determined

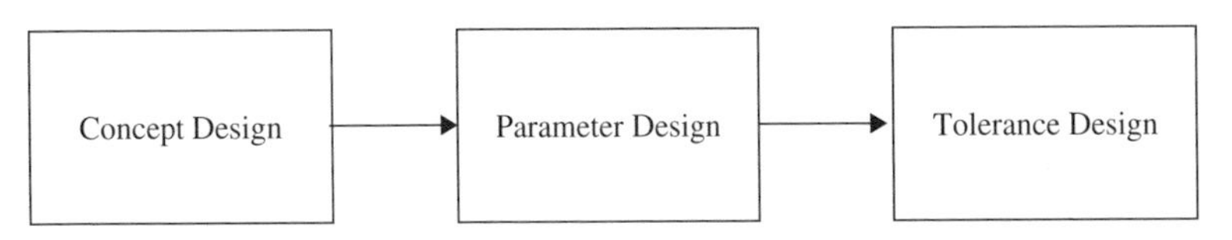

Figure 5.1 Taguchi's robust design.

in an experimental setup from a pool of economic alternatives. These alternatives assume the testing levels in search for the optimum.

A uni-FR solution entity can be classified as static or dynamic from a robustness perspective. A static entity has a fixed target value. The parameter design phase in the case of a static solution entity is to bring the FR mean, μ_{FR}, to the target, T_{FR}. On the other hand, the dynamic solution entity expresses a variable target, depending on customer intent. In this case, the operational vulnerability reduction phase (optimization phase) is carried over a range of useful customer applications called the *signal factor*. The signal factor can be used to set an FR to an intended value. For example, in the brake system of a vehicle, the signal factor is the force applied on the pedal.

Parameter design optimization requires classification of the FRs as *smaller-the-better* (e.g., minimize vibration, reduce friction), *larger-the-better* (e.g., increase strength), *nominal-the-best* (where keeping the product on a single performance objective is the main concern), and *dynamic* (where energy-related functional performance over a prescribed dynamic range of use is the perspective). The dynamic quality characteristic is the most general since it is carried out over a range of input signal. Taguchi characterizes dynamic systems based on the input–output relationship that exists for the *objective function* and can be described by physical or energy transfer expressions. Taguchi's objective function is synonymous with the value engineering concept of *purpose function*. The objective function includes the idea of perfection, called the *ideal function*. An ideal function is an input (signal)–output transformation of the uni-FR solution entity. In quality engineering, such a performance may be assessed by the degree to which a given entity deviates from its ideal functions.

Our approach here is to include the signal input among the p design parameters in the mathematical derivation; that is, the signal will be treated in a mathematical sense as a random variable as well as other DPs. In a practical sense, the numerical signal level is beyond the control of the designer. However, the range of application of the signal should be a premier design consideration.

When robustness cannot be assured by parameter design, we resort to the tolerance design phase. Tolerance design is the last phase of robust design. The practice is to upgrade or tighten the tolerances of some DPs so that quality loss can be reduced. However, tolerance tightening will usually add to the cost (Chapter 9). Our objective is thus to find the degree of tolerance of the DPs that will minimize the costs of both quality loss and tolerance control.

5.3 MATHEMATICAL RELATIONSHIP BETWEEN THE QUALITY LOSS FUNCTION AND AXIOMATIC MEASURES

The inspection schemes represent the heart of online quality control. Inspection schemes depend on the binary characterization of design parameters (i.e., being within or outside the specification limits). An entity is called *confirming* if an inspection finds that all of its DPs are within their respective specification limits; otherwise, it is *nonconfirming*. This binary representation of the acceptance criteria for a DP is not realistic since it characterizes equally entities that are marginally off specification limits and entities that are marginally within these limits. Taguchi proposed a continuous and better representation, the quality loss function (Taguchi and Wu, 1980). The loss function provides a better estimate of the monetary loss incurred by manufacturers and customers as an FR deviate from its targeted performance value, T. Determination of the target T implies the nominal-the-best and dynamic classifications.

A quality loss function can be interpreted as a means to translate variation and target adjustment to a monetary value. It allows the design teams to perform a detailed optimization of cost by relating engineering terminology to economical measures. In its quadratic form, a quality loss is determined by first finding the functional limits,[1] $T \pm \Delta\text{FR}$, of the relevant FR. The functional limits are the points at which the solution entity would fail (i.e., the point of unacceptable performance in approximately half of customer applications). In a sense, these represent performance levels that are equivalent to average customer tolerance. Kapur (1988) continued with this path of thinking and illustrated the derivation of specification limits using Taguchi's quality loss function. A quality loss is incurred due to deviation in the FRs from their intended targeted performance, T, caused by the NFs. Let L denote the quality loss function (QLF), taking the numerical value of the FR and the targeted value as arguments. By Taylor series expansion[2] at $\text{FR} = T$, we have

$$L(\text{FR}, T) = L(T, T) + \left.\frac{\partial L}{\partial \text{FR}}\right|_{\text{FR}=T} (\text{FR} - T) + \left.\frac{1}{2!}\frac{\partial^2 L}{\partial \text{FR}^2}\right|_{\text{FR}=T} (\text{FR} - T)^2 + \left.\frac{1}{3!}\frac{\partial^3 L}{\partial \text{FR}^3}\right|_{\text{FR}=T} (\text{FR} - T)^3 + \cdots \tag{5.1}$$

The target T is defined such that L is minimal at $\text{FR} = T \Rightarrow (\partial L/\partial \text{FR})_{\text{FR}=T} = 0$. In addition, the robustness theme implies that to minimize the quality loss, most entities should be delivering the FR at target (T) on a continuous basis or at least in the very near neighborhood of $\text{FR} = T$. In the latter case, the expansion quadratic term is the most significant term. This condition results in the approximation

$$L(\text{FR}, T) \cong K(\text{FR} - T_{\text{FR}})^2 \tag{5.2}$$

where $K = (1/2!)(\partial^2 L/\partial \text{FR}^2)_{\text{FR}=T}$. Let $\text{FR} \in [T_{\text{FR}} - \Delta\text{FR}, T_{\text{FR}} + \Delta\text{FR}]$, where T_{FR} is the target value and ΔFR is the functional deviation from the target. Let A_Δ be the quality loss incurred due to the symmetrical deviation, ΔFR; then

$$K = \frac{A_\Delta}{(\Delta\text{FR})^2} \tag{5.3}$$

[1]Functional limits or customer tolerance in robust design terminology is synonymous with the design range, DR, in axiomatic design approach terminology (see Section 2.4).

[2]The assumption here is that L is a higher-order continuous function such that derivatives exist, and is symmetrical around $\text{FR} = T$.

In Taguchi's tolerance design method, the quality loss coefficient K can be determined on the basis of losses in monetary terms by falling outside the *customer tolerance limits* (*design range*) instead of the specification limits generally used in process capability studies (e.g., *manufacturer limits*). The specification limits are most often associated with the DPs. Customer tolerance limits are used to estimate loss from the customer perspective or the quality loss to society, as proposed by Taguchi. Usually, customer tolerance is wider than manufacturer tolerance. In axiomatic quality both limits can be used to substitute for $T \pm \Delta\text{FR}$ in the derivation below. In this chapter we side with the design range limit terminology. Deviation from this practice will be noted where appropriate.

Let $f(\text{FR})$ be the probability density function (pdf) of the FR. Then via the expectation operator, E, we have

$$\begin{aligned} E[L(\text{FR}, T)] &= \int_{-\infty}^{\infty} K(\text{FR} - T_{\text{FR}})^2 f(\text{FR})\, d\text{FR} \\ &= K[\sigma_{\text{FR}}^2 + (\mu_{\text{FR}} - T_{\text{FR}})^2] \end{aligned} \tag{5.4}$$

Equation (5.4) is very fundamental to the axiomatic quality process optimization phase formulation in Chapter 9. Quality loss has two ingredients: loss incurred due to variability, σ_{FR}^2, and loss incurred due to mean deviation from target, $(\mu_{\text{FR}} - T_{\text{FR}})^2$. Usually, the second term is achieved by adjustment of the mean of a critically few DPs, a typical engineering problem. In addition, since complexity is a function of variability (Chapter 4), this equation can be used to bridge complexity and information measures to the quality loss function, a robust design measure. To do that we have to assume the probability density function of the functional requirement $f(\text{FR})$. Let FR $\sim$ Normal $(\mu_{\text{FR}}, \sigma_{\text{FR}}^2)$, and by substituting in (4.3), the continuous complexity h is given by

$$h = \ln\left\{(2\pi e)^{1/2}\sqrt{\frac{E[L(\text{FR}, T)]}{K} - (\mu_{\text{FR}} - T_{\text{FR}})^2}\right\} \quad \text{nats} \tag{5.5}$$

where $E[L(\text{FR}, T)] > K(\mu_{\text{FR}} - T)^2$. For a normally distributed FR, the significance of (5.5) is that we can quantify the information and complexity from the monetary value, and vice versa. We can now relate complexity to quality loss and cost in general. The logarithmic relationship implies that to reduce quality loss cost we need to reduce complexity; that is, complexity is not free. The effort resulting in complexity reduction, an operational vulnerability reduction, should be rewarded by quality loss reduction.

The derivation in (5.5) suits the nominal-the-best classification. Other uni-FR quality loss functions and mathematical forms for the other classifications may be found in Chen and Kapur (1989). The following forms of loss function were borrowed from that paper.

1. *Larger-the-better loss function.* For such functions as "increase strength" (FR = strength) or "multiply torque" (FR = torque), we would like a very

large target: ideally, $T_{FR} \rightarrow \infty$. The FR is bounded by a lower functional specifications limit, FR_l. The loss function is then given by

$$L(\text{FR}, T_{\text{FR}}) = \frac{K}{\text{FR}^2} \qquad \text{where} \quad \text{FR} \geq \text{FR}_l \tag{5.6}$$

Let μ_{FR} be the average FR numerical value of the system range (i.e., the average around which performance delivery is expected). Then by Taylor series expansion around FR = μ_{FR} we have

$$L(\text{FR}, T_{\text{FR}}) = K \left[\frac{1}{\mu_{\text{FR}}^2} - \frac{2(\text{FR} - \mu_{\text{FR}})}{\mu_{\text{FR}}^3} + \frac{3(\text{FR} - \mu_{\text{FR}})^2}{\mu_{\text{FR}}^4} - \cdots \right] \Bigg|_{\text{FR}=\mu_{\text{FR}}} \tag{5.7}$$

If the higher-order terms are negligibly small, we have

$$E[L(\text{FR}, T_{\text{FR}})] = K \left(\frac{1}{\mu_{\text{FR}}^2} + \frac{3}{\mu_{\text{FR}}^4} \sigma_{\text{FR}}^2 \right) \tag{5.8}$$

For a normally distributed FR, we have

$$h = \ln \left\{ (2\pi e)^{1/2} \left(\frac{2\pi \mu_{\text{FR}}^4 E[L(\text{FR}, T_{\text{FR}})]}{3K} - \frac{2\pi \mu_{\text{FR}}^2}{3} \right)^{1/2} \right\} \tag{5.9}$$

where $E[L(\text{FR}, T)] > K/\mu_{\text{FR}}^2$.

2. *Smaller-the-better loss function.* Functions such as "reduce noise" and "reduce wear" would preferably have zero as their target value. The loss function in this category and its expected values are given as

$$L(\text{FR}, T) = K \cdot \text{FR}^2 \tag{5.10}$$

and

$$E[L(\text{FR}, T)] = K(\sigma_{\text{FR}}^2 + \mu_{\text{FR}}^2) \tag{5.11}$$

respectively. The normal case is given as

$$h = \ln \left\{ (2\pi e)^{1/2} \sqrt{\frac{E[L(\text{FR}, T_{\text{FR}})]}{K} - \mu_{\text{FR}}^2} \right\} \tag{5.12}$$

where $E[L(\text{FR}, T_{\text{FR}})] > K\mu_{\text{FR}}^2$.

In the development above as well as in the next sections, the average loss can be estimated from a parameter design or even a tolerance design experiment by substituting the experiment variance S^2 and average $\overline{\text{FR}}$ as estimates for σ_{FR}^2 and μ_{FR} in (5.4), (5.8), and (5.11).

TABLE 5.1 Camera Spring Data

Machine	Data	S^2	Average FR
New	0.37, 0.41, 0.37, 043, 0.39, 0.35, 0.40, 0.36	0.0007	0.385
Old	0.55, 0.67, 0.70, 0.54, 0.41, 0.32, 0.46, 0.66	0.0184	0.539

Example 5.1 (Fowlkes and Creveling (1995)) A spring is used in the operation of a camera shutter. The spring manufacturing process has inherent variability, which is the source of the spring rate variance. Let T be the targeted numerical value of 0.5 oz/in. Let the functional limit be $T \pm 0.3$ oz/in. of the spring rate, where half the customers will consider the camera to be defective due to an improper shutter speed. The improper shutter speed results in improperly exposed pictures and thus customer dissatisfaction. The cost of repairing or replacing a camera that has an unacceptable spring rate is $A_\Delta = \$200$. Then, by using (5.3), we have

$$L(\text{FR}, T) = \frac{200}{0.3^2}(\text{FR} - 0.5)^2 = 2220(\text{FR} - 0.5)^2$$

There are two winding machines that are currently producing the springs: new and old. Eight springs produced by both machines were tested for the spring rate. The data are arranged in Table 5.1.

The new machine has a lower variance but a higher average; the old machine has an average that is close to the target of 0.5 oz/in. and a variance that is larger. By substituting S^2 and ave. FR as estimates for σ_{FR}^2 and μ_{FR}, respectively, we found that the quality loss for the new machine is \$3.08, whereas it is \$4.41 for the old machine. This situation among the machines amounts to 1.6723 nats of complexity.

5.4 MATHEMATICAL RELATIONSHIP BETWEEN THE QUALITY LOSS FUNCTION AND AXIOMATIC MEASURES OF HIGHER MODULARITY

Higher-order modularity design entities (e.g., a subsystem) have more than one FR. The array of FRs in a given mapping may be coupled with each other, which implies that the entity as a whole should be handled as one unit for quantification of complexity, quality loss, and then operational vulnerability reduction (optimization) afterward. For an uncoupled design and assuming statistical independence, the FRs can be treated independently. Each FR has its own quality loss and its own complexity degree. In this case, the losses are additive and the total loss function L is given by

$$L = L(\text{FRs}, T\text{s}) = \sum_{i=1}^{m} L(\text{FR}_i, T_i) \tag{5.13}$$

and the expected loss function of the solution entity is

$$E[L(\text{FRs}, T\text{s})] = \sum_{i=1}^{m} E[L(\text{FR}_i, T_i)] \tag{5.14}$$

Let $\mathbf{T} = \{T_1, T_2, \ldots, T_m\}^{\mathrm{T}}$ and $\mathbf{FR} = \{\mathrm{FR}_1, \mathrm{FR}_2, \ldots, \mathrm{FR}_m\}^{\mathrm{T}}$ be the target value array[3] and the FR array of the solution entity, respectively. Let **L(T,FR)** be the solution entity loss function. By expanding **L(T,FR)** at **FR = T**, we have

$$\mathbf{L(FR,T)} = \mathbf{L}|_{\mathbf{FR}=T} + \nabla\mathbf{L}|_{\mathbf{FR}=T}\{\mathbf{FR}-T\} + \tfrac{1}{2}\{\mathbf{FR}-T\}\mathbf{H}|_{\mathbf{FR}=T}\{\mathbf{FR}-\mathbf{T}\} + \cdots \tag{5.15}$$

where **H** is the Hessian[4] matrix. Again $\mathbf{L(FR},T)$ takes its minimum at $\{\mathbf{FR}\} = \{T\} \Rightarrow \nabla\mathbf{L}|_{\mathbf{FR}=T} = \mathbf{0}$. If the entity operates around the array $\{\mathbf{FR}\}=\{T\}$, the quadratic term dominates the expansion and we have

$$\mathbf{L(FR},T) \cong \tfrac{1}{2}\{\mathbf{FR}-T\}\mathbf{H}|_{\mathbf{FR}=T}\{\mathbf{FR}-\mathbf{T}\} \tag{5.16}$$

Let μ_{FR_i} and $\sigma^2_{\mathrm{FR}_i}$ be the mean and the variance of FR_i, $i = 1, \ldots, m$, respectively. The value expected for the total entity quality loss in (5.16) can be written as

$$\begin{aligned} E[L(\mathrm{FRs}, T\mathrm{s})] = {} & \sum_{i=1}^{m} \left.\frac{\partial^2 L}{\partial \mathrm{FR}_i^2}\right|_{\mathrm{FR}_i=T_i} [\sigma^2_{\mathrm{FR}_i} + (\mu_{\mathrm{FR}_i} - T_i)^2] \\ & + \sum_{i=1}^{m-1}\sum_{k=1+i}^{m} \left.\frac{\partial^2 L}{\partial \mathrm{FR}_i\, \partial \mathrm{FR}_k}\right|_{\substack{\mathrm{FR}_i=T_i \\ \mathrm{FR}_k=T_k}} [\mathrm{cov}(\mathrm{FR}_i, \mathrm{FR}_k) \\ & + (\mu_{\mathrm{FR}_i} - T_i)(\mu_{\mathrm{FR}_k} - T_k)] \end{aligned} \tag{5.17}$$

where "cov" denotes the statistical covariance operator.

Equation (5.17) is fundamental. It indicates that quality loss has three ingredients: FR individual variability, FR mean adjustment to target, and the covariance between the individual FRs. This should not be a surprise; that is, these ingredients are also components of complexity and information content measures. Check Sections 4.5 and 4.6: in particular, equation (4.32).

Let Y_{ij}, $i = 1, \ldots, m$; $j = 1, \ldots, m$, be a binary variable denoting the existence of the mapping between FR_i and DP_j (i.e., $Y_{ij} = 1$ if $\mathrm{FR}_i \rightarrow \mathrm{DP}_j$, 0 otherwise). Let μ_{DP_j} and $\sigma^2_{\mathrm{DP}_j}$ be the mean and the variance of the design parameter DP_j. We will assume that the DPs are uncorrelated[5] which also leads to the fact that the FRs are uncorrelated as a result. In addition, we employ the first-order error propagation formula throughout our derivation. The formula relates the variance of the FR, say FR_i, to the variance of its mapped-to DPs, which can be expressed as

[3]In axiomatic design terminology, we call this array the Ts.

[4]**H** is the second derivative matrix of L with entries $\partial^2 L/\partial \mathrm{FR}_i\ \partial \mathrm{FR}_j$. The entries in **H** can be found by regression (Johnson and Wichern, 1982).

[5]There is a difference between coupling and correlation or interaction. Coupling is certain and can be established by poor physical mapping practice. Correlation, on the other hand, is hypothesized to reflect an unknown statistical relationship between two quantities: say, two DPs that are correlated via noise factors. Correlation may happen as a result of packaging DPs (uncoupled or otherwise) in one physical entity, due to the intentional or unintentional employment of Corollary 2.3 (Section 2.5) with active noise factors.

$\sigma^2_{\mathrm{FR}_i} = \sum_{j=1}^{m} Y_{ij} \left(\partial \mathrm{FR}_i / \partial \mathrm{DP}_j\right)^2 \sigma^2_{\mathrm{DP}_j}$ (Ku, 1966). Under these assumptions, (5.17) can be approximated as

$$
\begin{aligned}
E[L(\mathrm{FRs}, T\mathrm{s})] \cong \sum_{i=1}^{m} \left.\frac{\partial^2 L}{\partial \mathrm{FR}_i^2}\right|_{\mathrm{FR}_i = T_i} & \left\{ \sum_{j=1}^{m} Y_{ij} \left(\frac{\partial \mathrm{FR}_i}{\partial \mathrm{DP}_j}\right)^2_{\mathrm{DP}_j = \mu \mathrm{DP}_j} \sigma^2_{\mathrm{DP}_j} \right. \\
& \left. + \left[\sum_{j=1}^{m} Y_{ij} \left(\frac{\partial \mathrm{FR}_i}{\partial \mathrm{DP}_j}\right)_{\mathrm{DP}_j = \mu \mathrm{DP}_j} \mu_{\mathrm{DP}_j} - T_i \right]^2 \right\} \\
& + \sum_{i=1}^{m-1} \sum_{k=1+i}^{m} \left.\frac{\partial^2 L}{\partial \mathrm{FR}_i \, \partial \mathrm{FR}_k}\right|_{\substack{\mathrm{FR}_i = T_i \\ \mathrm{FR}_k = T_k}} \left\{ \sum_{j=1}^{m} \left[Y_{ij} Y_{kj} \left(\frac{\partial \mathrm{FR}_i}{\partial \mathrm{DP}_j}\right)_{\mathrm{DP}_j = \mu \mathrm{DP}_j} \right. \right. \\
& \times \left(\frac{\partial \mathrm{FR}_k}{\partial \mathrm{DP}_j}\right)_{\mathrm{DP}_j = \mu \mathrm{DP}_j} \mu^2_{\mathrm{DP}_j} - \mu_{\mathrm{DP}_j} \left[Y_{ij} \left(\frac{\partial \mathrm{FR}_i}{\partial \mathrm{DP}_j}\right)_{\mathrm{DP}_j = \mu \mathrm{DP}_j} T_j \right. \\
& \left. \left. \left. + \; Y_{kj} \left(\frac{\partial \mathrm{FR}_k}{\partial \mathrm{DP}_j}\right)_{\mathrm{DP}_j = \mu \mathrm{DP}_j} T_i \right] \right] + T_i T_k \right\}
\end{aligned}
\tag{5.18}
$$

However, if we assume that the adjustment to target cost is insignificant following the steps of a uni-FR case, the loss due to adjustment can be ignored because

$$
\sum_{j=1}^{m} Y_{ij} \left(\frac{\partial \mathrm{FR}_i}{\partial \mathrm{DP}_j}\right)_{\mathrm{DP}_j = \mu \mathrm{DP}_j} \mu_{\mathrm{DP}_j} = T_i \quad \text{and} \quad \sum_{j=1}^{m} Y_{kj} \left(\frac{\partial \mathrm{FR}_k}{\partial \mathrm{DP}_j}\right)_{\mathrm{DP}_j = \mu \mathrm{DP}_j} \mu_{\mathrm{DP}_j} = T_k
$$

which gives

$$
\begin{aligned}
& \sum_{j=1}^{m} \left[Y_{ij} Y_{kj} \left(\frac{\partial \mathrm{FR}_i}{\partial \mathrm{DP}_j}\right)_{\mathrm{DP}_j = \mu \mathrm{DP}_j} \left(\frac{\partial \mathrm{FR}_k}{\partial \mathrm{DP}_j}\right)_{\mathrm{DP}_j = \mu \mathrm{DP}_j} \mu^2_{\mathrm{DP}_j} + T_i T_k \right] \\
& \quad = \sum_{j=1}^{m} \mu_{\mathrm{DP}_j} \left[Y_{ij} \left(\frac{\partial \mathrm{FR}_i}{\partial \mathrm{DP}_j}\right)_{\mathrm{DP}_j = \mu \mathrm{DP}_j} T_k + Y_{jk} \left(\frac{\partial \mathrm{FR}_k}{\partial \mathrm{DP}_j}\right)_{\mathrm{DP}_j = \mu \mathrm{DP}_j} T_i \right]
\end{aligned}
\tag{5.19}
$$

and the expected loss function is given by

$$
E[L(\mathrm{FRs}, T\mathrm{s})] \cong \sum_{i=1}^{m} \left.\frac{\partial^2 L}{\partial \mathrm{FR}_i^2}\right|_{\mathrm{FR}_i = T_i} \left[\sum_{j=1}^{m} Y_{ij} \left(\frac{\partial \mathrm{FR}_i}{\partial \mathrm{DP}_j}\right)^2_{\mathrm{DP}_j = \mu \mathrm{DP}_j} \sigma^2_{\mathrm{DP}_j} \right] \tag{5.20}
$$

To relate to complexity, we need to assume distribution of the DPs. A realistic case is that $\mathrm{DP}_j \sim \mathrm{Normal}(\mu_{\mathrm{DP}_j}, \sigma^2_{\mathrm{DP}_j})$. In this case h_j, the complexity of DP_j, is given as $h_j = \ln\sqrt{2\pi e}\,\sigma_{\mathrm{DP}_j}$ in (4.3). By substitution in (5.20), we have

$$E[L(\mathrm{FRs}, Ts)] \cong \sum_{i=1}^{m} \left.\frac{\partial^2 L}{\partial \mathrm{FR}_i^2}\right|_{\mathrm{FR}_i=T_i} \left[\sum_{j=1}^{m} Y_{ij} \left(\frac{\partial \mathrm{FR}_i}{\partial \mathrm{DP}_j}\right)^2_{\mathrm{DP}_j=\mu\mathrm{DP}_j} \frac{e^{2h_j-1}}{2\pi}\right] \tag{5.21}$$

There are two special cases of (5.21) that can be explored here: equal DP variance and equal sensitivity.

5.4.1 Equal Variance

In the case of equal variance, we have $\sum_{j=1}^{m} \sigma^2_{\mathrm{DP}_j} = m\sigma^2 \Rightarrow h_j = h, \forall j, j = 1, \ldots, m$. Therefore, the complexity h can be written as

$$h = \ln\sqrt{2\pi e}\frac{E[L(\mathrm{FRs}, Ts)]}{\sum_{i=1}^{m} \partial^2 L/\partial \mathrm{FR}_i^2|_{\mathrm{FR}_i=T_i}\left[\sum_{j=1}^{m} Y_{ij}\left(\partial \mathrm{FR}_i/\partial \mathrm{DP}_j\right)^2_{\mathrm{DP}_j=\mu\mathrm{DP}_j}\right]} \tag{5.22}$$

The conclusion drawn from (5.22) is important. For a given expected loss, the complexity contribution of the individual design parameters, DP_j, $j = 1, \ldots, m$, distributed normally, is inversely proportional to the adjusted sum of the square sensitivity of the FRs with respect to the mapped-to DPs evaluated at the target points. These sensitivities are the arguments of R and S, the independence measures defined in equations (3.4) and (3.5), respectively. The factor of adjustment is the second derivatives of the loss function with respect to the relevant FR at the target point. The sensitivity coefficients need to be maximized if the complexity due to variability of the DPs needs to be minimized. In the uncoupled case (i.e., $\sum_{j=1}^{m} Y_{ij} = 1, \forall\ \mathrm{FR}_i;\ i = 1, \ldots, m$) we have

$$h = \ln\sqrt{2\pi e}\frac{E[L(\mathrm{FRs}, Ts)]}{\sum_{i=1}^{m} \partial^2 L/\partial \mathrm{FR}_i^2|_{\mathrm{FR}_i=T_i}(\partial \mathrm{FR}_i/\partial \mathrm{DP}_j)^2_{\mathrm{DP}_j=\mu\mathrm{DP}j}}$$

which implies that the DP complexity due to variability, h, is

$$\propto \ln \frac{1}{\sum_{i=1}^{m} \partial^2 L/\partial \mathrm{FR}_i^2|_{\mathrm{FR}_i=T_i}(\partial \mathrm{FR}_i/\partial \mathrm{DP}_j)^2_{\mathrm{DP}_j=\mu\mathrm{DP}_j}}$$

the adjusted sum of the squares of the sensitivity coefficients at the target points. This is quite interesting, since the complexity due to sensitivity is a function of the logarithm of the sensitivity matrix. For uncoupled design, the determinant of the sensitivity matrix is given by $\prod_{i=1}^{m} |\partial \mathrm{FR}_i/\partial \mathrm{DP}_i| = \prod_{i=1}^{m} A'_{ii}$ (Section 4.5), the geometric mean of the diagonal elements, a term that also appears in equation (3.5), the mathematical definition of the semangularity coupling measure, S. The semangularity measure indicates the degree of parallelism between FRs and DPs. It is surprising to note that an effort to improve design independence will result in a complexity increment. Our findings here can be stated in the following theorem:

Theorem 5.1: Equal-Variance Case The total design complexity $h(\{\mathrm{FR}\})$ is proportional to the geometric mean of the sensitivity coefficients and inversely proportional to the adjusted sum of the squared sensitivity in an uncoupled–normally distributed design. The complexity of the functional requirement $h(\{\mathrm{FR}\})$ is given by

$$\begin{aligned} h(\{\mathrm{FR}\}) &= V + \ln \frac{\prod_{i=1}^{m} |\partial \mathrm{FR}_i/\partial \mathrm{DP}_i|}{\sum_{i=1}^{m} \partial^2 L/\partial \mathrm{FR}_i^2|_{\mathrm{FR}_i=T_i} (\partial \mathrm{FR}_i/\partial \mathrm{DP}_i)^2_{\mathrm{DP}_i=\mu_{\mathrm{DP}_i}}} \\ &= V + \ln \frac{\prod_{i=1}^{m} \sqrt{\sum_{i=1}^{m} \partial \mathrm{FR}_i/\partial \mathrm{DP}_i}}{\sum_{i=1}^{m} \partial^2 L/\partial \mathrm{FR}_i^2|_{\mathrm{FR}_i=T_i} (\partial \mathrm{FR}_i/\partial \mathrm{DP}_i)^2_{\mathrm{DP}_i=\mu_{\mathrm{DP}_i}}} \end{aligned} \tag{5.23}$$

where $V = \ln \sqrt{2\pi e}\ \mathrm{E}[(\mathrm{FRs}, T\mathrm{s})]$.

Proof Follow the argument above and use Theorem 4.1 and equation (3.5).

For the uncoupled design (i.e., $S = 1$), equation (5.23) can be written in two forms. In addition, for a quadratic loss function, we have $L(\mathrm{FR}_i, T_i) \cong K_i(\mathrm{FR}_i - T_i)^2$ as in (5.2), which gives the adjustment factor as $2K_i$. Hence, (5.23) can be written as

$$\begin{aligned} h &= V + \ln \frac{\prod_{i=1}^{m} |\partial \mathrm{FR}_i/\partial \mathrm{DP}_i|}{\sum_{i=1}^{m} 2K_i(\partial \mathrm{FR}_i/\partial \mathrm{DP}_i)^2} \\ &= V + \ln \frac{\prod_{i=1}^{m} |A'_{ii}|}{\sum_{i=1}^{m} 2K_i(A'_{ii})^2} \end{aligned} \tag{5.24}$$

We notice that if a polynomial loss function needs to be adopted, it has to be at least of second degree; otherwise, complexity will be undefined for normal sources of complexity.

5.4.2 Equal Sensitivity

In the uncoupled design case (i.e., $\sum_{k=1}^{m} Y_{ik} = 1, \forall i, i = 1, \ldots, m$), with an equal-sensitivity coefficient, (5.21) is written as

$$E[L(\mathrm{FRs}, T\mathrm{s})] \cong \sum_{i=1}^{m} \left. \frac{\partial^2 L}{\partial \mathrm{FR}_i^2} \right|_{\mathrm{FR}_i=T_i} \left(\frac{\partial \mathrm{FR}_i}{\partial \mathrm{DP}_i} \right)^2_{\mathrm{DP}_i=\mu_{\mathrm{DP}_i}} \sum_{j=1}^{m} \frac{e^{2h_j-1}}{2\pi} \tag{5.21}$$

$$\sum_{j=1}^{m} \frac{e^{2h_j-1}}{2\pi} = \frac{E[L(\mathrm{FRs}, T\mathrm{s})]}{\sum_{i=1}^{m} \partial^2 L/\partial \mathrm{FR}_i^2|_{\mathrm{FR}_i=T_i} (\partial \mathrm{FR}_i/\partial \mathrm{DP}_i)^2_{\mathrm{DP}_i=\mu_{\mathrm{DP}_i}}} \tag{5.22}$$

$$\sum_{j=1}^{m} (\sinh 2h_j + \cosh 2h_j) = 2\pi e \frac{E[L(\mathrm{FRs}, T\mathrm{s})]}{\sum_{i=1}^{m} \partial^2 L/\partial \mathrm{FR}_i^2|_{\mathrm{FR}_i=T_i} (\partial \mathrm{FR}_i/\partial \mathrm{DP}_i)^2_{\mathrm{DP}_i=\mu_{\mathrm{DP}_i}}} \tag{5.23}$$

and the total design complexity is given as

$$
\begin{aligned}
h(\{\mathrm{FR}\}) &= \sum_{i=1}^{m}(\sinh 2h_i + \cosh 2h_i) + \ln\left|\prod_{i=1}^{m}\left|\frac{\partial \mathrm{FR}_i}{\partial \mathrm{DP}_i}\right|\right| \\
&= 2\pi e \frac{E[L(\mathrm{FRs}, T\mathrm{s})]}{\sum_{i=1}^{m} \partial^2 L/\partial \mathrm{FR}_i^2|_{\mathrm{FR}_i=T_i}(\partial \mathrm{FR}_i/\partial \mathrm{DP}_i)^2_{\mathrm{DP}_i=\mu_{\mathrm{DP}_i}}} + \ln\left|\prod_{i=1}^{m}\left|\frac{\partial \mathrm{FR}_i}{\partial \mathrm{DP}_i}\right|\right|
\end{aligned}
\tag{5.24}
$$

In the derivation above, we generalized the quadratic loss function of the single FR case. We can do the same for smaller-the-better and larger-the-better requirement classifications.

Smaller-the-Better Quality Loss Function of Multiple FRs In this case, (5.10) needs to be generalized as

$$
L(\mathrm{FRs}, T\mathrm{s}) = \sum_{i=1}^{m-1}\sum_{k=1+i}^{m} K_{ik}\mathrm{FR}_i\mathrm{FR}_k \tag{5.25}
$$

and the expected value of loss is given by

$$
\begin{aligned}
E[L(\mathrm{FRs}, T\mathrm{s})] \cong{}& \sum_{i=1}^{m} \left.\frac{\partial^2 L}{\partial \mathrm{FR}_i}\right|_{\mathrm{FR}_i=T_i}(\sigma^2_{\mathrm{FR}_i} + \mu^2_{\mathrm{FR}_i}) \\
&+ \sum_{i=1}^{m-1}\sum_{k=1+i}^{m} \left.\frac{\partial^2 L}{\partial \mathrm{FR}_i \partial \mathrm{FR}_k}\right|_{\substack{\mathrm{FR}_i=T_i\\ \mathrm{FR}_k=T_k}} [\mathrm{cov}(\mathrm{FR}_i, \mathrm{FR}_k) + \mu_{\mathrm{FR}_i}\mu_{\mathrm{FR}_k}]
\end{aligned}
\tag{5.26}
$$

or in terms of the physical mapping for the uncorrelated DP case,

$$
\begin{aligned}
E[L(\mathrm{FRs}, T\mathrm{s})] \cong{}& \sum_{i=1}^{m} \left.\frac{\partial^2 L}{\partial \mathrm{FR}_i^2}\right|_{\mathrm{FR}_i=T_i} \left\{\sum_{k=1}^{m} Y_{ik}\left(\frac{\partial \mathrm{FR}_i}{\partial \mathrm{DP}_k}\right)^2_{\mathrm{DP}_k=\mu_k} \sigma^2_{\mathrm{DP}_k}\right. \\
&+ \left.\left[\sum_{k=1}^{m} Y_{ik}\left(\frac{\partial \mathrm{FR}_i}{\partial \mathrm{DP}_k}\right)_{\mathrm{DP}_k=\mu_k} \mu_{\mathrm{DP}_k}\right]^2\right\} \\
&+ \sum_{i=1}^{m-1}\sum_{k=1+i}^{m} \left.\frac{\partial^2 L}{\partial \mathrm{FR}_i \partial \mathrm{FR}_k}\right|_{\substack{\mathrm{FR}_i=T_i\\ \mathrm{FR}_k=T_k}} \left[\sum_{j=1}^{m} Y_{ij}\left(\frac{\partial \mathrm{FR}_i}{\partial \mathrm{DP}_j}\right)_{\mathrm{DP}_j=\mu_j} \mu_{\mathrm{DP}_j}\right] \\
&\times \left[\sum_{k=1}^{m} Y_{kj}\left(\frac{\partial \mathrm{FR}_k}{\partial \mathrm{DP}_j}\right)_{\mathrm{DP}_j=\mu_j} \mu_{\mathrm{DP}_j}\right]
\end{aligned}
\tag{5.27}
$$

In order to relate to complexity, we will again assume the case $\mathrm{DP}_j \sim$ Normal $(\mu_{\mathrm{DP}_j}, \sigma^2_{\mathrm{DP}_j})$. In this case, h_j, the complexity of DP_j, is given as $h_j = \ln \sqrt{2\pi e}\sigma_{\mathrm{DP}_j}$. Substitution in (5.27) yields

$$E[L(\mathrm{FRs}, T\mathrm{s})] \cong \sum_{i=1}^{m} \left.\frac{\partial^2 L}{\partial \mathrm{FR}_i^2}\right|_{\mathrm{FR}_i = T_i} \left\{ \sum_{j=1}^{m} Y_{ij} \left(\frac{\partial \mathrm{FR}_i}{\partial \mathrm{DP}_j}\right)^2_{\mathrm{DP}_j=\mu_j} \frac{e^{2h_j - 1}}{2\pi} + \left[\sum_{j=1}^{m} Y_{ij} \left(\frac{\partial \mathrm{FR}_i}{\partial \mathrm{DP}_j}\right)_{\mathrm{DP}_j=\mu_j} \mu_{\mathrm{DP}_j}\right]^2 \right\}$$
$$+ \sum_{i=1}^{m-1} \sum_{k=1+i}^{m} \left.\frac{\partial^2 L}{\partial \mathrm{FR}_i \partial \mathrm{FR}_k}\right|_{\substack{\mathrm{FR}_i = T_i \\ \mathrm{FR}_k = T_k}} \left[\sum_{j=1}^{m} Y_{ik} \left(\frac{\partial \mathrm{FR}_i}{\partial \mathrm{DP}_j}\right)_{\mathrm{DP}_j=\mu_j} \mu_{\mathrm{DP}_j}\right] \times \left[\sum_{j=1}^{m} Y_{kj} \left(\frac{\partial \mathrm{FR}_k}{\partial \mathrm{DP}_j}\right)_{\mathrm{DP}_j=\mu_j} \mu_{\mathrm{DP}_j}\right] \tag{5.28}$$

For the special case of equal variance, the complexity contribution of the individual DPs, h, can be written as in (5.29).

Larger-the-Better Quality Loss Function of Multiple FRs In this case, (5.6) needs to be generalized as in (5.30).

$$h = \ln \sqrt{2\pi e} \frac{E[L(\mathrm{FRs}, T\mathrm{s})] - \sum_{i=1}^{m} \left.\frac{\partial^2 L}{\partial \mathrm{FR}_i^2}\right|_{\mathrm{FR}_i=\mu_{\mathrm{FR}_i}} \left[\sum_{j=1}^{m} Y_{ij} \left(\frac{\partial \mathrm{FR}_i}{\partial \mathrm{DP}_j}\right)_{\mathrm{DP}_j=\mu_j} \mu_{\mathrm{DP}_j}\right]^2 - \sum_{i=1}^{m-1} \sum_{k=1+i}^{m} \left.\frac{\partial^2 L}{\partial \mathrm{FR}_i \partial \mathrm{FR}_k}\right|_{\substack{\mathrm{FR}_i=\mu_{\mathrm{FR}_i} \\ \mathrm{FR}_k=\mu_{\mathrm{FR}_k}}} \left[\sum_{j=1}^{m} Y_{ij} \left(\frac{\partial \mathrm{FR}_i}{\partial \mathrm{DP}_j}\right)_{\mathrm{DP}_j=\mu_j} \mu_{\mathrm{DP}_j}\right] \times \left[\sum_{j=1}^{m} Y_{kj} \left(\frac{\partial \mathrm{FR}_k}{\partial \mathrm{DP}_j}\right)_{\mathrm{DP}_j=\mu_j} \mu_{\mathrm{DP}_j}\right]}{\sum_{i=1}^{m} \partial^2 L / \partial \mathrm{FR}_i^2 |_{\mathrm{FR}_i=\mu_{\mathrm{FR}_i}} \left[\sum_{j=1}^{m} Y_{ij} (\partial \mathrm{FR}_i / \partial \mathrm{DP}_j)^2_{\mathrm{DP}_j=\mu_j}\right]} \tag{5.29}$$

$$L(\mathrm{FRs}, T\mathrm{s}) = \sum_{i=1}^{m} \sum_{k=1}^{i} \frac{K_{ik}}{\mathrm{FR}_i \mathrm{FR}_k} \tag{5.30}$$

The expected value of loss of the larger-the-better FRs is given by

$$E[L(\mathrm{FRs}, T\mathrm{s})] \cong \sum_{i=1}^{m-1} \sum_{k=1+i}^{m} \frac{\partial^2 L / \partial \mathrm{FR}_i \partial \mathrm{FR}_k \Big|_{\substack{\mathrm{FR}_i=\mu_{\mathrm{FR}_i} \\ \mathrm{FR}_k=\mu_{\mathrm{FR}_k}}}}{\mu_{\mathrm{FR}_i} \mu_{\mathrm{FR}_k}}$$

$$+\sum_{i=1}^{m}\left[\frac{3(\partial^2 L/\partial \mathrm{FR}_i^2)\big|_{\mathrm{FR}_i=\mu_{\mathrm{FR}_i}}}{\mu_{\mathrm{FR}_i}^4}+\sum_{\substack{k=1\\k\neq i}}^{m}\frac{\partial^2 L/\partial \mathrm{FR}_i\partial \mathrm{FR}_k\big|_{\substack{\mathrm{FR}_i=\mu_{\mathrm{FR}_i}\\ \mathrm{FR}_k=\mu_{\mathrm{FR}_k}}}}{\mu_{\mathrm{FR}_i}^3\mu_{\mathrm{FR}_k}^3}\right]\sigma_{\mathrm{FR}_i}^2$$

$$+\sum_{i=1}^{m-1}\sum_{k=1+i}^{m}\frac{\partial^2 L/\partial \mathrm{FR}_i\partial \mathrm{FR}_k\big|_{\substack{\mathrm{FR}_i=\mu_{\mathrm{FR}_i}\\ \mathrm{FR}_k=\mu_{\mathrm{FR}_k}}}}{\mu_{\mathrm{FR}_i}^2\mu_{\mathrm{FR}_k}^2}[\mathrm{cov}(\mathrm{FR}_i,\mathrm{FR}_k)] \tag{5.31}$$

Equation (5.31) is generated by Taylor series expansion of (5.30) at $\mathrm{FR}_i = \mu_{\mathrm{FR}_i}$. Third- and higher-order terms are ignored. If the DPs are uncorrelated, we have

$$\begin{aligned}E[L(\mathrm{FRs}, T\mathrm{s})] &\cong \sum_{i=1}^{m-1}\sum_{k=1+i}^{m}\frac{\partial^2 L/\partial \mathrm{FR}_i\partial \mathrm{FR}_k\big|_{\substack{\mathrm{FR}_i=\mu_{\mathrm{FR}_i}\\ \mathrm{FR}_k=\mu_{\mathrm{FR}_k}}}}{\mu_{\mathrm{FR}_i}\mu_{\mathrm{FR}_k}}\\ &+\sum_{i=1}^{m}\left[\frac{3(\partial^2 L/\partial \mathrm{FR}_i^2)\big|_{\mathrm{FR}_i=\mu_{\mathrm{FR}_i}}}{\mu_{\mathrm{FR}_i}^4}+\sum_{\substack{k=1\\k\neq i}}^{m}\frac{\partial^2 L/\partial \mathrm{FR}_i\partial \mathrm{FR}_k\big|_{\substack{\mathrm{FR}_i=\mu_{\mathrm{FR}_i}\\ \mathrm{FR}_k=\mu_{\mathrm{FR}_k}}}}{\mu_{\mathrm{FR}_i}^3\mu_{\mathrm{FR}_k}^3}\right]\sigma_{\mathrm{FR}_i}^2\end{aligned} \tag{5.32}$$

and with the physical mapping and the use of propagation error formula, we have

$$\begin{aligned}&E[L(\mathrm{FRs}, T\mathrm{s})]\\ &\cong \sum_{i=1}^{m-1}\sum_{k=1+i}^{m}\frac{\partial^2 L/\partial \mathrm{FR}_i\partial \mathrm{FR}_k\big|_{\substack{\mathrm{FR}_i=\mu_{\mathrm{FR}_i}\\ \mathrm{FR}_k=\mu_{\mathrm{FR}_k}}}}{\left[\sum_{j=1}^{m} Y_{ij}(\partial \mathrm{FR}_i/\partial \mathrm{DP}_j)\mu_{\mathrm{DP}_j}\right]\left[\sum_{j=1}^{m} Y_{kj}(\partial \mathrm{FR}_k/\partial \mathrm{DP}_j)\mu_{\mathrm{DP}_j}\right]}\\ &+\sum_{i=1}^{m}\left\{\left[\frac{3(\partial^2 L/\partial \mathrm{FR}_i^2)\big|_{\mathrm{FR}_i=\mu_{\mathrm{FR}_i}}}{\left[\sum_{j=1}^{m} Y_{ij}(\partial \mathrm{FR}_i/\partial \mathrm{DP}_j)_{\mathrm{DP}_j=\mu_j}\mu_{\mathrm{DP}_j}\right]^4}\right.\right.\\ &\left.\left.+\sum_{k=1+i}^{m}\frac{\partial^2 L/\partial \mathrm{FR}_i\partial \mathrm{FR}_k\big|_{\substack{\mathrm{FR}_i=\mu_{\mathrm{FR}_i}\\ \mathrm{FR}_k=\mu_{\mathrm{FR}_k}}}}{\begin{array}{c}\left[\sum_{j=1}^{m} Y_{ij}(\partial \mathrm{FR}_i/\partial \mathrm{DP}_j)_{\mathrm{DP}_j=\mu_j}\mu_{\mathrm{DP}_j}\right]^3\\ \left[\sum_{j=1}^{m} Y_{kj}(\partial \mathrm{FR}_k/\partial \mathrm{DP}_j)_{\mathrm{DP}_j=\mu_j}\mu_{\mathrm{DP}_j}\right]\end{array}}\right]\left[\sum_{j=1}^{m} Y_{ij}\left(\frac{\partial \mathrm{FR}_i}{\partial \mathrm{DP}_j}\right)^2_{\mathrm{DP}_j=\mu_j}\sigma_{\mathrm{DP}_j}^2\right]\right\}\end{aligned} \tag{5.33}$$

In the equal-normal-variance case, the individual DPs complexity contribution, h, is

$$h \cong \ln \sqrt{2\pi e}\,\frac{E[L(\mathrm{FRs}, Ts)] - \sum_{i=1}^{m-1}\sum_{k=1+i}^{mi}\dfrac{\partial^2 L/\partial \mathrm{FR}_i \partial \mathrm{FR}_k\big|_{\substack{\mathrm{FR}_i=\mu_{\mathrm{FR}_i}\\ \mathrm{FR}_k=\mu_{\mathrm{FR}_k}}}}{\left(\sum_{j=1}^{m} Y_{ij}(\partial \mathrm{FR}_i/\partial \mathrm{DP}_j)_{\mathrm{DP}_j=\mu_j}\mu_{\mathrm{DP}_j}\right)\left(\sum_{j=1}^{m} Y_{kj}(\partial \mathrm{FR}_k/\partial \mathrm{DP}_j)_{\mathrm{DP}_j=\mu_j}\mu_{\mathrm{DP}_j}\right)}}{\sum_{i=1}^{m}\left[\dfrac{3(\partial^2 L/\partial \mathrm{FR}_i^2)\big|_{\mathrm{FR}_i=\mu_{\mathrm{FR}_i}}}{\left[\sum_{j=1}^{m} Y_{ij}(\partial \mathrm{FR}_i/\partial \mathrm{DP}_j)\mu_{\mathrm{DP}_j}\right]^4} + \sum_{i=1+i}^{m}\dfrac{\partial^2 L/\partial \mathrm{FR}_i \partial \mathrm{FR}_j\big|_{\substack{\mathrm{FR}_i=\mu_{\mathrm{FR}_i}\\ \mathrm{FR}_j=\mu_{\mathrm{FR}_j}}}}{\left[\sum_{j=1}^{m} Y_{ij}(\partial \mathrm{FR}_i/\partial \mathrm{DP}_j)_{\mathrm{DP}_j=\mu_j}\mu_{\mathrm{DP}_j}\right]^3\left[\sum_{j=1}^{m} Y_{kj}(\partial \mathrm{FR}_k/\partial \mathrm{DP}_j)_{\mathrm{DP}_j=\mu_j}\mu_{\mathrm{DP}_j}\right]}\right] \times \left[\sum_{j=1}^{m} Y_{ij}\left(\dfrac{\partial \mathrm{FR}_i}{\partial \mathrm{DP}_j}\right)^2_{\mathrm{DP}_j=\mu_j}\right]} \tag{5.34}$$

Again the individual complexity is inversely proportional to the second derivatives, which in the case of larger-the better is adjusted by the brackets of mean terms as a result of nonlinearity.

A modular solution entity may have a combination of the three FR robust design classifications: nominal-the-best, smaller-the-better, and larger-the-better. The appropriate equations applicable to each category need to be used in the hierarchical level of interest. In the derivation above, the assumptions include the uncorrelated DPs with equal variance normally distributed DPs.

The uncorrelation assumption needs to be relaxed if the DPs are functions of the NFs and are hosted physically in one entity. The presence of correlation posts a new source of complexity because we need to know, in addition to the inherent variation of the individual DPs as well as the sensitivities, how they correlate. For example, the covariance term in (5.17) can be written as

$$\mathrm{cov}\left(\sum_{j=1}^{m} Y_{ij}\left(\frac{\partial \mathrm{FR}_i}{\partial \mathrm{DP}_j}\right)_{\mathrm{DP}_k=\mu_k}\mathrm{DP}_k, \sum_{l=1}^{m} Y_{kl}\left(\frac{\partial \mathrm{FR}_k}{\partial \mathrm{DP}_l}\right)_{\mathrm{DP}_l=\mu_l}\mathrm{DP}_l\right)_{\substack{Y_{ij}=Y_{kl}=1\\ k\neq l}}$$

Let L_{corr} and h_{corr} be the quality loss and the complexity due to correlation, respectively. Then we have

$$E[L_{\mathrm{corr}}(\mathrm{FRs}, Ts)] = \sum_{i=1}^{m-1}\sum_{j=1+i}^{m}\frac{\partial^2 L}{\partial \mathrm{FR}_i \partial \mathrm{FR}_k}\bigg|_{\substack{\mathrm{FR}_i=T_i\\ \mathrm{FR}_k=T_k}}$$

$$\times \operatorname{cov}\left(\sum_{j=1}^{m} Y_{ij}\left(\frac{\partial \mathrm{FR}_i}{\partial \mathrm{DP}_j}\right)_{\mathrm{DP}_j=\mu_j} \mathrm{DP}_k, \sum_{l=1}^{m} Y_{kl}\left(\frac{\partial \mathrm{FR}_k}{\partial \mathrm{DP}_l}\right)_{\mathrm{DP}_l=\mu_l} \mathrm{DP}_l\right)_{\substack{Y_{ij}=Y_{kl}=1\\ k\neq l}}$$

$$= \sum_{i=1}^{m-1}\sum_{k=1+i}^{m} \left.\frac{\partial^2 L}{\partial \mathrm{FR}_i \partial \mathrm{FR}_k}\right|_{\substack{\mathrm{FR}_i=T_i\\ \mathrm{FR}_k=T_k}} \sum_{j=1}^{m-1}\sum_{l=j+1}^{m} Y_{ij}Y_{kl}$$

$$\times \left(\frac{\partial \mathrm{FR}_i}{\partial \mathrm{DP}_j}\right)_{\mathrm{DP}_j=\mu_j} \left(\frac{\partial \mathrm{FR}_k}{\partial \mathrm{DP}_l}\right)_{\mathrm{DP}_l=\mu_l} \operatorname{cov}(\mathrm{DP}_k, \mathrm{DP}_l)$$

$$= \sum_{i=1}^{m-1}\sum_{k=1+i}^{m} \left.\frac{\partial^2 L}{\partial \mathrm{FR}_i \partial \mathrm{FR}_k}\right|_{\substack{\mathrm{FR}_i=T_i\\ \mathrm{FR}_k=T_k}} \sum_{\substack{j=1\\ k\neq l}}^{m-1}\sum_{l=j+1}^{m} Y_{ij}Y_{kl}$$

$$\times \left(\frac{\partial \mathrm{FR}_i}{\partial \mathrm{DP}_j}\right)_{\mathrm{DP}_j=\mu_j} \left(\frac{\partial \mathrm{FR}_k}{\partial \mathrm{DP}_l}\right)_{\mathrm{DP}_l=\mu_l} \rho_{jl}\sigma_{\mathrm{DP}_j}\sigma_{\mathrm{DP}_l} \tag{5.35}$$

where $E[L_{\mathrm{corr}}(\mathrm{FRs}, T\mathrm{s})]$ is a fraction of $E[L(\mathrm{FRs}, T\mathrm{s})]$ and ρ_{jl} is the correlation coefficient between the two relevant DPs.

The correlation term in (5.35) is not the only correlation contribution in (5.17). If the DPs are correlated, the error propagation formula of FR_i is given by

$$\sigma^2_{\mathrm{FR}_i} \cong \sum_{j=1}^{m}\sum_{l=1}^{m} \left(\frac{\partial \mathrm{FR}_i}{\partial \mathrm{DP}_j}\right)_{\substack{\mathrm{DP}_l=\mu_{\mathrm{DP}_l}\\ \mathrm{DP}_j=\mu_{\mathrm{DP}_j}}} \left(\frac{\partial \mathrm{FR}_i}{\partial \mathrm{DP}_l}\right)_{\substack{\mathrm{DP}_j=\mu_{\mathrm{DP}_j}\\ \mathrm{DP}_l=\mu_{\mathrm{DP}_l}}} \operatorname{cov}(\mathrm{DP}_j, \mathrm{DP}_l) \tag{5.36}$$

and hence (5.17) can be written as

$$E[L(\mathrm{FRs}, T\mathrm{s})]$$

$$= \sum_{i=1}^{m} \left.\frac{\partial^2 L}{\partial \mathrm{FR}_i^2}\right|_{\mathrm{FR}_i=T_i} \left\{ \sum_{j=1}^{m}\sum_{l=1}^{m} Y_{ij}Y_{il} \left(\frac{\partial \mathrm{FR}_i}{\partial \mathrm{DP}_j}\right)_{\substack{\mathrm{DP}_j=\mu_{\mathrm{DP}_j}\\ \mathrm{DP}_l=\mu_{\mathrm{DP}_l}}} \left(\frac{\partial \mathrm{FR}_i}{\partial \mathrm{DP}_l}\right)_{\substack{\mathrm{DP}_j=\mu_{\mathrm{DP}_j}\\ \mathrm{DP}_l=\mu_{\mathrm{DP}_l}}} \right.$$

$$\left. \times \operatorname{cov}(\mathrm{DP}_j, \mathrm{DP}_l) + \left[\sum_{j=1}^{m} Y_{ij}\left(\frac{\partial \mathrm{FR}_i}{\partial \mathrm{DP}_j}\right)_{\mathrm{DP}_j=\mu_j} \mu_{\mathrm{DP}_j} - T_i\right]^2 \right\}$$

$$+ \sum_{i=1}^{m-1}\sum_{k=1+i}^{m} \left.\frac{\partial^2 L}{\partial \mathrm{FR}_i \partial \mathrm{FR}_k}\right|_{\substack{\mathrm{FR}_i=T_i\\ \mathrm{FR}_k=T_k}} \sum_{\substack{j=1\\ k\neq l}}^{m-1}\sum_{l=j+1}^{m} Y_{ij}Y_{kl}$$

$$\times \left(\frac{\partial \mathrm{FR}_i}{\partial \mathrm{DP}_k}\right)_{\substack{\mathrm{DP}_j=\mu_{\mathrm{DP}_j}\\ \mathrm{DP}_l=\mu_{\mathrm{DP}_l}}} \left(\frac{\partial \mathrm{FR}_k}{\partial \mathrm{DP}_l}\right)_{\substack{\mathrm{DP}_j=\mu_{\mathrm{DP}_j}\\ \mathrm{DP}_l=\mu_{\mathrm{DP}_l}}} \operatorname{cov}(\mathrm{DP}_j, \mathrm{DP}_l)$$

$$+\sum_{i=1}^{m-1}\sum_{k=1+i}^{m}\left.\frac{\partial^2 L}{\partial \mathrm{FR}_i \partial \mathrm{FR}_k}\right|_{\substack{\mathrm{FR}_i=T_i\\ \mathrm{FR}_k=T_k}}$$

$$\times\left\{\sum_{j=1}^{m}\left[Y_{ij}Y_{kj}\left(\frac{\partial \mathrm{FR}_i}{\partial \mathrm{DP}_j}\right)_{\substack{\mathrm{DP}_j=\mu_{\mathrm{DP}_j}\\ \mathrm{DP}_l=\mu_{\mathrm{DP}_l}}}\left(\frac{\partial \mathrm{FR}_k}{\partial \mathrm{DP}_j}\right)_{\substack{\mathrm{DP}_j=\mu_{\mathrm{DP}_j}\\ \mathrm{DP}_l=\mu_{\mathrm{DP}_l}}}\mu^2_{\mathrm{DP}_j}-\mu_{\mathrm{DP}_j}\right.\right.$$

$$\left.\left.\times\left[Y_{ij}\left(\frac{\partial \mathrm{FR}_i}{\partial \mathrm{DP}_j}\right)_{\substack{\mathrm{DP}_j=\mu_{\mathrm{DP}_j}\\ \mathrm{DP}_l=\mu_{\mathrm{DP}_l}}}T_k+Y_{kj}\left(\frac{\partial \mathrm{FR}_j}{\partial \mathrm{DP}_j}\right)_{\substack{\mathrm{DP}_j=\mu_{\mathrm{DP}_j}\\ \mathrm{DP}_l=\mu_{\mathrm{DP}_l}}}T_i\right]\right]+T_iT_k\right\}$$

(5.37)

The term $\mathrm{cov}(\mathrm{DP}_j, \mathrm{DP}_l)$ represents the statistical correlation term, which is explored in Section 4.6. Substituting (4.48) into (5.35), we have

$$E[L_{\mathrm{corr}}(\mathrm{FRs}, T\mathrm{s})]=\sum_{i=1}^{m-1}\sum_{k=1+i}^{m}\left.\frac{\partial^2 L}{\partial \mathrm{FR}_i \partial \mathrm{FR}_k}\right|_{\substack{\mathrm{FR}_i=T_i\\ \mathrm{FR}_k=T_k}}\sum_{\substack{j=1\\ k\neq l}}^{m-1}\sum_{l=j+1}^{m}Y_{ij}Y_{kl}$$
$$\times\left(\frac{\partial \mathrm{FR}_i}{\partial \mathrm{DP}_j}\right)_{\mathrm{DP}_j=\mu_j}\left(\frac{\partial \mathrm{FR}_k}{\partial \mathrm{DP}_l}\right)_{\mathrm{DP}_l=\mu_l}\rho_{jl}\frac{e^{h^{kl}_{\mathrm{corr}}-1}}{2\pi\sqrt{1-\rho^2_{jl}}} \tag{5.38}$$

The correlation due to the complexity term in (5.38) can be substituted in equations (5.17), (5.26), and (5.31) using the same procedure as above or after manipulating the equivalent version of (5.17) that appears in (5.38).

If the correlation coefficients are equal for all (j, l) combinations, such that $\rho_{jl}=\rho, \forall k; l, j=1,\ldots,m$, the correlation complexity can be written as

$$h_{\mathrm{corr}}=\ln\frac{2\pi e E[L_{\mathrm{corr}}(\mathrm{FRs}, T\mathrm{s})]\sqrt{1-\rho^2}}{\sum_{i=1}^{m-1}\sum_{k=1+i}^{m}\left.\partial^2 L/\partial \mathrm{FR}_i \partial \mathrm{FR}_k\right|_{\substack{\mathrm{FR}_i=T_i\\ \mathrm{FR}_k=T_k}}\sum_{\substack{j=1\\ k\neq l}}^{m-1}\sum_{l=j+1}^{m}Y_{ij}Y_{kl}(\partial \mathrm{FR}_i/\partial \mathrm{DP}_j)_{\mathrm{DP}_j=\mu_j}(\partial \mathrm{FR}_k/\partial \mathrm{DP}_l)_{\mathrm{DP}_l=\mu_l}} \tag{5.39}$$

Equation (5.39) has the same form as (5.22), but due to correlation, the complexity was adjusted using the correlation coefficient terms.

5.5 ESTIMATION OF THE EXPECTED LOSS FUNCTION

It is obvious that in the derivations of previous sections, complexity measures require an estimation of the expected quality loss to be available. In uni-FR cases, $E[L(\mathrm{FR}_i, T_i)]$ can be estimated from the data obtained from the parameter design experiment or by sampling. The experimental estimated loss for smaller-the-better, larger-the-better, and nominal-the-best are described by

$$E[L(\mathrm{FR}_i, T_i)] = K\left(S_i^2 + \overline{\mathrm{FR}}_i^2\right) \tag{5.40}$$

$$E[L(\mathrm{FR}_i, T_i)] = K\left(\frac{1}{\overline{\mathrm{FR}}_i^2} + \frac{3S_i^2}{\overline{\mathrm{FR}}_i^2}\right) \tag{5.41}$$

$$E[L(\mathrm{FR}_i, T_i)] = K\left[S_i^2 + \left(\overline{\mathrm{FR}}_i - T_i\right)^2\right] \tag{5.42}$$

respectively, where $S_i^2 = (1/N)\sum_{n=1}^{N}\left(\mathrm{FR}_{i,n} - \overline{\mathrm{FR}}_i\right)^2$ and $\overline{\mathrm{FR}}_i$ are the FR_i experimental variance and average, respectively. The equations above are generated by substituting the experiment variance S^2 and average $\overline{\mathrm{FR}}$ as estimates for σ_{FR}^2 and μ_{FR} in equations (5.4), (5.8), and (5.11). These expected loss estimates were defined by Taguchi and Wu (1980) and share the general form $E[L(\mathrm{FR}, T)] = K(\mathrm{MSD})$, where MSD is the mean-squared deviation.

For higher-order modular entities, the expected loss can be obtained from a series of parameter design experiments. The experimental series should be conducted according to the sequence revealed by the design mapping within the modular hierarchy of interest.

5.6 MATHEMATICAL RELATIONSHIP BETWEEN THE SIGNAL-TO-NOISE RATIO AND AXIOMATIC MEASURES

Our goal in this section is to connect complexity to the SN ratio for the various quality engineering classifications of the FRs. We will also extend our treatment for normal sources of variation in this section. Other distributions may warrant different results than those derived for normally distributed FRs. The normal distribution is selected because of several of its attractive attributes, including its universal representation of many physical phenomena as well as for its appealing logarithmic form.

Taguchi and Wu (1980) define the average quality loss as $L = K(\mathrm{MSD})$. In addition, they define SN ratios for the larger-the-better and smaller-the-better FRs, which are given in equations(5.40) and (5.41) for larger-the-better and smaller-the-better, respectively:

$$\mathrm{SN} = -10\log_{10}\left(\frac{1}{N}\sum_{n=1}^{N}\frac{1}{\mathrm{FR}_i^2}\right) \tag{5.43}$$

$$\mathrm{SN} = -10\log_{10}\left(\frac{1}{N}\sum_{n=1}^{N}\mathrm{FR}_i^2\right) \tag{5.44}$$

The constant N represents the number of units (that have FR_i as their functional requirement) measured in an experiment or in a sample. Experiments are conducted and the FR measurements are collected. For higher-order modules, parameter optimization to reduce variability should follow the sequence identified in the design matrix. Substituting equations (5.43) and (5.44) in equations (5.6) and (5.10), we have

$$E[L(\text{FR}_i, T_i)] = K\left(10^{-\text{SN}/10}\right) \tag{5.45}$$

where $E[L(\text{FR}_i, T_i)]$ is the average quality loss per unit. It is obvious that as SN increases, the expected loss per unit decreases. The expected loss value in (5.45) can be substituted in (5.9) for the larger-the-better case to give

$$h_{\text{FR}_i} = \ln\left[\sqrt{2\pi e}\left(\frac{\mu_{\text{FR}_i}^4 e^{\ln 10^{-0.1}\text{SN}} - \mu_{\text{FR}_i}^2}{3}\right)^{1/2}\right] \tag{5.46}$$

where $\text{SN} \geq \ln 10^{-0.1}\mu_{\text{FR}}^2$. It is obvious that complexity depends on both the mean and SN. The mean is a function of the sensitivity as well as the mean of the mapped to DPs. For smaller-the-better, we have

$$h_{\text{FR}_i} = \ln\left[\sqrt{2\pi e}(e^{\ln 10^{-0.1}\text{SN}} - \mu_{\text{FR}_i}^2)^{1/2}\right] \tag{5.47}$$

where $\text{SN} \geq (\ln 10^{-0.1})^{-1}\ln\left(1/\mu_{\text{FR}}^2\right)$. For the nominal-the-best case, the SN is defined as

$$\text{SN} = -10\log_{10}\frac{\sigma_{\text{FR}_i}^2}{\mu_{\text{FR}_i}^2} \tag{5.48}$$

However, for normal sources of complexity, we have $\sigma_{\text{FR}_i}^2 = e^{2h_{\text{FR}_i}-1}/2\pi$. By substitution we have

$$h_{\text{FR}_i} = \ln(\sqrt{2\pi e}\ \mu_{\text{FR}_i}) + (0.5\ln 10^{-0.1})\text{SN} \tag{5.49}$$

which is a linear relationship. Notice that the intercept adds to the complexity since it is a function of the mean. The nominal-the-best category of FRs that have their mean value less than $1/\sqrt{2\pi e}$ will experience a negative intercept since the logarithmic argument in the intercept is less than unity.

5.7 SUMMARY

The concept of quality in engineering design was introduced in Section 5.1. Reliability testing and online quality control are no longer effective as the quality engineering method for today's design engineer.

In Section 5.2, robust design, founded by Genichi Taguchi, was described as an important method to enhance quality through improving product development and manufacturing process design. It must be recognized that contemporary engineering methodologies have contributed greatly to the successful development of modern sophisticated systems and products. Growth in global competition and emerging market demands drive the need to maximize the efficiency and effectiveness of the product development process. Customer satisfaction is emerging as a critical demand, requiring manufacturers to develop high-quality, reliable products quickly. The axiomatic quality process integrates key elements of reliability engineering and robust engineering to achieve the aforementioned targets.

In Section 5.3 we established the groundwork for an innovative concept that strategically links robust design (quality engineering) and axiomatic design, thus enabling several concepts of the axiomatic quality process. Robust design minimizes quality loss by systematic selection of design parameter settings through application of the quality loss function, design of experiments, statistics, and optimization methodologies. Parameter design requires that variability first be minimized and then the target (mean) be adjusted. In tolerance design, the design specifications are completed by determining allowable control factor (DP) variability to achieve quality requirements and minimize costs. Taguchi's approach is to desensitize the design solution to environmental, manufacturing, deterioration, and usage conditions. Noise factors consist of both uncontrollable factors and factors that although controllable in principle, are controlled only at considerable effort/cost and are therefore considered as noise factors in the design process. Methods for addressing the effects of noise factors were introduced. The three phases of Taguchi's philosophy—concept design, parameter design, and tolerance design—were described. Parameter design establishes the smaller-the-better, larger-the-better, and nominal-the-best classifications of functional requirements. Static and dynamic solutions are characterized from a robustness perspective, based on the nature of the physics of the target. The ideal function, defined as a pure transfer of energy between input and output, has monetary implications. A camera shutter spring case study was presented to describe loss function and complexity calculations.

In a discussion of the mathematical relationship between quality loss function, signal-to-noise ratio, and axiomatic design measures, in particular complexity, in lieu of information content, measures of higher modularity were derived and findings presented in several theorems. An extensive mathematical proof was offered in Section 5.4. Equations for estimating the expected loss function smaller-the-better, larger-the-better, and nominal-the-best cases were presented in Section 5.5. The mathematical relationship between the signal-to-noise ratio and axiomatic measures was presented in Section 5.6 for the smaller-the-better, larger-the-better, and nominal-the-best classifications.

CHAPTER 6

AXIOMATIC QUALITY AND RELIABILITY PROCESS

6.1 INTRODUCTION

As discussed in previous chapters, there is much vulnerability induced in current design and manufacturing engineering practices, which often leads to the creation of systematic quality issues in the designed entity. Design vulnerability can be categorized as follows in the light of our discussion in Chapter 1: (1) *conceptual vulnerability*, in particular coupling, leading to lack of conceptual robustness of the fundamental design concepts [this category is associated with design analysis activities and results from violation of the independence axiom (Chapters 3 and 4)] and (2) *operational vulnerability*, leading to lack of operational robustness postlaunch over the design life cycle. Operational robustness is enabled by conceptual robustness (i.e., absence of conceptual vulnerability). Operational vulnerability results when the system is subjected to noise factor effects, such as customer use or abuse, material degradation, and piece-to-piece manufacturing variation (see Chapter 1).

The objective of this chapter is to present the *axiomatic quality process*, which is tasked with providing solution methods to the two major categories of vulnerabilities listed above. The process is pieced together from concepts borrowed from the axiomatic design method, quality and reliability engineering, six-sigma, and

Axiomatic Quality: Integrating Axiomatic Design with Six-Sigma, Reliability, and Quality Engineering, by Basem Said El-Haik
ISBN 0-471-68273-X

the author's own research and experience. Axiomatic quality has three phases, as depicted in Figure 6.1:

1. Customer attributes-to-functional requirements mapping
2. Conceptual design for capability (CDFC) phase
3. Optimization phase
   a. Parameter optimization phase
   b. Tolerance optimization phase
   c. Axiomatic reliability phase

Axiomatic quality is focused on providing a solution framework for design vulnerabilities in order to produce healthy conceptual entities with six-sigma quality potential. In phase 1 of the axiomatic quality process, the mapping from the customer attributes domain to the functional requirements (FR) domain is conducted using quality function deployment (QFD) over two stages. The first QFD stage represents a mapping from the raw customer attributes to substitute quality characteristics called critical-to-satisfaction (CTS) characteristics. The second stage is a mapping from the CTSs to the FRs.

The conceptual design for capability (CDFC) is the following phase, where the six-sigma conceptual potential of the FRs is set. The CDFC phase presents a systematic approach for establishing the capability at the conceptual level in the designed entity by reducing the coupling vulnerability of the design functional

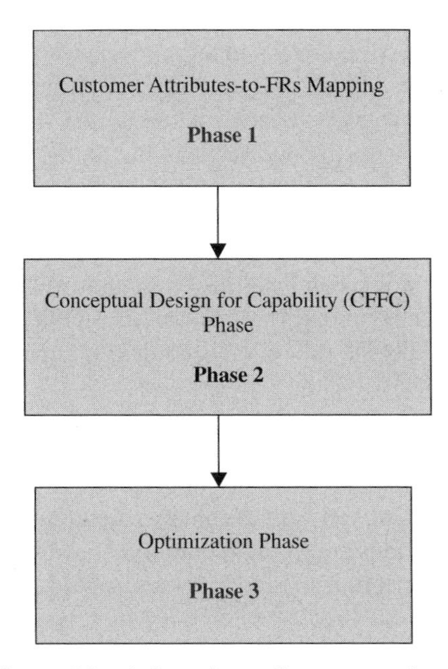

Figure 6.1 Axiomatic quality process phases.

requirements, the array **{FR}**. The equation **{FR}** = **[A]{DP}** is used, where **{FR}** is the array of FRs, **[A]** the design matrix, and **{DP}** the array of design parameters. Section 6.4 and Chapter 8 complement each other. We packaged most of the CDFC phase in this chapter, leaving the special steps of the CDFC phase to be explored in Chapter 8. This division is made to have Chapter 6 comprise an overall axiomatic quality process chapter with occasional reference to complementary chapters when needed.

The *optimization phase* is concerned with the operational vulnerabilities and is introduced in this chapter. It provides an approach that improves the robustness of the product or a system by defining settings and tolerances for the design parameters and the process variables, thus limiting both coupling- and complexity-induced vulnerabilities (due to axiom violation) and achieving these objectives at the minimum possible cost. It is by controlling the DPs that both optimization mechanisms sought for every functional requirement—the variation reduction, σ^2_{FR}, and adjustment to target, $\mu_{FR} \Rightarrow T$—can be achieved.

Optimization in the context of axiomatic quality has two tracks: empirical testing and mathematical programming (Chapter 9). Nonlinear optimization models are derived with an objective function pieced from design vulnerability and economic measures.

6.2 AXIOMATIC QUALITY PROCESS[1]

The basic objective of axiomatic quality when adopted up front is to design a system right the first time, by eliminating or reducing conceptual and operational vulnerabilities. This objective is very much aligned with several current design for six-sigma approaches. In essence, it requires analytical means to achieve and sustain it. Recently, many companies adopting the six-sigma philosophy are devising in-house approaches for DFSS. It is the author's perception that most of these approaches are geared toward packaging several statistical techniques with some complexity covariant beyond the traditional six-sigma (e.g., DMAIC) approach. Unfortunately, this practice is usually coupled with minimal caution regarding a statistical tool's compatibility with the problems of interest. This practice does not guarantee the achievement of six-sigma capability in the entity designed. In effect, the potential of a well-thought-out approach is compromised by this simplistic conduct.

The axiomatic quality process, including axiomatic reliability (Chapter 10), recognizes the peculiarities of the development process, guides the creativity of the design team, has generic applicability in both design and manufacturing arenas, and provides analytical tools and techniques that eliminate or reduce design vulnerabilities at the various stages of Figure 1.8. For example, some activities in initial stages are prescriptive and may include activities that cannot be solved

[1]The material presented in Chapter 6 uses physical mapping, **{FR}** = **[A]{DP}**, as a illustration vehicle. The axiomatic quality process is equally applicable for process mapping, **{DP}** = **[B]{PV}**.

by statistical means (e.g., concept analysis and synthesis). Systematic preventive design methods such as axiomatic design are more suitable in these stages.

The axiomatic design method, when integrated with quality concepts, forms the backbone of the axiomatic quality process. This process is a balanced approach that possesses a set of tools and concepts that covers a range of applications, from design analysis, synthesis, and evaluation to statistical and optimization techniques. The axiomatic quality process has a rich toolbox that applies to many situations encountered within the product development cycle.

The axiomatic quality process map, presented in Section 6.3, distinguishes between two types of design projects:

1. *Creative design* (design at the "white paper" stage), where conceptual tools can be used extensively by employing the axiomatic quality CDFC phase.
2. *Incremental design assignment* (design from a datum system), where at least one baseline (datum) design exists, with possibly some wealth of relevant data readily available. The degree of deviation of the pursed system from a datum is the key factor in deciding on the usefulness of such data.

In both cases, the axiomatic quality process should be carried over several sequential phases: customer attributes–to-FRs mapping, the CDFC phase, followed by an optimization phase. The first phase is concerned with mapping the customer attributes to actionable and technical functional requirements. The second phase is concerned with concept generation, analysis, synthesis, selection, and evaluation, such that a healthy system can be achieved. That is, the CDFC phase enables the objective of establishing a six-sigma capability or potential in the optimization phase of the design entity, via an axiomatic treatment. The optimization phase follows directly and is concerned with formulating six-sigma capability as an optimization problem where DP settings and tolerances of the design parameters and process variables are determined to minimize design vulnerability, quality loss, and other measures. Such optimization represents axiomatic quality operational robustness components.

6.2.1 Why the Axiomatic Quality Process?

Most current quality and reliability methods are empirical in nature. They represent the best thinking of the design community, which unfortunately, lacks a scientific design base while relying on subjective engineering judgment. When a company suffers from detrimental loss of customer satisfaction in certain design FRs, engineering judgment or statistical problem solving may not be sufficient to obtain a satisfactory solution, especially for problems of conceptual root. Therefore, a systemic strategy of axiomatic quality is needed that anticipates such problems and provides means for solution at both the conceptual and operational levels.

It is a fact that design mappings do exist (Figure 6.2), whether or not the design team is aware of them. When the three mappings follow the design axioms, a

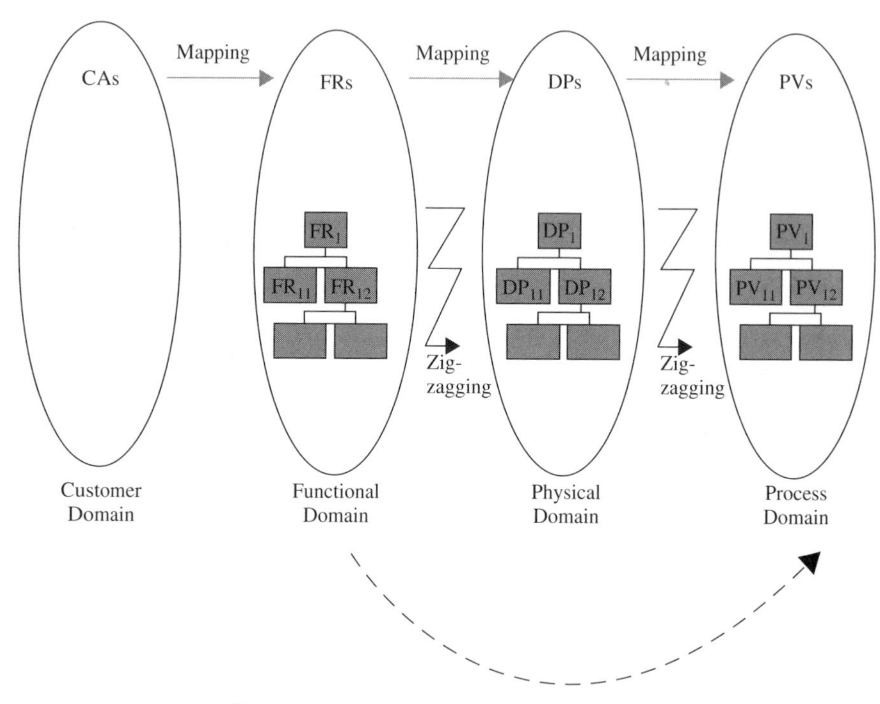

Figure 6.2 Design mappings and domains.

potential conceptual robustness of the design entity is created. However, the type of design objective for the project of interest, whether creative or incremental, is key to deciding whether to modify existing mappings of the datum design or to develop new ones.

Several design and problem-solving methodologies concentrate on finding solutions in the manufacturing environment for a problematic functional requirement. However, they do not employ design mappings or design principles to obtain or validate such a solution. Most often, a solution is obtained without regard to vulnerabilities at the conceptual level of the system. It is typical simply to use the PVs to improve a problematic FR quality. The conceptual framework of current six-sigma implementation, the DMAIC approach, usually bypasses the physical domain (the dashed mapping in Figure 6.2), ignoring the DPs' potential contribution to an overall solution.

Two major disadvantages can be highlighted in the context of bypassing the physical domain. The first disadvantage happens when the team overlooks the need for design changes when the traditional adjustment of the PVs is not alone sufficient to solve the problem, indicating limited solution capability in manufacturing. The risk in this scenario occurs when the problem solver, usually a black belt team, introduces a major manufacturing change (i.e., alters the PV array) when it is not necessary. The second disadvantage is concerned with the ignorance of coupling vulnerability. This would impair the achievement of six-sigma capability (as discussed in Chapter 8). Ignorance of coupling may introduce new symptoms in other FRs when the solution to the original problem is institutionalized.

On the other hand, employing a subset of the PVs for a solution is usually cheaper than pursuing the DPs, since the latter involves both design change and process changes, while the former is limited to process changes only. The adjustment of process variables may or may not produce a satisfactory solution to the problem at hand, depending on the sensitivities in both physical and process mappings. In either case, solutions can be implemented using changes induced in the chosen mapping independent variables, the codomain variables. The independent variables may be the DPs or the PVs, according to the mapping of interest and where the solution is sought. A *change* can be soft or hard. *Soft changes* involve adjusting the means to new target values (nominals) within the specified tolerances, changing the tolerance ranges, or both. *Hard changes* involve eliminating or adding DPs or PVs in the relevant mapping. In a manufacturing environment, soft process changes can be carried out by parametric adjustment in the DP domain, within the tolerances permitted, whereas hard changes may require **{PV}** array alteration. This alteration may result in some manufacturing process or cell redesign to enable the addition, modification, or deletion of process variables. Such activity needs to be conducted with the axiomatic quality CDFC phase in mind. Design changes to reduce or eliminate a detrimental effect on a certain functional requirement may call for hard changes in both the design entity and its manufacturing processes when soft changes cannot produce the result desired.

Mathematically, let the concerned FRs be expressed using $\{\mathbf{FR}\} = [\mathbf{A}]\{\mathbf{DP}\}$, where $\{\mathbf{DP}\}$ is an array of mapped-to DPs of size p. Upon differentiation, the matrix $[\mathbf{A}']$ is the sensitivity design matrix with entries $A'_{ij} = \partial \mathrm{FR}_i / \partial \mathrm{DP}_j, i = 1, \ldots, m; j = 1, \ldots, p$. Let each DP in the array $\{\mathbf{DP}\}$ be written in transfer function format as $\mathrm{DP}_j = g(\mathbf{PV}_k)$, where PV_k, $k = 1, \ldots, n$, is an array of process variables that are mapped-to DP_j. Soft changes may be implemented using sensitivities in physical and process mappings. Using the chain differentiation rule, we have

$$\frac{\partial \mathrm{FR}_i}{\partial \mathrm{PV}_k} = \frac{\partial \mathrm{FR}_i}{\partial \mathrm{DP}_j} \frac{\partial \mathrm{DP}_j}{\partial \mathrm{PV}_k} \tag{6.1}$$

The first term in (6.1) represents a design change and the second represent a process change. An efficient axiomatic quality strategy should utilize both terms if all potential improvements to a six-sigma solution need to be accomplished.

The axiomatic quality process is based on several guidelines. First, a design concept needs to be approached from a total perspective, depending on the extensive use of design axioms, quality engineering, and six-sigma concepts. Second, the objective is to "design it right the first time," with consistent development performance at the conceptual and operational levels. Third, design optimization should consider vulnerability and economic measures.

In the following sections we present the axiomatic quality process. We focus on each of the axiomatic quality process phases, highlighting possible interfaces with engineering design and quality engineering practices. Essential steps of the CDFC phases are covered in Chapter 8. This chapter can serve as an axiomatic quality process summary, with reference to the detailed chapters where needed. Special attention is given to the first design mapping, from the CA domain to the FR domain, the first phase of the process.

6.2.2 Axiomatic Quality Process Map

A road map of the axiomatic quality process is presented in Figure 6.3. The process is about taking the steps necessary to develop solution entities with unprecedented customer delight for its total life. This process is based on the theoretical frameworks introduced in earlier chapters and developed next.

The objective of this chapter is to mold the theoretical development in a comprehensive implementable sequence. We would like to think of the theoretical development achieved, as applied to Figure 6.3, as the guiding rules for design decision making. As such, we consolidate the advantages of both the algorithmic and prescriptive design approaches. The axiomatic quality process is a design process composed of several steps, which utilizes the applicable design, quality,

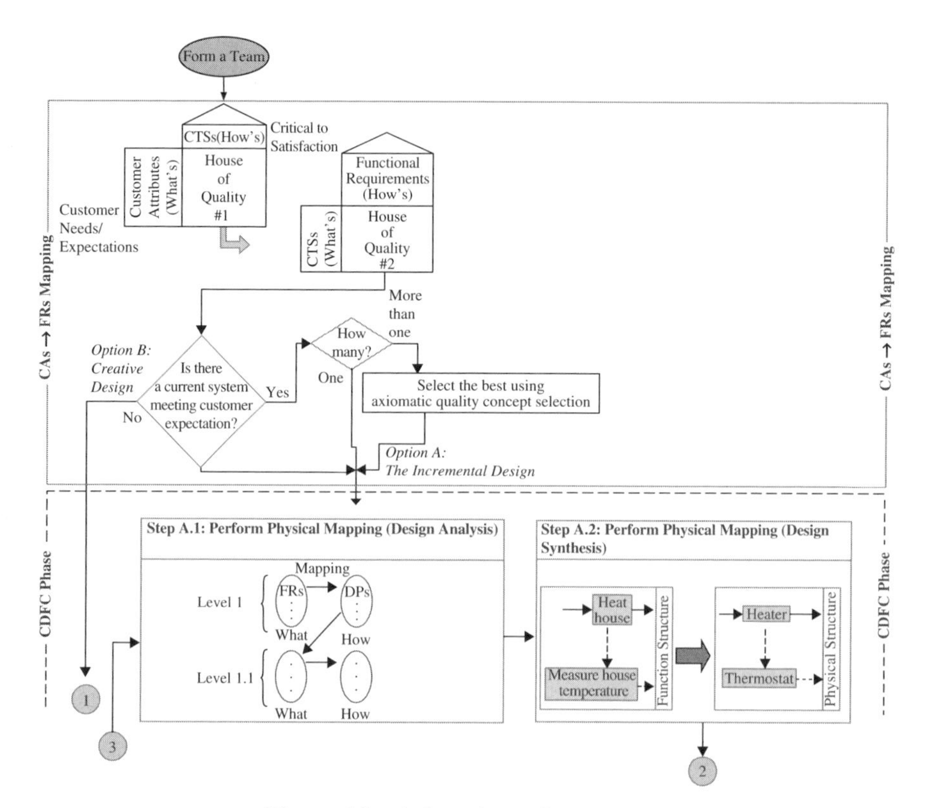

Figure 6.3 Axiomatic quality process.

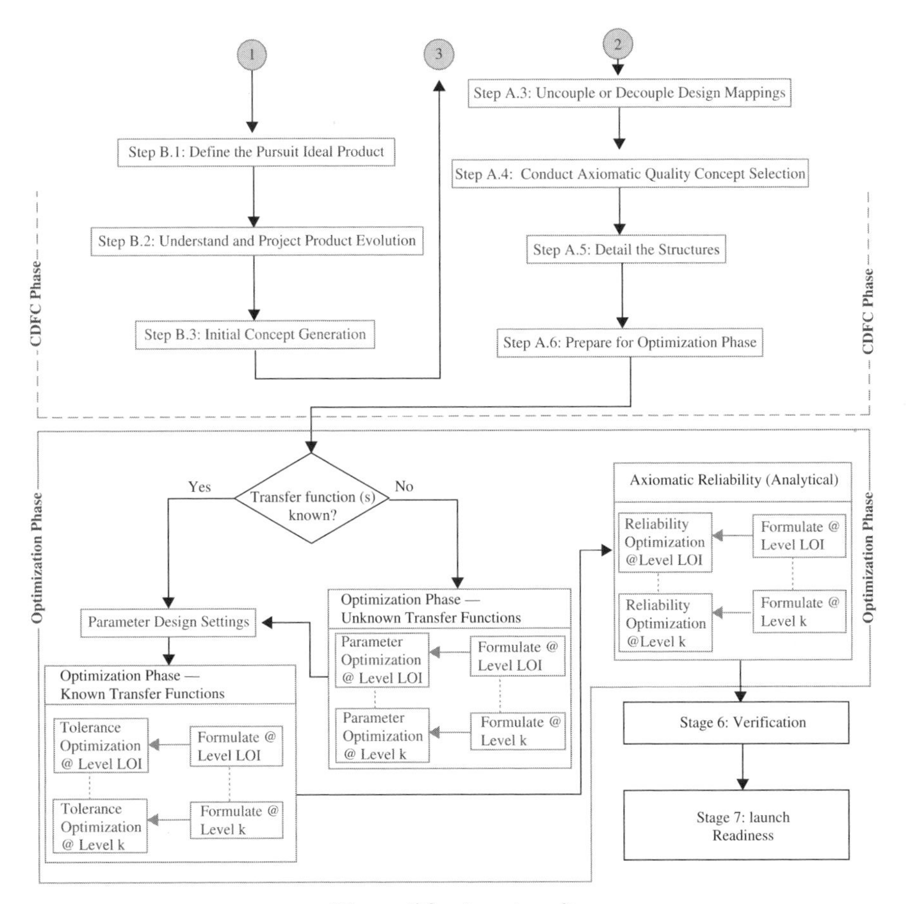

Figure 6.3 (*continued*)

and reliability engineering concepts within each step. We discuss these steps and place emphasis on the cross-functional design team.

A well-developed team has the potential to conceive winning solutions. The growing synergy that arises from ever-increasing numbers of high-performance teams accelerates improvement throughout the design organization. The investment for small, up-front investments in team performance can be enormous. Continuous vigilance at improving and measuring team performance throughout a product cycle will be rewarded with ever-increasing capability and commitment to delivering winning products.

Implementation is, of course, the key procedure. The impact of any process depends on the effectiveness of implementation (i.e., how well the methods, tools, and concepts are practiced by design teams). Intensity and constancy of purpose beyond the norm are required to implement and improve the process. In any type of race, those who go fastest win. Rapid implementation of new race plans and new processes, plus commitment, training, and practice, characterize winning teams.

6.2.3 Axiomatic Quality Design Team

The success of product development activities depends on the performance of this team. Special effort may be necessary to create a multifunctional team that collaborates to achieve a shared design vision. Roles, responsibilities, members, and resources are best defined up front, collaboratively, by all team members. It is very important to "get it right up front," to avoid costly downstream mistakes, problems, and delays. Once the team is established, however, it is just as important to maintain the team to improve its performance continuously. This first step is therefore an ongoing effort throughout the development cycle of design analysis, synthesis, formulation, manufacturing, and production.

The primary challenge for a design organization is to learn and improve faster than the competitor. Lagging competitors must go faster to catch up. Leading competitors must go faster to stay in front. A synergetic design team should learn rapidly, not only about what needs to be done but also about how to do it—how to implement change pervasively when needed.

This first step in team building is to create an environment of teamwork. One thing the team leader will eventually learn is that team members have very different abilities, motivations, and personalities. For example, some team members will be pioneers and others will not want to serve. If the leader allows the latter behavior, such members will become dead weight and a source of frustration. The team leader must not let this happen. When team members do not comply, it is not entirely their fault. If the leader wants the team to succeed, he or she has to accept that they must manage others actively. One of the first things the leader should do as a team is to make sure that every member knows every other member beyond simply an introduction. It is important to get an idea of what each person is good at and what resources each person can bring to the project.

One thing to realize is that when teams are new, each member is wondering about his or her identity within the team. Identity is a combination of personality, competencies, behavior, and position in the organization chart. The team leader needs to push for another dimension of identity, that is, belonging to the same team, with the axiomatic quality project as the task on hand. In addition to the explicit project phased activities, what are the real project goals? A useful exercise is to create a project charter, with a vision statement, among themselves and with the project stakeholders. The charter is basically a contract that says what the team is about, what their objectives are, what they are ultimately trying to accomplish, where to get resources, and what kinds of benefits will be gained as a return on their investment on closing the project. The best charters are usually those that synthesize from each member's input. A vision statement may also be useful. Each member should figure out separately what he or she thinks the team should accomplish, and then together, see if there are any common elements out of which they can build a single coherent vision to which each person can commit. The reason that it is helpful to use common elements of member input is to capitalize on the common direction and to motivate the team going forward.

Teamwork fosters culture transformation and instills execution and pride. It is difficult for teams to succeed without a leader, who should be equipped with several leadership qualities acquired by experience and through training. It is a fact that there will be team functions that need to be performed, and he or she can do all of them or can split the job up among pioneer thinkers within his or her team. One key function is that of facilitator. The leader will call meetings, keep members on track, and pay attention to team dynamics. As a facilitator, the leader makes sure that the team focuses on the project, engages participation from all members, prevents personal attacks, suggests alternative procedures when the team is stalled, and summarizes and clarifies the team's decisions. In doing so, the leader should stay neutral until the data start speaking and should stop meetings from running too long, even if they are going well, or people will try to avoid coming next time. Another key function is that of liaison: serving as liaison between the team and the project stakeholders for most of the work in progress. Finally, there is the project management function. As a manger of the design project, the leader organizes the project plan and sees that it is implemented. He or she needs to be able to break a project task down into scoped and bounded activities, with crisp deliverables to be handed out to team members as assignments. The leader has to be able to budget time and resources and get members to execute their assignments at the right time.

Team meetings can be very useful if done right. One simple thing that helps a lot is having an updated agenda. Having a written agenda, the leader will make it useful for the team to steer things back to the project activities and assignments, the compass. The design teams emerge and grow through systematic efforts to foster continuous learning, shared direction, interrelationships, and a balance between intrinsic motivators (a desire that comes from within) and extrinsic motivators (a desire stimulated by external actions).

6.3 CUSTOMER ATTRIBUTES-TO-FRs MAPPING: UNDERSTANDING THE VOICE OF THE CUSTOMER

As depicted in Figure 6.3, the axiomatic quality process requires some input to formulate the design problem objective and scope correctly. Correct formulation of the design project objective ranges from modification (incremental design) to totally new design (creative design). Before leaping to the CDFC phase of the axiomatic quality process, customer requirements need to be studied thoroughly. This cannot happen without a well-architected cross-functional design team tasked with the design problem. We emphasize the customer-to-FRs mapping, the first phase of the axiomatic quality process, as the initial necessary work of any design problem if we are to reach optimum conclusions.

The identification of key customer design wants is a good lead into how the "voice of the customer" is collected and analyzed. A major step is listening to the customer captures wants and needs through focus groups, interviews, councils, field trials and observations, surveys, or any other form of customer engagement. The design team needs to analyze the customer's communications and assign satisfaction performance ratings to design (product and product attributes) using the quality function deployment (QFD) method, depicted in Figure 6.4.

QFD is best viewed as a planning and problem-solving tool that relates a list of wants and needs of customers to product technical functional requirements, although in more than one stage. With the application of QFD, possible relationships are explored between characteristics expressed by customers and substitute quality requirements expressed by the design team (Cohen, 1988). The substitute quality characteristics are called *critical to satisfaction* in the six-sigma context. In the QFD methodology the customer defines the product using his own expressions, which usually do not carry any actionable engineering terminology. The voice of the customer can be discounted into a list of needs used later as input to a relationship diagram, which is called QFDs *house of quality* (HOQ). Full QFD activity expands over four stages. Yang and El-Haik (2003) discussed the role of QFD within the broad perspective of design for six-sigma (DFSS).

Following are brief descriptions of each room of the HOQ.

Customer Attributes (CAs) These are obtained from the voice of customer from such sources as surveys, interviews, focus groups, showrooms, claim data,

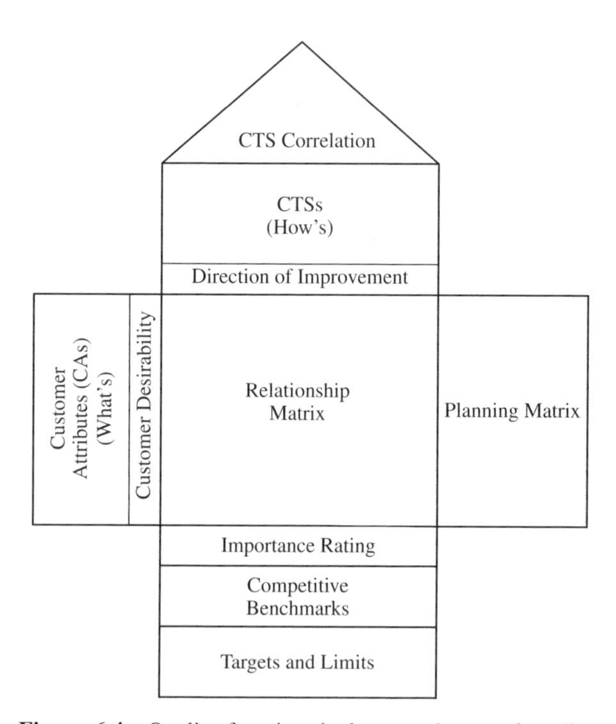

Figure 6.4 Quality function deployment house of quality.

warranty, and promotion campaigns. Usually, customers use fuzzy expressions in characterizing their needs, with several dimensions to be satisfied simultaneously. Affinity and tree diagrams may be used to complete the list of needs. Most of these CAs are very general ideas that require more detailed definition. For example, customers often say that they look for something "stylish" or "cool' when they purchase a product. "Cool" may be a very desirable feature, but since it is interpreted differently by different people, it cannot be acted upon directly. Legal and safety requirements or other internal wants are considered extensions of the CAs for the voice of the business. The CAs can be characterized using the Kano model.

Critical-to-Satisfaction (CTS) A CTS design feature is derived by the design team to respond to the CAs. Each of the initial CAs needs operational definitions. The objective is to determine a set of CTSs with which CAs can be materialized. A key logical question to ask is: How can the CAs be delivered? The answer is in effect a translation of customer expectations into design criteria such as speed, torque, and time to delivery. For each customer attribute (CA) there should be one or more CTSs that describe the means of attaining customer satisfaction. For example, a "cool car" can be achieved through body style (different and new), seat design, leg room, lower noise, harshness, and vibration requirements. At this stage only overall characteristics that can be measured and controlled need to be determined. These substitute for customer needs and expectations and are traditionally known as substitute quality characteristics (SQCs). In this book we adopt the (CTS) terminology aligned with six-sigma in naming the SQCs. Teams should define the CTSs in a solution-neutral environment and not be restricted by listing specific parts and processes. Just itemize the means (the CTSs) whereby the list of CAs can be realized. The 1-to-1 relationships between the CAs and the CTSs are not part of the real world, and many CTSs will relate to many customer wants.

Relationship Matrix The process of relating CAs to CTSs often becomes complicated by the absence of 1-to-1 relationships, as some of the CTSs affect more than one CA. In many situations, they affect one another adversely. CTSs that could have an adverse effect on another customer want are important. For example, "cool" and "powerful" are two of the CAs that a customer would want in a vehicle. The CTSs that support "cool" are speed, lower noise, roominess, and seat design requirements, among others. These CTSs will also have some effect on the "powerful" as well. A relationship is created in the HOQ between the CTSs as columns and the CAs in the rows. The relationship in every (CA, CTS) cell can be displayed by placing a symbol representing the cause-and-effect relationship strength in that cell.

After determining the relationship strength of each (CA, CTS) cell, the design team should take the time to review the relationship matrix. For example, blank rows or columns indicate gaps in either team's understanding or deficiency in fulfilling customer attributes. A blank row shows a need to develop a CTS for the

CA in that row, indicating a potentially unsatisfied customer attribute. When a blank column exists, one of the CTSs does not affect any of the CAs. Delivering that CTS may require a new CA that has not been identified, or it might be a waste. The relationship matrix gives the design team the opportunity to revisit their work, leading to better planning and therefore better results.

What is needed is a way of determining to what extent the CTS at the head of the column contributes to meeting customer attributes at the left of the row. This is a subjective weighing of the possible cause–effect relationships. We utilize such subjectivity in formulating the axiomatic quality concept selection method in Chapter 7.

To rank-order the CTS and customer features, we multiply the numerical value of the symbol representing the relationship by the *customer desirability index*. When summed over all the customer features in the CA array, this product provides a measure of the relative importance of such CTSs to the design and is used as a planning index to allocate resources and to compare the strength, importance, and interactions of various relationships. This importance rating is called the *technical importance rating*.

Importance Ratings Importance ratings are a relative measure indicating the importance of each CA or CTS to the design. In QFD, there are two importance ratings:

1. *The customer desirability index*. This is obtained from the voice of customer activities such as surveys and clinics, and is usually rated on a scale from 1 (not important) to 5 (extremely important).
2. *Technical importance rating*. This is calculated as follows:
   a. By convention, each symbol in the relationship matrix receives a value representing the strength in the (CA, CTS) cell.
   b. These values are then multiplied by the customer desirability index, resulting in a numerical value for the symbol in the matrix.
   c. The technical importance rating for each CTS can then be found by adding together the values of the products in each column.

   Technical importance ratings have no physical interpretation; their value lies in their ranking relative to one another. They are utilized to determine which CTSs have priority and should receive the most resource allocation. In doing so, the design team should use the technical importance rating in their project objectives as a compass coupled with such other factors as difficulty, innovation, cost, reliability, timing, and all other measures.

Planning Matrix This task includes comparisons of competitive performance and identification of a benchmark in the context of ability to meet specific customer needs. It is also used as a tool to set goals for improvement using a ratio of performance (goal rating/current rating). Hauser and Clausing (1988) view this matrix as a perceptual map in trying to answer the following question: How can we change the existing product or develop a new one to reflect customer intent given that the customer is more biased toward certain features? The product of customer value, the targeted improvement ratio for the raw (feature), and the sales point constitutes a measure of how the raw feature affects sales and will provide a weighted measure of the relative importance of this customer feature to be considered by the team. See Chapter 7 for more insights into this room use for concept selection.

CTS Correlation (the Roof of HOQ) Each cell in the roof is a measure of the possible correlation of two different CTS characteristics. Use of this information improves a team's ability to develop a systems perspective for the various CTSs under consideration.

Designing and manufacturing activities involve many trade-off decisions, due primarily to the intentional or unintentional violation of design axioms. The correlation matrix is one of the more commonly used optional extensions over the original QFD, developed originally in Japan. Traditionally, the major task of the correlation matrix is to make trade-off decisions by identifying the qualitative correlations among the various CTSs. This is a very important function in the QFD because CTSs are most often *coupled*. For example, a matrix contains "quality" and "cost," in which the design team is looking to decrease cost, but any traditional improvement in this aspect (e.g., upgrading material) may have a negative effect on the quality. This is called a *negative correlation* in QFD literature and must be identified so that a trade-off can be addressed appropriately. Trade-offs are customarily accomplished by revising the long-term objectives (CTS settings). These revisions are called *realistic objectives*. Using the negative correlation example discussed previously, in order to resolve the conflict between cost and quality, a cost objective would be changed to a realistic objective.

In a correlation matrix, once again, symbols are used for ease of reference to indicate the various levels of correlation. In a coupled design scenario, both positive and negative interactions may result. If one CTS supports another CTS directly, a positive correlation is produced. Correlations and coupling can be resolved only through conceptual methods such as TRIZ (Chapter 8) and axiomatic design (Chapters 2 and 3). Otherwise, a coupled design results and trade-offs are inevitable, leading to compromising customer satisfaction with design physics.

Many coupling situations are the result of a conflict between design intent and the laws of physics. In many cases, the laws of physics win, due primarily to the ignorance of design teams as to design mapping and other axiomatic concepts. In several transactional (service-type) design projects, coupling situations may have to be resolved by high-level management because departmental and sectional functional lines are being crossed.

Targets or CTS Settings For every CTS shown on the relationship matrix, a setting should be determined. The goal here is to quantify the customers' needs and expectations and create a target for the design team. The settings also create a basis for assessing success. For this reason, CTSs should be measurable. It is necessary to review the CTSs and develop a means of quantification. Target orientation to provide a visual indication of target type is usually optional but highly recommended in light of Chapter 5.[2] In addition, the tolerance around targets needs to be identified based on the company's marketing strategy and (comparison) with the best-in-class competitor. This will enable determination of the FR specification limits, $T \pm \Delta\text{FR}$.

Competitive Assessments or Benchmarking Competitive assessments are used to compare the competitors' design with the team's design. There are two types of competitive assessments:

1. *Customer competitive assessment.* This is documented in the planning matrix. A voice of the customer (VOC) analysis is used to rate the CAs of the various designs in a particular segment of the market.
2. *Technical competitive assessment.* This is at the bottom of the relationships matrix. It rates CTSs against competitor CTSs from a technical perspective.

Both assessments should be aligned, and a conflict or mismatch between them indicates a failure to understand the VOC by the team. In a case such as this, the team needs to revisit the CTS array, check their understanding, and contrast that understanding with the VOC data. Further research may be needed. The team may add new CTSs that reflect the customer's perceptions to fill any gaps discovered. Any unexpected items that violate conventional wisdom should be noted for future reference. Situations such as this can be resolved only by having the design team involved in comparing competitive designs. In this way, the team that is responsible for designing customer attributes will interpret exactly what those wants are.

In defining their wants and delights, customers use some vague and fuzzy terms that are difficult to interpret or attribute to specific engineering terminology. Since they are technical terms, however, the design team determines FRs as implied by customers. As a result, uncertainty or linguistic inexactness in the VOC may lead to inaccurate or incorrect interpretation and therefore unsatisfactory results. Linguistically, customers use the three types of inexactness—generality, ambiguity, and vagueness terms—in describing what they want. The generality happens where a word applies to a multiplicity of objects in the relevant fields of design. For example, the object "seat" in the function "manufacture seat" can apply to things differing in size, shape, and material. Ambiguity happens due to the association of a finite number of alternative meanings having the same phonetic form. Customers may also use vague linguistic terms where no precise boundaries to the meaning of a word can be definited.

[2]We are referring to the classification of CTSs and FRs as smaller-the-better, larger-the-better, or nominal-the-best.

To overcome the linguistic inexactness barrier, we propose two QFD stages in the first mapping of axiomatic quality to reach a firm FR definition systematically, as depicted in Figure 6.5. QFD stage 1 translates customer needs and expectations into the CTSs. Subsequently, the CTSs must be converted into functional requirements in QFD stage 2. QFD stage 2 is a new HOQ on which the CTSs and their target values from QFD stage 1 occupy most of the room still available in the HOQ (Figure 6.5). The CTSs and their settings for each matrix are deployed progressively as "what's" on the charts or matrices that represent the next stage (see, e.g., Figure 6.4). Once QFD stage 2 is completed, the functional requirements can be cascaded using the axiomatic design *zigzagging process*. The zigzagging approach provides more insight into the design when the axiomatic quality CDFC phase is employed.

6.3.1 QFD Stage I

Customer attributes (the "what's" in Figure 6.4) are usually obtained from market research and customer engagement techniques. Market research is gathered in two ways: through indirect information (e.g., surveys, questionnaires, competitive benchmarking, projections, consumer labs, trade journals, the media) and through direct customer engagement, including current, potential, and competitors' customers (e.g., interviews, focus groups, customer councils, field observations and trials, and any other means appropriate).

In the context of Figure 6.2, CAs are potential benefits that the customer could receive from the design and are characterized by qualitative and quantitative data. Each attribute is ranked according to its relative importance to the customer.

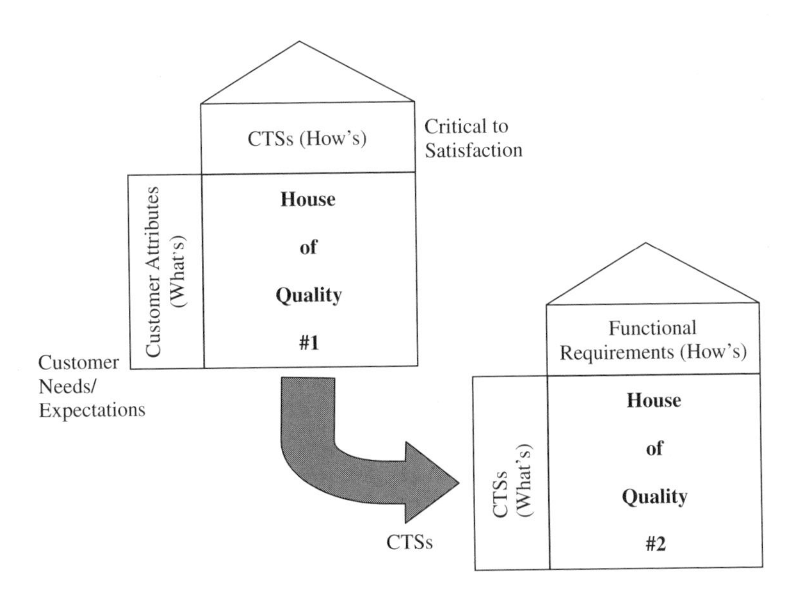

Figure 6.5 Axiomatic design for CA-to-FR mapping.

This ranking is based on the customer's satisfaction with similar design entities featuring that attribute (incremental redesign case). A model recommended for data characterization was developed by Robert Klein (Cohen, 1995).

An understanding by the design team of customer expectations (wants, needs) and "delights" is a prerequisite to further development and is, therefore, the most important action prior to beginning physical and process mapping. The fulfillment of these expectations and the provision of exciting delights will lead to satisfaction. This satisfaction will ultimately determine what products and products the customer is going to endorse and buy. In doing so, the design team needs to identify constraints that limit the delivery of such satisfaction. Constraints present opportunities to exceed expectations and create delighters. The identification of customer expectations is a vital step for the development of six-sigma products that customer will buy in preference to those of competitors. Noriaki Kano, a Japanese consultant, has developed a model relating design characteristics to customer satisfaction (Cohen, 1995). This model (see Figure 6.6) divides characteristics into categories, each of which affects customers differently: dissatisfiers, satisfiers, and delighters.

Dissatisfiers are also known as basic, "must-be," or expected characteristics and can be defined as a characteristic that a customer takes for granted and that causes dissatisfaction when it is missing. *Satisfiers* are known as performance, one-dimensional, or straight-line characteristics and are defined as things the customer wants and expects—the more the better. *Delighters* are those features that exceed competitive offerings in creating pleasant, unexpected surprises. Not all customer satisfaction attributes are equal in importance. Some are more important than others to customers in subtly different ways. For example, dissatisfiers

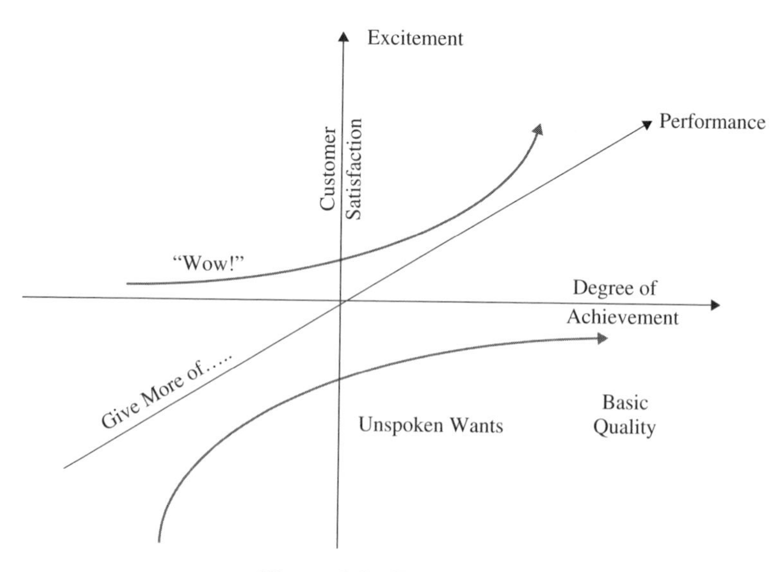

Figure 6.6 Kano model.

may not matter when they are met, but subtract from overall design satisfaction when they are not delivered.

When customers interact with the team, delights often surface that neither would have conceived independently. Another source of delighters may emerge from team creativity, as some features have the unintended result of becoming delights in the eyes of customers. Any design feature that fills a latent or hidden need is a delight, and with time, becomes a want. Delighters can be sought in areas of weakness, competition benchmarking; and technical, social, and strategic design innovation.

The design team should conduct a customer evaluation study. This is difficult to do in creative design situations (see Figure 6.3). Customer evaluations is conducted to assess how well the current or proposed design delivers on the needs and desires. The method used most frequently for this evaluation is to ask the customer (e.g., via a clinic or survey) how well the design entity is meeting each expectation. To outperform the competition, the team must also understand the evaluation and performance strategies of their toughest competition. In the planning matrix of the QFD method depicted in Figure 6.4, the team has the opportunity to grasp and compare, side by side, how well the current, proposed, or competitive design solutions are in delivering on customer needs.

The objective of the planning matrix evaluation is to broaden the team's strategic choices for setting customer performance goals. For example, armed with meaningful customer desires, the team could aim their efforts at either the strengths or weaknesses of best-in-class competitors. In another choice, the team might explore other innovative avenues to gaining competitive advantages.

The array of customer attributes should include all customer and regulatory requirements and all social and environmental expectations. It is necessary to understand requirement and prioritization similarities and differences in order to understand what can be standardized and what needs to be tailored.

Customer attributes (i.e., QFD stage 1 "what's") and social, environmental, and other company wants can be refined in matrix form for each market segment identified. The customer importance rating is the main driver for assigning priorities from both customer and corporate perspectives, as obtained through direct or indirect engagement forms with the customer.

6.3.2 QFD Stage 2

At this point it is important to inject additional QFD terminology into the section. The CTS is an array of design features derived by the design team to answer the CA array. The CTS array is the "what's" room in this QFD stage. Each CTS needs an operational definition. The objective is to determine a set of FRs, the "how's" in this case, with which the CTSs can be helped to materialize. The answering activity translates customer expectations into requirements such as speed, torque, and time to delivery. For each CTS, there should be one or more FRs that describe a means of attaining customer satisfaction.

Only overall CTSs that can be measured and controlled need to be used. Similar to QFD stage I, relationships between technical CTS and FR arrays are often used to prioritize CTSs filling the relationship matrix of QFD stage 2. For each CTS, the design team has to assign a value that reflects the extent to which the FRs defined contribute to meeting it. This value, along with the importance index of the CTS, establishes the contribution of the FRs to overall satisfaction and can be used for prioritization.

The analysis of the relationships of FRs and CTSs allows a comparison with other indirect information, which needs to be understood before prioritization can be finalized. The new information from the planning matrix in the QFD needs to be contrasted with the design available information (if any) to ensure that reasons for modification are understood.

The purpose of the CA-to-FR mapping phase of axiomatic quality is to define the design in terms of customer expectations, benchmark projections, institutional knowledge, and interfaces with other systems, and to translate this information into technical functional requirement targets and specifications. This will facilitate the start of physical mapping in the CDFC phase. This step will provide a starting basis for the logical question pair (what? how?) employed to define the design structures.

A major reason for customer dissatisfaction is that design specifications do not link adequately to customer use of the product or product. Often, the specifications were written after the design was completed, or the specifications may be outdated. This reality may be attributed to the current planned design practices, which do not allocate activities and resources in areas of importance to customers and waste resources by spending too much time in activities that provide marginal value, a gap that is nicely filled by the QFD stages in axiomatic quality first mapping (Figure 6.3). The target and tolerance settings activities in QFD stage 2 will also be stressed at the beginning of the design physical and process mappings.

The approach is to spent time up-front understanding customer expectations and delights together with corporate and regulatory wants. This understanding is then translated into FRs with design specifications (tolerances) over two stages of QFD, which then cascaded to all levels of design hierarchy. The power of first gaining complete understanding of requirements and then translating them into specifications was highlighted by Pugh (1991).

It is a good practice to set FR definition guidelines to minimize existing inexactness as well as to reduce any confusion the team already has. The following guidelines prove to be useful (Shiba et al., 1993):

- Avoid attribute description, such as binary (0–1) requirements. Instead, use ratio-scale variable requirements with defined measurement systems and units of measurement.
- FRs should be defined in a solution-neutral environment. Avoid statements of solution (e.g., "design is made of x ..."); instead, use "design fulfills x ...."
- Avoid auxiliary verbs such as "must" or "have to"; instead, say "design needs to be...."

- Avoid intangible concepts, instead, use concrete terms ("add more sugar" rather than "tasty").
- Avoid statements in nonpositive form (e.g., "design does not . . ."); instead, "design performs x . . .."
- Avoid abstract words such as "reliable" and "durable"; instead, "design withstands x environmental conditions. . .."
- Avoid premature detail (e.g., "design dimensions are $a'' \times b''$ . . .").
- Use this function definition template: active verb + noun + qualifier.

6.4 CONCEPTUAL DESIGN FOR CAPABILITY PHASE

This section serves as a core of the CDFC phase. Chapter 8 is a natural extension to clarify some of the concepts briefed here. This section and Chapter 8 will give a complete picture of the CDFC phase.

The CA-to-FR mapping discussed in Section 6.3 begins by considering high-level customer attributes of the design. These are the true attributes, which define what the customer would like if the design entity were ideal. This consideration of a product from a customer perspective must address the requirements from higher-level systems, internal customers (such as manufacturing/production, assembly, product, packaging, etc.), external customers, and regulatory legislation. True attributes are not directly operational in the world of design teams. For this reason it is necessary to relate customer attributes to the CTSs and then to functional requirements (Section 6.3) that may readily be measured and when properly targeted will substitute or assure performance to the true quality of the attributes (Cohen, 1995).

In performing the physical and process mappings of the axiomatic design methodology, the design team may start developing a testing matrix for validation and keep updating it as more details are achieved. They need to create tests that cover all customer attributes and eliminate unnecessary and redundant tests (i.e., testing a hidden factory).

6.4.1 Define FR Specification Target Values and Allowable Tolerances

This step is conducted utilizing historical targets, and variation provides an initial source of information, if any. Not all FR targets and allowable variation can be estimated at this step. However, it is very beneficial to start this activity to gauge the gap of what is already known and what needs to be known. Such a gap is wider for creative design projects than those that are incremental. Competitive benchmarking, use profiles, and testing are useful tools to aid the design team in understanding customer use and competitive performance.

On the other hand, it is also important to understand competition trends. The trend appreciation is vital because the team should set the design targets to beat what the competition will release, not what they have in the market now. Based on the information above, the design team selects the appropriate test target and allowable variation for each test. This selection is based on the team's understanding of the relationship matrix in QFD stage 1 and 2 studies so that the appropriate values may be chosen to satisfy design targets. Usually, targets may be modified on the light of customer studies. This involves verifying the target and variation with actual customers. On some occasions, surrogates might be pieced together to measure customer reaction; on others, a meeting with internal customers may be necessary. Targets are tuned, and trade-off decisions are refined after assessing customer reaction. The preliminary specification may now be written. The design team will select tests for the verification and in-process (ongoing) testing.

The action suggested here are prerequisite to proceeding in the right path according to the design project classification (i.e., incremental or creative) (Figure 6.3). The design team then proceeds to check the availability of datum solutions that address the array of FRs. The team will study the datum entities against the functional requirements generated by stage 2 QFD to check if at least one solution, a design entity, exists that is the approximate physical translation of the functional requirements. A "yes" answer (Figure 6.3) implies that the entity selected is a *derivative* of baseline design (within a narrow deviation from the baseline), and design changes (hard and/or soft) may be enough to achieve the design described by the customer. The team may declare the project as an incremental design problem and work toward improvements in order to satisfy customer requirements, progressing from the datum design as a starting point. Assuming an incremental design scenario with several baselines, the best entity could be selected using the axiomatic quality concept selection method in Chapter 7. In the absence of datum entities, the only option is the creative design, which requires more conceptualizing work and therefore extensive employment of innovative methods.

The objective of specifying target and tolerances of the FRs and then to the DPs is to verify choices for the functional solution entity elements and interfaces. Once the targets and tolerances have been determined, the physical mapping activity begins. This activity employs the zigzagging method (Chapter 3) to reveal design hierarchy together with design matrixes and coupling assessment at each hierarchal level. It should also witness extensive activities of the axiomatic design concepts utilization.

6.5 OPTION A: CONCEPTUAL DESIGN FOR THE CAPABILITY PHASE OF AN INCREMENTAL DESIGN

An *incremental design* is a design that can be achieved by hard or soft changes. In the context of axiomatic quality process, hard changes imply eliminating or adding DPs or PVs in the concerned mapping without altering the original set of FRs. Hard changes imply a redesign (incremental) cycle. Soft changes imply adjusting the target values within the tolerances specified, changing the tolerance ranges, or both. Soft process changes can be achieved by adjusting the DPs within the permitted tolerances, while hard changes may require PV array alteration. The objective is to keep transferred variation to the FRs within its specified tolerances in the context of Figure 1.5. A creative design case occurs when there is a need to alter the FR vector by adding or deleting functional requirements.

What needs to be done in this phase is to make sure that technical characteristics, the FRs, important to customers are optimized in the development of the product. Efficiency can be gained by having the entire design team plan the engineering tasks up front. Activities that should be considered include, but are not limited to, functional structuring, zigzagging and mapping, analytical transfer function development, preparation for tests for development/discovery, and preparation for optimization. The team needs to map the relationship of each activity to product and process characteristics, the DPs and the PVs. A matrix is created to compare the list of brainstormed design activities to the list of implementable requirements. Relationships are designated as no relationship, weak relationship, medium relationship, and strong relationship. In addition, the list of activities may be improved by both adding steps to cover missing or weak areas of the matrix (to improve the development of FRs that are covered weakly in the original list) and to delete redundant activities that are already covered sufficiently by other planned activities. The team may consider combining activities for synergy. For example, robust design studies with transfer function optimization can be conducted using computer simulations. Once the engineering activities are planned, workload and timing can be established and resources allocated.

Recommended thoughts in this context are:

- Understand timing, resource constraints, and checkpoints. The workload cannot be established until the timing is thoroughly understood. Constraints with respect to capital equipment and budget also need to be understood, as well as the requirements for various checkpoints.
- Estimate the workload associated with the activity. The engineering workload is now estimated based on the planned engineering activities with the required timing. The team needs to map out a sequence of planned events; determine which events must be done in series and which in parallel; and identify the critical path.
- Allocate resources. Resources for the various activities are then allocated.

6.5.1 Step A.1: Perform the Physical Mapping (Design Analysis)

The primary intention of physical mapping is to fulfill the FRs through the zigzagging process presented in Chapter 3. In each hierarchical design level, the array of FRs is a discrete list of independent requirements by definition. Considering the independence axiom, the design team would like to have a one-to-one relationship between FRs and DPs in all design mappings across the overall design hierarchy. Ideally, we would want a square diagonal matrix per Theorem 2.4 (Section 2.5). Several concepts of the axiomatic design method presented in Chapters 2 and 3 need to be employed in this step, including zigzagging, uncoupling or decoupling, and complexity reduction.

6.5.2 Step A.2: Perform the Physical Mapping (Design Synthesis)

The synthesis activity is conducted using the following steps:

1. Capitalize on the functional mappings and design hierarchy performed earlier (revisit Chapters 2 and 3).

2. Search for design parameters (DPs) to fulfill the functional requirements in all design hierarchal levels. As the team zigs to the physical domain from the functional domain, there may be several conceptual solution or DPs for each FR. As design hierarchy is established through zigzagging, possible alternatives, multiple DP choices, can be brainstormed for an FR. For example, the manufacturing function "build assembly" can be satisfied by DP_1 = "fasteners," DP_2 = "welding," or DP_3 = "casting process."

It is necessary to discuss and compare various conceptual solutions for each FR. The design conceptual solutions, as well as the methods of combining them, will enable the architect of the physical structure. In this step, the design activities decide both the feasibility and compatibility of alternative solutions by narrowing down the *theoretically* possible solutions to *practically* possible solutions of the preliminary structure.

3. Combine the design parameters into overall physical structure alternatives. The development of the overall design through a combination of solution alternatives for FRs can be identified by a matrix technique called a *morphological matrix*, where functions are listed in the rows and the DPs (components, subsystem, physical effect, physical fields, etc.) are laid down in the columns, as depicted in Figure 6.7. The functions need to be grouped according to the type of exciting input signal (energy type, material type, control type).

Mathematically, let i be the index of FRs, $i = 1, 2, \ldots, m$; j be the index of DPs, $k = 1, 2, \ldots, K$; F_i be the set of potential (alternative) DPs of FR_i with cardinality N_i, and $\mathbf{F}$ be the union set of the overall unique potential DPs. Identification of the k alternative solutions according to the FR_i may be facilitated by use of the morphological approach of Zwickey (1984).

The morphological matrix is a process to generate technically and economically feasible design concepts starting with a refined FR definition obtained from QFD HOQ stage 2. The feasible permutations, some percentage of $\prod_{i=1}^{m} N_i$ total permutations, of the individual solution entity in the structure via the synthesis activity needs to be identified, thus broadening the selection options. Connecting all possible solutions using arrows in Figure 6.7 identifies a feasible synthesized system or overall solution: for example, two possible solution variants (systems) that can be identified in Figure 6.7 that are considered synthesis feasible in two modules (the dashed boxes). Alternatives are then winnowed to a reasonable

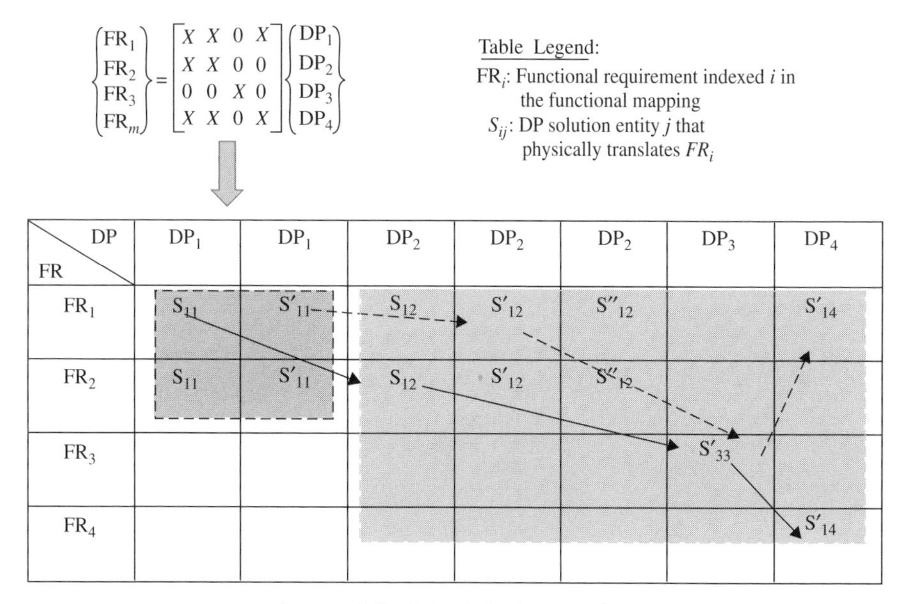

$$\begin{Bmatrix} FR_1 \\ FR_2 \\ FR_3 \\ FR_m \end{Bmatrix} = \begin{bmatrix} X & X & 0 & X \\ X & X & 0 & 0 \\ 0 & 0 & X & 0 \\ X & X & 0 & X \end{bmatrix} \begin{Bmatrix} DP_1 \\ DP_2 \\ DP_3 \\ DP_4 \end{Bmatrix}$$

Figure 6.7 Morphological matrix.

number. That is, the team needs to concentrate on promising combinations of connected arrows (i.e., solution entities). The challenge here is to ensure feasibility; that is, physical and geometrical compatibility do exist, as well as the smooth flow of energy, material, and information subject to design constraints.

4. For each feasible design entity (set of connected arrows in Figure 6.7) that is of interest to them, the team needs to develop the corresponding design structure. A structure is a graphical depiction of design intent broken down into task elements in the form of a block diagram. A physical structure is the embodiment of the design mappings (i.e., fulfills all FRs defined in terms of energy, material, and information flows). In forming the structure, the following is suggested (Pahl and Beitz, 1988):

- Energy FRs (indicated by ——→)
- Material FRs (indicated by ——→)
- Information FRs (indicated by - - -→)

An example of physical structure is depicted in Figure 6.8, which highlights the house heating system example. Notice the way the functional requirements were defined, as well as adoption of the arrow legend.

A structure is a visual description of a design physical mapping and is denoted mathematically as ψ. The design is first defined in terms of its high-level functional requirements, design parameters, and mappings, which are refined progressively to lower hierarchical levels using the axiomatic design zigzagging process.

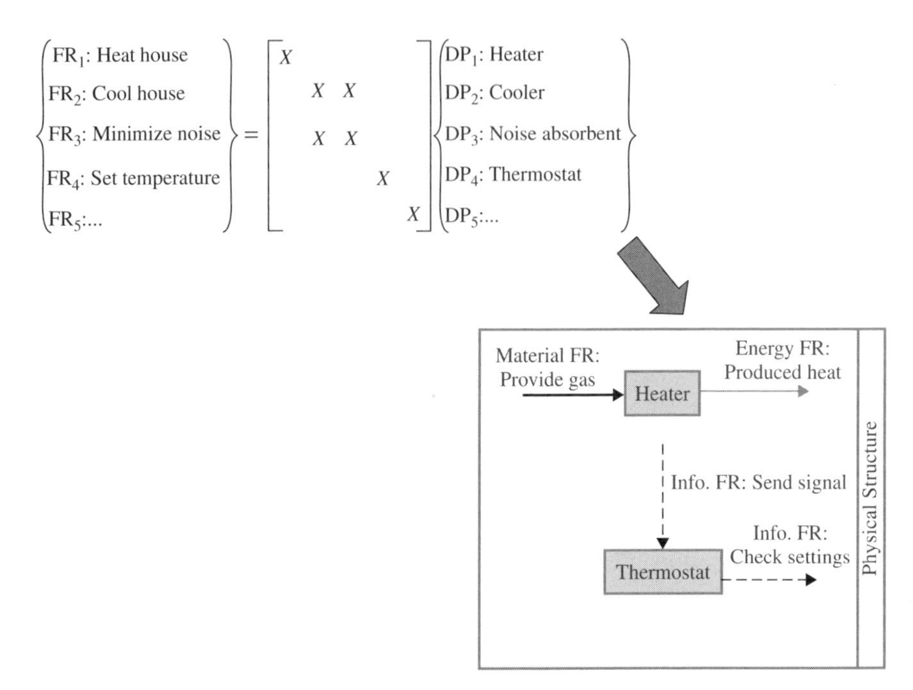

Figure 6.8 Physical structure example.

The advantages of this approach are design synthesis and modular groupings, verification, and testing. The latter can be addressed from the modularity point of view (i.e., from bigger to smaller modules); each level is integrated from the top down; customer requirements are met; and vulnerabilities are considered at the outset, not as an afterthought.

5. Write or rewrite the design matrix and classify the design accordingly. Check for coupling. In the case of a coupled design, the first option is to solve the coupling (i.e., elimination by applying axiomatic design theorems and corollaries). If this is not feasible, we need to use the vulnerability reduction algorithm in Chapter 9. Use the design matrices as a design guide for subsequent design activities to maximize controllability (adjustability) and minimize complexity

6. Pool the population of feasible physical structures and firm into solid solution alternatives by using creativity tools (e.g., TRIZ). TRIZ is a combination of methods, tools, and a way of thinking. In the context of axiomatic quality, TRIZ can be used in concept generation and as a coupling vulnerability solving tool, as discussed in Chapter 8. In essence, when two functional requirements are coupled, TRIZ may suggest different design parameters to uncouple them, resulting in decoupled or uncoupled design. At a very high level, TRIZ is a systematic study of excellence, a philosophy with five key elements: ideality, functionality, resource, contradictions (synonymous to coupling), and evolution.

6.5.3 Step A.3: Uncouple or Decouple the Design Mappings

The objective of this step is to reduce to the farthest possible limit the coupling conceptual vulnerability that appears in coupled design matrices through the hierarchy. Coupling is represented by unwanted off-diagonal elements of a design matrix. The team should consider coupling existence as opportunities for improvement and innovation. It is a very rich source for customer delighters. Therefore, this step witnesses an extensive use of independence axiom theorems and corollaries as well as innovative thinking using TRIZ. This step is covered briefly in this section and is presented in Chapter 8 in more detail.

6.5.4 Step A.4: Conduct Axiomatic Quality Concept Selection

The objective here is to select the best concept from the pool generated so far. We would like to select the best structure to be advanced in the axiomatic quality process stream of development. We will use mathematical programming formulations in Chapter 7, which will enable the design team to select robust concepts based on the information axiom.

Design analysis and physical synthesis are the premier activities performed in this step of the axiomatic quality process. The general process of selecting a winning structure flows very much like Pugh concept generation. However, Pugh (1991) does not differentiate distinctly between functional decomposition (partitioning) and concept creation. At this stage of activity, it is necessary to loop forward to perform the process mapping of axiomatic design (Figure 6.3). Success dictates a concurrent engineering team with heavy participation from manufacturing and production operations.

This step produces convergence on the best solution entity and includes iterative substeps that need to be performed with discipline and rigor. The following first-run sequence of suggested steps facilitates convergence to the best concept in the spirit of Pugh matrix selection (1991):

1. List product attributes, the FRs, in the matrix rows.
2. List alternative solution entities in columns. Entities are defined by performing the physical mapping. Avoid coupling through creative selection of the DPs. If complete elimination is not possible, weaken the coupling by proper selection of the DPs according to design theorems listed in Section 2.5 and Chapter 8 formulations. Coupling vulnerability represents technical bottlenecks and conflicts that need to be resolved or at least minimized.
3. Choose a datum (baseline) from the alternatives generated with which all other concepts are to be compared from the alternative entities. *Score* concepts versus defined selection criteria.
4. Perform trade-off studies to generate more healthy alternatives. Look at the negatives. What is needed in the design to reverse the unsatisfactory FR performance (relative to the datum)? Will the improvement reverse one or more of the existing higher scores (relative to the datum)? If possible,

introduce the modified solution entity into the matrix and retain the original solution entity in the matrix for reference purposes. Eliminate truly weak concepts from the matrix. This will reduce the matrix size. See if strong concepts begin to emerge from the matrix. If it appears that there is an overall uniformity of strength, this may be an indication of mixed interpretation by the team. Uniformity of one or more of the concepts may suggest that they are subsets of the others (i.e., they are not distinct). In this case, the matrix cannot make a distinction where none exists.

5. Having scored the concepts relative to the datum, sum the ranks across all criteria. These scores must not be treated as absolute, as they are for guidance only and as such must not be summed algebraically. That is, the positive (relative to datum) scores are summed together, the negative scores are summed together, and so on. Certain concepts will exhibit relative strengths, while others will demonstrate relative vulnerability.
6. Select the best using the integer programming formulations of Chapter 7.

Before the beginning of the optimization phase, the design needs to pass the conceptual vulnerability reduction in the CDFC, as depicted in Figure 6.3. The objective is to reduce design vulnerabilities by determining the settings of the DPs and PVs that optimize conceptual vulnerabilities. The specification limits should be set around these setting values, which collectively determine the optimization space boundaries. The analytical and conceptual vulnerability reduction techniques within the context of the axiomatic quality process are discussed in Chapter 8.

The team needs to seek big ideas for competitive advantage and customer delight, challenge conventional functional structures with innovative ideas, capitalize on new technologies (DPs), forecast preliminary performance of solution entity elements in the presence of noise factors (operational vulnerabilities), interfaces, and conceptual vulnerabilities. In doing so, the team seeks to minimize and simplify the physical structure according to the information axiom. When creating alternatives, the team leader should foster an environment in which "out-of-the-box" thinking and brainstorming are encouraged. When winnowing ideas, a more structured, disciplined environment is fostered. Iterate back and forth between expansion and contraction of ideas. Improving functionality is the guiding principle in the search for robust structures.

6.5.5 Step A.5: Detail the Structures

Adding available information (not added so far) and identifying the gaps where further information is needed are steps to detailed design structure. For example, adding noise factors with the (energy, material, information) classification discussed earlier for the FRs and noise factors is a very important detailing step. Effects of noise factors are added here as other variables contributing to the structure, as arrows pointing to the relevant structure elements. This step, coupled with the transfer function equation definition, with FR specifications, at the level of impact (LOI) within the design hierarchy, is vital in structure detailing. The LOI is the level at which corporate and team knowledge enable transfer function identification and detailing by writing the respective transfer function equations. Usually, these are the lowest-level design matrices (i.e., the last design matrixes defined using the zigzagging process). Above this level is just layers of abstractions obtained through the zigzagging approach and documented in the mappings of interest. Figure 6.9 depicts the LOI levels of the zigzagging design matrix tree. Notice that not all levels of impact are the lowest hierarchical levels.

The greatest potential here is to enable the optimization phase by utilizing this step as a first perspective of the transfer function detailing. Definition of the transfer function via the mappings is central to selecting and understanding good structural (solution entity) choices.

A detailed structure determines the opportunity to capture the "maximum potential for customer satisfaction" defined in CA-to-FR mapping, the first phase of the axiomatic quality process. The purpose of structural detailing is to establish a framework that enables subsequent axiomatic quality steps to realize this maximum potential. In axiomatic quality, the synthesis of a physical structure is conducted by grouping design parameters into a number of modular elements, which collectively meet customer attributes. Structural modularity, a synthesis step, and mapping, an analysis step, start with the creative, heuristic process of defining functions through employment of zigzagging process logical questions. Structural definition proceeds by identifying conceptual vulnerabilities. A structure is the foundation for analytical or experimental optimization phase for cost, quality, and performance. Because structure definition segments design activities, the design team is able to gain an understanding of how to optimize the functional requirements and structural elements, as well as how to gain an understanding of interfaces, couplings, and measurements. The wrong choice of structure can seldom, if ever, be recovered in subsequent development stages.

The design team needs to check the specification against the structure and add to the structure the noise factors from the environment, customer use, and manufacturing variation. The objective of specifying targets and tolerances of the FRs is to ensure delivery of the customer attributes and to verify structure modular choices. The specifications for the DPs and PVs are set to deliver the targeted FR performance.

6.5.6 Step A.6: Prepare for the Optimization Phase of the Structure Selected

A design physical structure is developed by grouping DPs into modular elements. In the optimization phase, several venues may be followed, depending on the quality of detailing activities that occurred prior to this step, in particular, the transfer function determination at the LOI. We will entertain nonlinear optimization formulations as well as traditional parameter design experimentation to obtain valid transfer functions in Chapter 9.

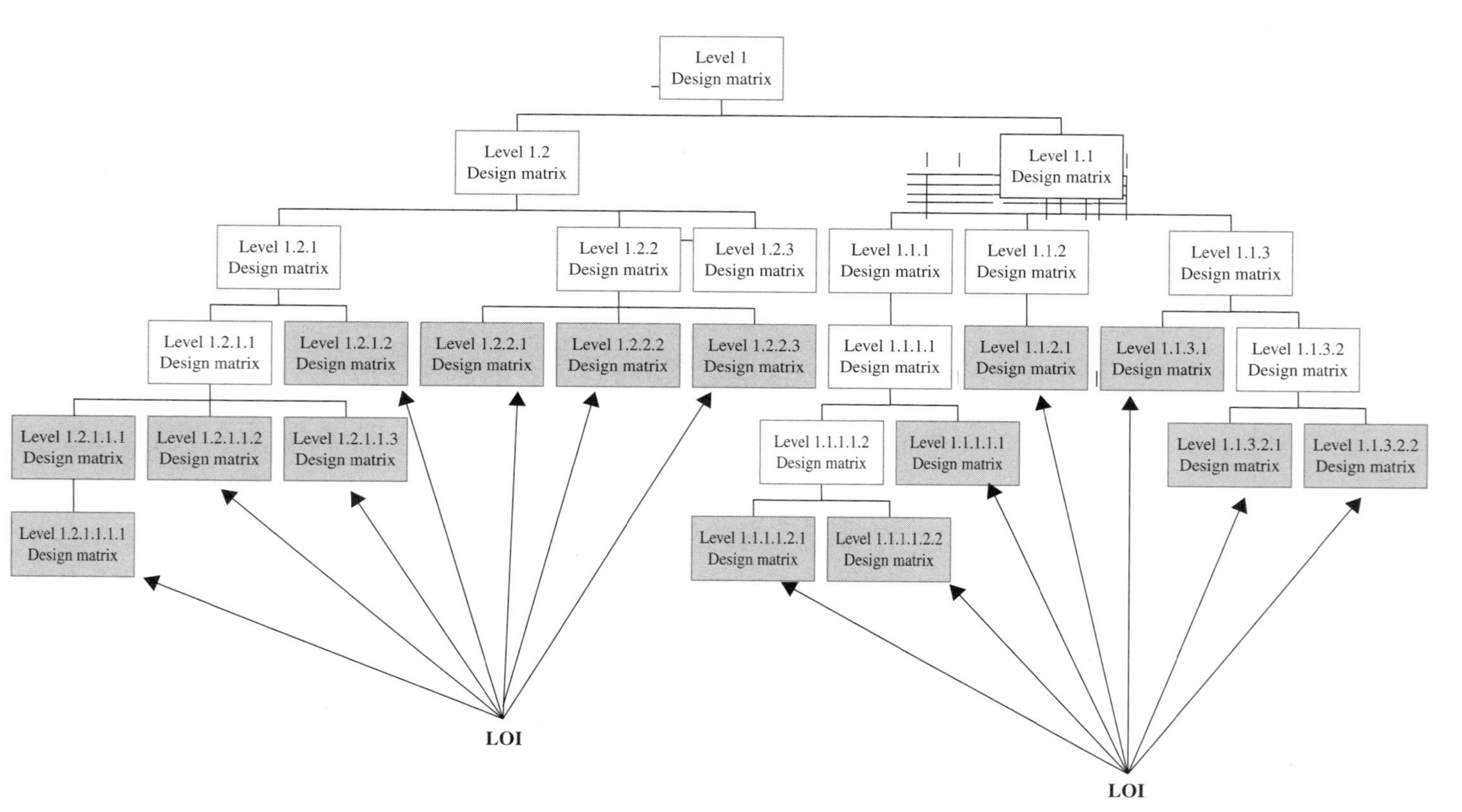

Figure 6.9 Design matrix tree level of impact.

On the experimentation side and from Taguchi's perspective on dynamic systems, an exploration of the solution entity function should precede the experimentation. The *ideal function* is the desired, customer-focused response that would be found if the solution entity element accomplished its intended function perfectly. Solution entity functions may be static or dynamic: constant, linear, or nonlinear response. The transfer function maps the domain requirements (FRs) to the codomain parameters (DPs). The DPs control FR delivery of the solution entity at the initiation of a user's signal (command). The signal may be direct or indirect (e.g., by outputs from other design elements). Inputs may have constant (static) or variable (dynamic) values. The FRs may assume constant, linear, or nonlinear relationships to a range of input values in accordance with the design team's current knowledge of the functional requirements. The team needs to identify where transfer functions are available or missing (at levels on or below the level of impact). Missing transfer functions may be derived or obtained empirically by experimentation.

6.6 OPTION B: CONCEPTUAL DESIGN FOR THE CAPABILITY PHASE OF A CREATIVE DESIGN

Engineering design can be defined as a set of intellectual actions with the objective of creating optimized solution entities (systems) to meet certain needs: however, within some conflicting constraints. Design starts as an idea, a thought, or a concept that matures when the right circumstances of technical capability and economic feasibility are available. Good design depends on a solid foundation of mathematics, laws of physics, design team experience, consideration of societal needs and requirements, and so on. Of special importance are the customer needs, which change continuously at a rapid pace. Such factors produce a complex environment of product design, making the achievement of an optimized solution entity difficult to realize.

Creative engineering design projects include most of the life-cycle stages presented in Chapter 1. A project starts with an understanding of customer attributes, followed by formal definition of the design problem, analysis and synthesis activities, and verification prior to launch. True, such stages sound very similar to an incremental design case. An assessment, however, indicates more concentration of conceptual and creativity tools prior to stage 1 of Figure 1.8, as reflected by several additional steps adding to the effectiveness of the axiomatic quality process.

6.6.1 Step B.1: Define the Pursuit Ideal Product

The definition required here is to turn the knowledge gained from continuous monitoring of consumer attributes, wants trends, competitive benchmarking, and customer satisfiers and dissatisfiers into a preliminary definition of an ideal product. This will help identify areas for further research and dedicated effort. The product should be described from a customer viewpoint (both external and internal) and should provide the first insight into what a good product or process could look like. Concept models and design studies are good sources for evaluating consumer appeal and areas of likes or dislikes.

When groups of customers interact with a design team, delights are often created which neither group would have conceived independently. However, not all innovative ideas come from customer engagement. Delights are unique and may emerge from a number of sources. Many new delights come from engineering creativity, product evolution, known coupling opportunities, and historical issues. Some features have the unintended result of becoming delights in the eyes of customers. Anything that fills a latent or hidden need is a delight, and with time, becomes a want. There are many ways to create innovative delights. Tools such as brainstorming and TRIZ (a technique for systematic innovation developed by Genrich Altshuller of the former Soviet Union) are very helpful (see Chapter 8).

6.6.2 Step B.2: Understand and Project Product Evolution

According to TRIZ, solution entity evolution follows certain basic patterns of development. Evolutionary trends in the functional performance (FRs) of a certain design can be plotted over time and have been found to evolve in a manner that resembles an S curve (Figure 6.10). To predict logical next stages of concept development and to form opinions regarding the limitations of the current technologies, the design team can use such knowledge available to them. The team needs to list historical breakthroughs in technology according to an FR and compare the development of a solution entity (concept) with generic evolution. The team will want to relate technological breakthroughs with evolutionary solution entity improvements. The nature of improvements will assist the team in identifying the stage of development within the particular solution entity, the concept

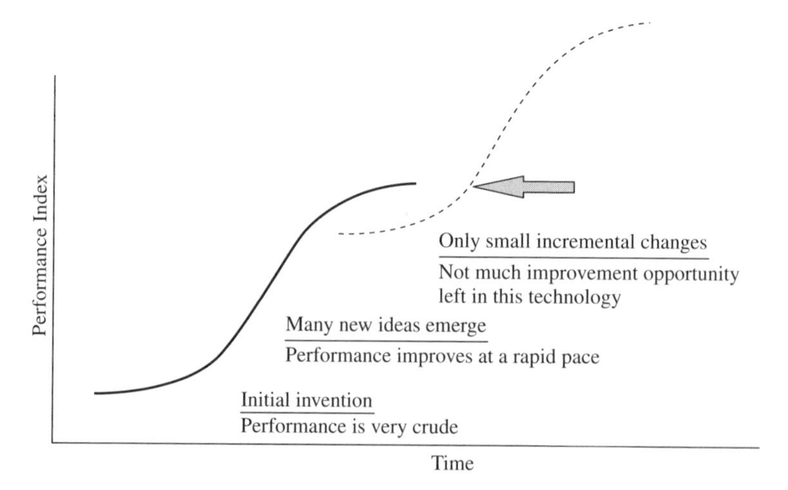

Figure 6.10 S curve of evolution.

under study. The pertinent information can be accessed through literature and patent searches, together with benchmarking of competitive and noncompetitive corporations. In addition, brainstorming of how TRIZ generic evolutionary principles apply to this solution entity element is particularly useful. This activity involves the study of established paths of technical evolution to anticipate future solution entity evolution.

The compatibility of generic evolutionary principles to product applications needs to be evaluated. This activity involves the study of established paths of technical evolution to anticipate future solution entity evolution. The *theory of inventive problem solving* (TIPS, also known as TRIZ) is a valuable methodology for gaining understanding and making projections about technical evolution (Altshuller, 1980; Dovoino, 1993; Tsurikov, 1993).

6.6.3 Step B.3: Initial Concept Generation

The team may start with the FR vector and employ a zigzagging process similar to Section 6.5.1 of the incremental design scenario (Figure 6.3). The rest of the creative design option loops back to incremental design steps beginning with step A.1. The TRIZ methodology provides ample opportunities to supplement the physical structure of the design as well as a solution pool to resolve coupling vulnerability. This will provide an initial structure that needs to be detailed with any available information. To provide discipline in the process of analysis, synthesis, and selection of the solution entities, it is recommended that a morphological matrix be developed that will enable the evaluation of alternative concepts against the FRs defined in the first phase of the axiomatic quality process. Besides the FRs, the entire team may decide additional criteria of selection.

Information from target determination should be cascaded to the solution entity hierarchical levels under development and translated into measurement criteria. A targeted product cost should be established at the outset and checked against those of similar products. A life-cycle costing approach is also suggested to offset short-term thinking and to minimize economic suboptimization.

A selection matrix may take a shape similar to Figure 6.5 or and other format the team sees appropriate for expressing ideas, and the criteria for evaluating these ideas, in a visual and user-friendly fashion. It will be used to justify that the best DP choice has been made and to justify why certain DP solution entities are inferior and deserve to be discarded. Feasibility and compatibility of the DPs when synthesized together should be checked with great caution. Available documentation, if any, for the physical structure of some of the mappings should be provided in preparation for including them in the evaluation matrix.

A matrix for organizing concepts should be constructed. Solution entity alternatives should be characterized with sufficient clarity to ensure that all team members understand intent. Entities should be augmented with schematics, the structure, and other appropriate descriptors. Word descriptions and models should be utilized for additional clarity. All concepts should be characterized at same level of detail. Alternatives should be titled and numbered for ease of reference.

Alternatives are formed by the synthesis activity. Synthesis in this case means selecting a feasible structure. Since a functional requirement may be mapped physically to possibly different DPs, not all of them are feasibly synthesized with the DPs of the rest of the FRs. The techniques useful in idea generation and synthesis include analogy, brainstorming, inversion, combination, and evolution.

The purpose of the concept generation step is to generate technical solutions (physical structure) to meet the requirements derived from the customer attributes. This step also involves solution evaluation and selection. Concept design methods apply to the entire solution entity (as a process continuum) and should be applied at every hierarchical level of the product and process design. The conceptual stage of any design is concerned with the synthesis of separate elements into a connected whole. This step provides a systematic way of generating and selecting the best product and process design to satisfy customer, corporate, and regulatory requirements. This method of concept creation minimizes constraints to creative thinking. During this step of design, creativity is critically important and should be leveraged to the fullest to produce a competitive advantage. This concept generation and selection method also helps to prevent conceptually weak designs from going into production. A design that exhibits conceptual vulnerability have limited potential for improvement in the use environment. Pugh (1991) stated: "The wrong choice of concept in a given design situation can rarely, if ever, be recouped by brilliant detail design." The solution entities need to be generated and selected (Chapter 7). The healthy generation practice objective is to eliminate functional coupling via creativity approaches, including TRIZ.

6.7 AXIOMATIC QUALITY OPTIMIZATION PHASE

The major steps within the axiomatic quality optimization phase are depicted in Figure 6.3. The purpose of the optimization step in the development process is to minimize the operational vulnerability of variability from intended or ideal design through analytical or experimental optimization techniques. The optimization phase is enabled by minimizing conceptual vulnerabilities that were resolved in the CDFC phase. The analytical techniques involve nonlinear programming formulations. The experimental optimization techniques include exposing product and process requirements to representative sources of variation (noise factors) while testing. The decision as to which optimization stream to be adopted depends on the availability of the transfer functions in the design mappings after step A.5 (Figure 6.3). A transfer function, a mathematical model, may be developed that will predict the optimum combination of design parameters and parameter target settings. This activity enables simultaneous evaluation of many designs or process parameters for their improvement potential. This step facilitates efficiency in design development by stressing the development of sound measurement strategies, which are based on the measurement of functional requirements. On the empirical optimization side, this phase involves a systematic approach to anticipating downstream sources of product and process noise. The optimization phase takes maximum advantage of the cost and quality performance advantage which exists for preventing problems (during the early stages of product/process design).

Optimization (in the context of axiomatic quality) means that the team must seek to identify the best expression of a design (product and process) that is the lowest total cost solution to the customer-driven product design specification. "Good enough" or "within specifications" are not concepts that are compatible with the optimization phase. The target is six-sigma. To analyze a system's robustness and adjustability, unique metrics such as quality loss function (Chapters 1 and 5) are available which make it possible for engineers to use powerful experimental methods to optimize a product's insensitivity to noise factors, the sources of variability.

The objective of the engineering team is to develop products and processes that function as intended under a wide range of conditions for the duration of a product's design life. Optimization techniques are systematic activities that extract the best functional performance from design parameters under development and produce product performance that is minimally affected by noise at the six-sigma level. The optimum design is the one that will provide the best functional performance for the system with the least amount of conceptual vulnerability while reaching insensitivity to uncontrollable sources of variation, thus minimizing operational vulnerability. The analytical and empirical optimization phase details are deferred to Chapter 9. In the following sections we explore the deployment stream of the axiomatic quality process.

6.8 AXIOMATIC QUALITY PROCESS DEPLOYMENT

The extent to which the axiomatic quality process produces results is affected directly by the plan with which it is deployed. In this section we present a high-level perspective of a sound plan by outlining the critical elements of successful deployment. We must point out up front that a successful initiative is the result of key contributions from people at all levels and functions of the company concerned with product development. In short, successful initiatives require buy-in, commitment, and support from officers, executives, and management staff before and while operational and process-level employees execute design projects.

This top-down approach is critical to the success of such a process and can be linked to a design for six-sigma (DFSS) deployment program. In essence, axiomatic quality and DFSS can be used interchangeably (see Yang and El-Haik, 2003). DFSS is a disciplined methodology that embeds customer expectations into the design, applies the transfer function approach to ensure that customer expectations are met, predicts design performance prior to the pilot phase, builds performance measurement systems (score cards) into the design to ensure effective ongoing process management, leverages a common language for design, and uses tollgate reviews to ensure accountability (Chapter 1).

Although *black belts*, axiomatic quality or DFSS project team leaders, are the focal point for executing projects and generating cash from process improvements, their success is linked inextricably to the way that leaders and managers establish the design culture, create motivation, allocate goals, institute plans, set procedures, initialize systems, select projects, control resources, and maintain recognition and rewards.

Benchmarking the six-sigma initiative in several successful deployments, we can conclude that a top-down deployment approach will work for axiomatic quality deployment as well. This conclusion reflects the critical importance of securing and cascading the buy-in from the top leadership level. The black belts and green belts are the focused force of deployment under the guidance of the master black belts and champions. Success is measured by increased revenue and customer satisfaction and the cash flow generated both long and short term (soft and hard), a project at a time. Axiomatic quality projects should diligently be scoped and aligned to company objectives with in a prioritization scheme. Deployment benefits cannot be harvested without a sound strategy with the long-term vision of establishing the design culture desired. In the short term, deployment success depends on motivation, management commitment, project selection and scoping, an institutionalized reward and recognition system, and optimized resource allocation. This section contains a brief description for use by the axiomatic quality or DFSS deployment team.

We are categorizing the deployment process, in terms of evolution time, into three phases:

1. The predeployment phase to build the infrastructure
2. The deployment phase, where most activities will happen
3. The postdeployment phase, where sustainment needs to be accomplished

Predeployment is a phase representing the period of time when a design leadership team lays the groundwork and prepares the company for implementation, ensures the alignment of their individual deployment plans, and creates synergy and heightened performance.

The first step in an effective deployment starts with the product development top leadership of the deployment company. It is at this level that the team tasked with deployment works with the senior product development executives in developing a strategy and plan for deployment that is designed for success. The initiative marketing and culture selling should come from the top.

Initially, it is advisable that select senior leadership as a team meet jointly with the deployment team in an off-site location (with limited distractions). The meeting should comprise a balanced mix of strategic thinking, axiomatic quality and DFSS high-level education, interaction, and hands-on planning. On the education side, overviews of the process concepts and a demonstration of methods, improvement measures, and management controls will be very useful. Specifically, the following should be a minimum set of objectives for this launch meeting:

- To experience the application of some of the tools during the meeting.
- To brainstorm a deployment strategy and a corresponding deployment plan with high first-time-through capability.

- To understand the organizational infrastructure requirements for deployment.
- To set financial and cultural goals, targets, and limits for the initiative.
- To discuss project pipeline and black belt resources in all phases of deployment.
- To design a mechanism for tracking the progress of the initiative, establishing a robust financial management and reporting system for the initiative.

The first step taken by the deployment leader is to establish a deployment team to develop strategies and oversee deployment. With the help of the deployment team, the leader is responsible for designing, managing, and delivering successful deployment of the axiomatic quality initiative throughout the company, locally and globally. He or she needs to work with human resources personnel to develop policy to ensure that the initiative becomes integrated into the culture, which may include integration with internal leadership development programs, career planning for black belts and deployment champions, a reward and recognition program, and progress reporting to the senior leadership team. In addition, the deployment leader needs to provide training, communication (as a single point of contact to the initiative), and infrastructure support to ensure consistent deployment.

Deployment structure is not limited to the deployment team overseeing deployment both strategically and tactically, but should include extended operatives such as project champions, functional area deployment champions, process and design owners where solutions will be implemented, and master black belts (MBBs) who mentor and coach the black belts. All should have very crisp roles and responsibilities with defined objectives. A premier deployment objective can be that the black belts are used as a task force to improve customer satisfaction, company image, and other strategic long-term objectives. To achieve these objectives, the deploying company should establish a deployment structure formed from directors, the centralized deployment team overseeing deployment, and MBBs with defined roles and responsibilities for long- and short-term planning. The structure can take the form of a council with a definite meeting schedule.

To jump-start the deployment process, training is usually outsourced in the first year or two into deployment. The deployment team needs to devise a qualifying scheme for training vendors once their strategy is finalized and approved by the senior leadership of the company. Specific training session content for executive leadership, champions, and black belts should be planned with the heavy participation by selected vendors. This facilitates a coordinated effort, allowing better management of the training schedule and prompter service. Attendance is required during training sessions. To get the full benefit of the training course, each attendee needs to be present for all of the material that is presented. Each training course should be developed carefully and condensed into the shortest possible period by the vendor. Missing any part of a course will result in a diminished understanding of the topics covered and, as a result, may severely delay the progression of projects.

Deployment is the period of time when champions are trained and select initial black belt projects, as well as when the initial wave of black belts are trained and complete projects that yield significant operational benefit, both soft and hard. Additionally, this phase includes the following assignment of the deployment team:

- Reiterate to key personnel their responsibilities at critical points in the deployment plan.
- Reinforce the commitment among project champions and black belts to execute selected design projects aggressively. Mobilize and empower both populations to carry out their respective roles and responsibilities effectively.
- Recognize exemplary performance in execution and in culture with the project champion and black belt populations.
- Build information packets for project champions and black belts that contain administrative, logistical, and other information they need to execute their responsibilities at given points in time.
- Document and publicize successful projects and the positive consequences for the company and its employees.
- Document and distribute project savings data by business unit, product, or other appropriate area of focus.
- Hold events or meetings with all core and support design employees in locations where leadership is present and involved and where such topics are covered.

The black belts as project leaders will implement the DFSS or axiomatic quality process and tools on projects aligned with the business objectives. They lead projects, institutionalize a timely project plan, determine appropriate tool use, perform analyses, and act as the central point of contact for their projects. Training for black belts includes detailed information about the concept, methodology, and tools. Depending on the curriculum, the duration is usually hands-on. Black belts will come with a training focused scoped project that has ample opportunity for tool application to foster learning while delivering deployment objectives. The time period between training sessions will be spent on gathering data, forming and training their teams, and applying concepts and tools where necessary. Axiomatic quality concepts and tools flavored by some soft skills are the core of the curriculum. Of course, training and deployment will be in synchrony with the development process already adopted by the deploying company.

The postdeployment phase, the period of time when subsequent waves of black belts are trained, is characterized by the synergy buildup to critical mass and when additional elements of deployment are implemented and integrated. The purpose is to determine factors toward keeping and expanding the momentum of axiomatic quality deployment to be sustainable.

This book covers the axiomatic quality methodology, which exhibits the merging of many tools at both the conceptual and analytical levels and penetrates dimensions such as mapping, optimization, and selection by integrating tools, principles, and concepts. This vision of the process is a core competency in a company's overall technology strategy to accomplish its goals. An evolutionary strategy that moves the deployment of the axiomatic quality method toward the ideal culture is discussed. In the strategy we have identified the critical elements, needed decisions, and deployment concerns.

The literature suggests that more innovative methods fail immediately after initial deployment than at any stage. Useful innovation attempts that are challenged by cultural change are not terminated directly, but are allowed to fade slowly and silently. A major reason for the failure of technically feasible innovations is the inability of leadership to commit to an integrated, effective, cost-justified, and evolutionary program for sustainability that is consistent with the company's mission. The deployment parallels in many aspects the technical innovation challenges from a cultural perspective. The axiomatic quality initiative is particularly vulnerable if they are too narrowly conceived, built upon only one major success mechanism, or lack fit to the larger organizational objectives. The tentative top-down deployment approach has been working where the top leadership support should be the significant driver. However, this approach can be strengthened when built around such mechanisms as the superiority of axiomatic quality as a design process and its attractiveness to designers who want to become more proficient at their jobs.

Although there are needs to customize a deployment strategy, it should not be rigid. The strategy should be flexible enough to meet expected improvements. The deployment strategy itself should be axiomatically driven and robust to anticipated changes. It should be insensitive to expected swings in the financial health of company and should be attuned to the company's objectives on a continuous basis.

The strategy should consistently build coherent linkages between axiomatic quality and daily design business. For example, engineers and architectures need to see how all of the principles and tools fit together, complement one another, and build toward a coherent whole process. Axiomatic quality needs to be seen, initially, as an important part, if not the central core, of an overall effort to increase technical flexibility.

The process of design can be improved by constant deployment of the axiomatic quality process, which begins from different premises: namely, the principles of design. The design axioms are central to the conceptual component of axiomatic quality. In a sustainability strategy, the following attributes would be persistent and pervasive features:

- Developing a deployment measurement system that tracks the critical-to-deployment requirements and failure modes and implements corrective actions
- Continuing improvement in the effectiveness of axiomatic quality deployment by benchmarking other successful DFSS deployment elsewhere

- Enhancing control over the company's objectives via selected axiomatic quality projects that really move the needle
- Extending involvement of all levels and functions
- Embedding the process into the everyday operations of company

The prospectus for sustaining success will improve if the strategy yields a consistent day-to-day emphasis on recognizing that axiomatic quality represents a cultural change and a paradigm shift. Several deployments found it very useful to extend their initiative to key suppliers. Some call these project *intraprojects* when they span different areas, functions, and business domains. This ultimately will lead to integrating the axiomatic quality process as a superior design approach within the program management system and to aligning the issues of funding, timing, and reviews to the embedded philosophy.

What about culture? What we are finding powerful in design cultural transformation is the premise that the company results wanted is the culture wanted. Leadership must first identify objectives that the company must achieve. These objectives must be defined carefully so that the other elements, such as employee's beliefs, behaviors, and actions, support them. A company has certain initiatives and actions that they must maintain to achieve the new results. But to achieve six-sigma results using axiomatic quality, certain things must be stopped (e.g., old processes) while others must be started (e.g., deployment). These changes will cause a behavioral shift that people must make in order for the new cultural transition to evolve. True behavior change will not occur, let alone last, unless there is an accompanying change in leadership and deployment team belief. Beliefs are powerful in that they dictate action plans that produce desired results. Successful deployment benchmarking (initially) and experiences (later) determine the beliefs, and beliefs motivate actions, so ultimately leaders must create experiences that foster beliefs in people. The bottom line is that for axiomatic culture to be achieved, the design department cannot operate with an old set of actions, beliefs, and experiences; otherwise, the results it gets will be the results that it is obtaining currently. Experiences, beliefs, and actions—these have to change.

The biggest impact on the culture of a company is the initiative of the founders themselves, starting from the top. The new culture is just maintained by the employees once the transition is complete; they keep it alive. Leadership sets up structures (deployment team) and processes (deployment plan) that consciously perpetuate the culture. A new culture means a new identity and a new direction, the six-sigma way.

CHAPTER 7

AXIOMATIC QUALITY PROCESS CONCEPT SELECTION PROCESS

7.1 INTRODUCTION

The concept selection problem is to select the "best" conceptual design solution entity from a pool of feasible alternatives, usually by the conclusion of the preliminary design stage of Figure 1.8. The identification of good selection criteria is a key for successful design release. In this chapter, the concept selection problem is formulated as an integer programming problem. Complexity, value, cost, and customer satisfaction are criteria used to derive the selection program objective function. The mathematical form of the proposed objective function can be obtained conveniently by borrowing from concepts from a portfolio of methods that includes QFD, axiomatic design, and value engineering. The function is then employed into our integer programming formulation, which is expanded to include technical synthesis feasibility (morphological matrix, introduced in Chapter 6) and assembly feasibility as constraints. The proposed formulation is sufficiently robust to adapt design situations with deterministic (crisp) assessment or fuzzy assessment.

As discussed earlier, the goal of engineering design is to create the design entities that satisfy the needs and delights of customers. The design team's creativity, experience, and scientific knowledge are essential for developing good design entities. For design situations that start with a pool of several baselines prior to the CDFC phase (Figure 7.1) or while processing step A.4 of the CDFC phase as depicted in Figure 6.3, usually more than one conceptual entity will

Axiomatic Quality: Integrating Axiomatic Design with Six-Sigma, Reliability, and Quality Engineering, by Basem Said El-Haik
ISBN 0-471-68273-X

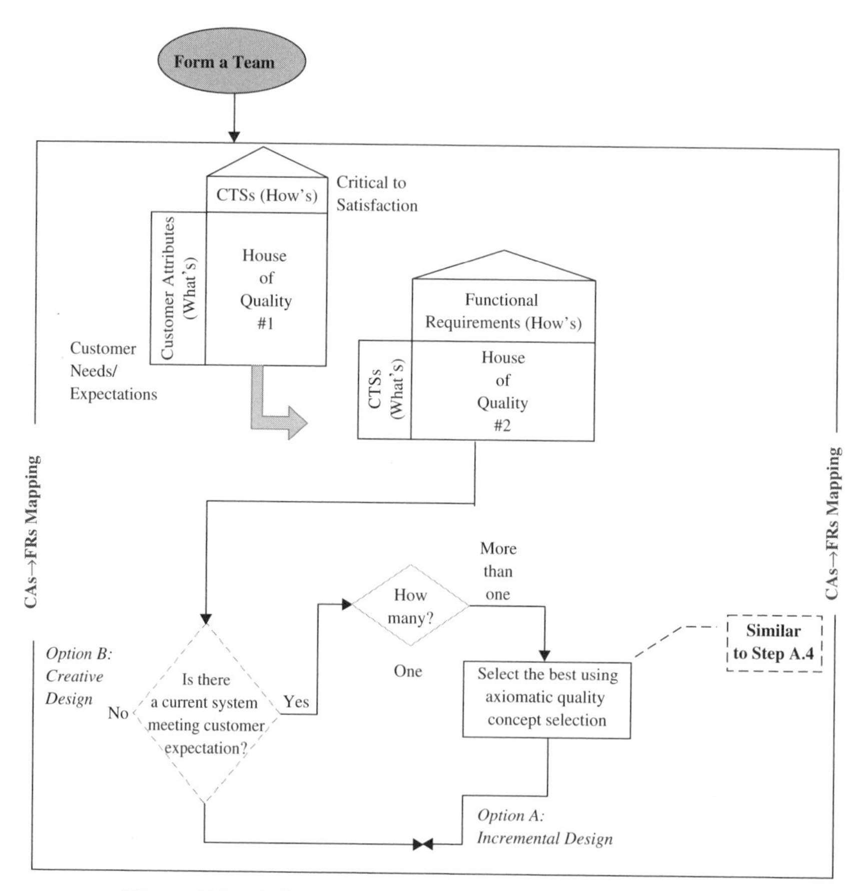

Figure 7.1 Axiomatic quality concept selection opportunity.

be conceived. The concept selection problem is to select the *best* design entity that not only satisfies the customer requirements but also outperforms the other alternative solutions based on a set of selection criteria. The selection problem involves the following three major steps: (1) identification of the selection criteria, (2) ranking (scoring) of different design entities against the selection criteria, and (3) identification of the "best" (optimum) entity. The selection problem is trivial when only one criterion is used. The best conceptual entity is the one that scores favorably in the ranking. However, the problem becomes more complex when multiple criteria are involved in the absence of credible and realistic data. The selection problem may become judgmental and exposed to design team bias, as ranking will be driven to favor some predetermined conceptual entity. The bias problem can be eliminated by the systematic employment of a disciplined selection process. Process creditability and robustness are greatly enhanced when coupled with axiomatic design as a scientific-based design method.

In this chapter we propose to formulate the selection problem as an integer programming problem with two principal selection criteria: customer satisfaction and design complexity. The choice of design complexity as a selection criterion stems from the information axiom application (Suh, 1990). In addition, the formulation proposed is built around the generic conceptual framework of QFD, presented in Chapter 6.

This chapter starts by formulating the design's technical, manufacturing, and assembly feasibilities as a requirement for selection problem formulation. This is very important because it forces the design team to think in terms of the physical structure rather than the design mapping alone. That is, we are selecting a design that has embodiment abilities. Therefore, the concept of design module feasibility will be examined first to enable the formulation. Crisp and fuzzy optimization formulations of the concept selection problem are presented as nonlinear integer programming problems.

7.2 DESIGN FEASIBILITY IN AXIOMATIC QUALITY

7.2.1 Modules

In the axiomatic quality process, a product is made up of several components, subsystems, or other hierarchical modules, called collectively the *physical structure*. Depending on coupling vulnerability and interrelationships with modules within the design environment but outside the scope of the project, a design mapping cannot always be hosted by one module, and hence there is no one-to-one mapping between a design mapping and a hierarchical module. This is true for complex systems of higher modularity. For low-modularity design projects, a one-to-one mapping between the design mappings and the modules in the physical structure has a higher existence possibility. Nevertheless, a module is designed to deliver an array of FRs. The physical entity of a module is a set of DPs grouped together in the form of a product with some hierarchical classification (component, subsystem, system, and super system).

Let m_d be the number of the FRs in the dth module and D be the number of modules in the design team project. Then the total number of independent functional requirements, m, in the project should satisfy $\sum_{d=1}^{D} m_d = M$. Assuming probabilistic independence, for any module the information content of the module can be defined as

$$\begin{aligned} H_{\text{module}_d} &= -\log_v(\text{Prob}_{\text{module}_d}) \\ &= -\log_v(\text{Prob}_1 \times \text{Prob}_2 \times \cdots \times \text{Prob}_{m_d}) \\ &= -\sum_{j=1}^{m_d} \log_v(\text{Prob}_j) \end{aligned} \tag{7.1}$$

where Prob_i is the probability of success of the FR indexed i, $i = 1, \ldots, m_d$, and H represents entropy, as discussed in Chapter 4. In a given mapping, each FR can be viewed as a stand-alone information source, or equivalently, a complexity source. Assuming statistical independence, the probability of success is multiplicative.

Equation (7.1) takes the following equation as an average form:

$$H_{\text{module}_d} = -\sum_{i=1}^{m_d} \text{Prob}_i \log_v(\text{Prob}_i) \qquad i = 1, \ldots, m_d$$

$$= -\sum_{d=1}^{D}\sum_{i=1}^{m_d} \text{Prob}_i \log_v(\text{Prob}_i) \qquad i = 1, \ldots, m_d; \quad d = 1, \ldots, D \tag{7.2}$$

which can be generalized to quantify the information content for the whole product, where the summation is taken up to D, adding one module at a time. The entropy described in (7.1) or (7.2) can be used as an assessment for the new design by substituting for respective probabilities from baseline (datum) data. The entropy values serve as an expected performance index for the newly developed alternatives (concepts). The smaller entropy, H, indicates a lower degree of complexity. It is obvious that *maximizing the probability of success can reduce H and hence overall design complexity* (see Chapter 4).

A practical and cost-effective way to maximize the probability of success, and therefore reduce the design complexity, is accomplished by using standard reusable design modules or DPs if available. In addition, there are two major advantages of using standard modules. First, design teams do not have to reinvent what has already existed, therefore saving valuable design resources. Second, the use of standard modules will improve the quality and reliability levels.

7.2.2 Design Technical (Morphological) Feasibility

Clearly, a system or product is made of a number of modules. Each module is pieced from a set of DPs, which is a key design decision to be taken in step A.4 of the axiomatic quality process. A design parameter set forming a design module may has elements that belong to different design mappings in several hierarchical levels. Feasibility thinking should guard this synthesis activity. Design feasibility here has two aspects: first, *technical morphological feasibility*; that is, the set of design parameters constituting the module should be able to deliver the FRs hosted in the same the module. Second, the module itself should be feasible as to assembly and manufacturing, which means that the constituent module DPs can be synthesized in a manner that enables the use of manufacturing and assembly processes. Processes that do not require additional investment should be considered to meet any cost constraints already imposed. In doing so, the design team should not limit their synthesis to the intuitive assumption of hardware-based elements, but rather as generic physical instances that can be materialized by software and fields entities as well.

The mathematical formulation of *technical feasibility* of the dth module is as follows. Let i be the index of FRs, $i = 1, 2, \ldots, m_d$; j be the index of DPs forming the module, $k = 1, 2, \ldots, K$; F_i be the set of potential (alternative) DPs of the functional requirement FR_i with cardinality N_i; and $\mathbf{F}$ be the union set of the overall unique potential DPs. The reader is encouraged to revisit step A.3 of the axiomatic quality process in Chapter 6.

Example 7.1 A given design problem has the following arrays of design functional requirements, $m_d = 3$, where the arrow denotes the possible mapping between the FR domain and the DP domain.

$$\mathbf{FR} = \{\text{FR}_1, \text{FR}_2, \text{FR}_3\}$$

$$\text{FR}_1 \rightarrow F_1 = \{\text{DP}_1, \text{DP}_2\} \qquad \text{with } N_1 = 2$$

$$\text{FR}_2 \rightarrow F_2 = \{\text{DP}_1, \text{DP}_3\} \qquad \text{with } N_2 = 2$$

$$\text{FR}_3 \rightarrow F_3 = \{\text{DP}_1, \text{DP}_4, \text{DP}_5\} \qquad \text{with } N_3 = 3$$

For example, $\text{FR}_1 \rightarrow F_1 = \{\text{DP}_1, \text{DP}_2\}$ means that FR_1 can be performed by either DP_1 or DP_2. It is also assumed that there is no duplication of DPs in each of the synthesized modules. For example, DP_1 can be used to deliver all FR_1, FR_2, and FR_3 in Example 7.1. When we select this option, the module will have only one DP_1. In general, to furnish each module, the union set $\mathbf{F} = \cup_i F_i$, the set of unique potential DPs, will be selected by dropping overlaps and its cardinality $K \leq N_1 \times N_2 \times \cdots \times N_m$, with equality satisfied only when there is no overlapping DPs among the union set.

Consider the case of a single DP that serves more than one FR, say FR_a and FR_b. In this case we have $F_a \cap F_b \neq \phi$, which implies that the coupling vulnerability may be created due to poor selection of the DPs during design synthesis activities. In other words, a coupling-free (independent) design solution can be achieved when $F_a \cap F_b = \phi$ for $a = 1, 2, \ldots, m_d - 1; b = 2, 3, \ldots, m_d$, or for a sufficient subset of $\mathbf{F}$ that covers the FRs.

In last example, the set $\mathbf{F} = \{\text{DP}_1, \text{DP}_2, \text{DP}_3, \text{DP}_4, \text{DP}_5\}$ and $K = 5$ $(\leq 2 \times 2 \times 3 = 12)$. In addition, the mapping process can be coded mathematically via the variable $T_{ik} = 1$ if $\text{FR}_i \rightarrow \text{DP}_k$ and 0 otherwise. This binary variable is different from the variable Y_{ik} introduced in Chapter 5, that is, denoting an already synthesized module described by the exhibited mapping. The technology binary T_{ik} represents a mathematical formulation for the morphological matrix introduced in Chapter 6 of all possibly synthesizable sets of solutions (with no DP overlaps). A morphological technology matrix $T_{m_d \times K}$ can be defined as

$$\mathbf{T} = \begin{array}{c} \begin{matrix} \text{DP}_1 & \text{DP}_2 & \text{DP}_3 & \text{DP}_4 & \text{DP}_5 \end{matrix} \\ \begin{bmatrix} 1 & 1 & 0 & 0 & 0 \\ 1 & 0 & 1 & 0 & 0 \\ 1 & 0 & 0 & 1 & 1 \end{bmatrix} \end{array} \begin{matrix} \leftarrow \text{FR}_1 \\ \leftarrow \text{FR}_2 \\ \leftarrow \text{FR}_3 \end{matrix}$$

There are 12 solution combinations in this example. Assume that not all of them satisfy the independence axiom, which is only satisfied by two overall solutions (S_1, S_2), each of which is a subset of possibly synthesizable solutions:

$$\mathbf{T}_{S_1} = \begin{array}{c} \begin{matrix} \text{DP}_2 & \text{DP}_3 & \text{DP}_4 \end{matrix} \\ \begin{bmatrix} 1 & 0 & 0 \\ 0 & 1 & 0 \\ 0 & 0 & 1 \end{bmatrix} \end{array} \begin{matrix} \text{FR}_1 \\ \text{FR}_2 \\ \text{FR}_3 \end{matrix} \quad \text{and} \quad \mathbf{T}_{S_2} = \begin{array}{c} \begin{matrix} \text{DP}_2 & \text{DP}_3 & \text{DP}_5 \end{matrix} \\ \begin{bmatrix} 1 & 0 & 0 \\ 0 & 1 & 0 \\ 0 & 0 & 1 \end{bmatrix} \end{array} \begin{matrix} \text{FR}_1 \\ \text{FR}_2 \\ \text{FR}_3 \end{matrix}$$

Notice that the solutions are formed by column selection from the technology matrix and coupling was avoided by dropping the DP_1 column.

A mathematical formulation of manufacturing and assembly feasibility is presented next. The manufacturing or production feasibility can be formulated following the steps presented here. The assembly feasibility should be tested in the physical domain among the DPs themselves in a given module subsequent to the technology determination. The binary characterization variable $Z_{kl} = 1$ if $DP_k \rightarrow DP_l$ and 0 otherwise denotes the assembly feasibility between pairs of DPs. The unity value, $Z_{kl} = 1$, indicates that DP_k and DP_l can be assembled together. Hence, a 0–1 assembly matrix $\mathbf{Z_{K \times K}}$ can be constructed as follows. Clearly, the $\mathbf{Z}$ matrix is symmetrical, (i.e., $z_{kl} = z_{lk}$):

$$\mathbf{Z} = \begin{array}{c} \begin{array}{ccccc} DP_1 & DP_2 & DP_3 & DP_4 & DP_5 \end{array} \\ \left[\begin{array}{ccccc} 1 & 1 & 1 & 1 & 1 \\ 1 & 1 & 1 & 0 & 0 \\ 1 & 1 & 1 & 1 & 0 \\ 1 & 0 & 1 & 1 & 0 \\ 1 & 0 & 0 & 0 & 1 \end{array}\right] \end{array} \begin{array}{c} \\ DP_1 \\ DP_2 \\ DP_3 \\ DP_4 \\ DP_5 \end{array}$$

From $\mathbf{Z}$ we can construct the following assembly matrixes for the S_1 and S_2 solutions:

$$\mathbf{Z}_{S_1} = \begin{array}{c} \begin{array}{ccc} DP_2 & DP_3 & DP_4 \end{array} \\ \left[\begin{array}{ccc} 1 & 1 & 0 \\ 1 & 1 & 1 \\ 0 & 1 & 1 \end{array}\right] \end{array} \quad \text{and} \quad \mathbf{Z}_{S_2} = \begin{array}{c} \begin{array}{ccc} DP_2 & DP_3 & DP_5 \end{array} \\ \left[\begin{array}{ccc} 1 & 0 & 0 \\ 0 & 1 & 0 \\ 0 & 0 & 1 \end{array}\right] \end{array}$$

Assembly feasibility is depicted by the off-diagonal elements. Assembly feasibility need not be confused with coupling when Corollary 2.3 (Section 2.5.1) is employed. It is obvious that S_1 is assembly feasible in only two DPs, DP_3 and DP_4, and that S_2 is not assembly feasible at all. Also, S_1 is a coupling-free solution. Under binding technological and/or constraints, it is sometimes inevitable to trade independence with feasibility, as is the case with solutions S_3 and S_4:

$$\mathbf{T}_{S_3} = \begin{array}{c} \begin{array}{ccc} DP_1 & DP_2 & DP_3 \end{array} \\ \left[\begin{array}{ccc} 1 & 1 & 0 \\ 1 & 0 & 1 \\ 1 & 0 & 0 \end{array}\right] \end{array} \begin{array}{c} \\ FR_1 \\ FR_2 \\ FR_3 \end{array} \quad \text{and} \quad \mathbf{Z}_{S_3} = \begin{array}{c} \begin{array}{ccc} DP_1 & DP_2 & DP_3 \end{array} \\ \left[\begin{array}{ccc} 1 & 1 & 1 \\ 1 & 1 & 0 \\ 1 & 0 & 1 \end{array}\right] \end{array}$$

$$\mathbf{T}_{S_4} = \begin{array}{c} \begin{array}{ccc} DP_1 & DP_3 & DP_4 \end{array} \\ \left[\begin{array}{ccc} 1 & 0 & 0 \\ 1 & 1 & 0 \\ 1 & 0 & 1 \end{array}\right] \end{array} \begin{array}{c} \\ FR_1 \\ FR_2 \\ FR_3 \end{array} \quad \text{and} \quad \mathbf{Z}_{S_4} = \begin{array}{c} \begin{array}{ccc} DP_1 & DP_3 & DP_4 \end{array} \\ \left[\begin{array}{ccc} 1 & 1 & 1 \\ 1 & 1 & 1 \\ 1 & 1 & 1 \end{array}\right] \end{array}$$

Note that S_3 is assembly feasible because DP_3 and DP_2 are assembly feasible with DP_1, an indirect synthesis link. There are no assembly-related restrictions on S_4, and hopefully, in the spirit of Corollary 2.3, the degree of coupling will not increase. Once the feasibility criteria are satisfied, the designs degree of coupling can be quantified using semangularity and reangularity, the axiomatic measures (Suh, 1990).

7.3 CONCEPT SELECTION PROBLEM

As the physical mapping process is performed, it is possible that an FR may be mapped to many alternative physical entities (a DP or a group of DPs), each having its own manufacturing processes, material variability, geometrical tolerance, and other physical attributes. A DP is a complexity source and hence an information source. In the mapping of interest, we would like to select the "best" DPs (or PVs) that satisfy the FRs (or DPs) with the maximum customer satisfaction and minimal design vulnerabilities. To achieve this objective, the concept selection problem is formulated by using the framework of QFD (Chapter 6) and the axiomatic design principles. This formulation is unique in the axiomatic quality process context.

In the QFD planning matrix introduced in Figure 6.4, the product of customer *attribute value* (AV), targeted *improvement ratio* (IR) for a customer attribute (a row, a "what"), and *sales point* (SP) provide a weighted measure of the relative importance of this customer feature, where SP is a measure of how the raw feature affects sales (see Figure 7.2). The product is denoted AW (*attribute weight*). The other relative measure is the subjective cause–effect weight in the relationship matrix (Figure 6.4), the weight W, that a CTS or an FR (a column in the respective QFD stage, a "how") may play in satisfying a customer attribute. The weight W_{ij} gives a measure of the strength of the FR_i relationship to the attribute CA_j. The summation of W_{ij} in each column is denoted here as FW (functional requirement weight), which gives a measure about how much this FR is related to the overall customer attributes. The product of FR weight (FW) times the raw weight (AW) and summing over all the rows (customer attributes) on the right of the house of quality provides a measure of the relative importance of that functional requirement to overall customer satisfaction. For example, $\sum_{j=1}^{J} W_{ij} AW_j$, $j = 1, \ldots, J$, is a measure of the customer perceived satisfaction index for FR_i (assuming that J is the number of customer attributes in QFD). In addition to customer satisfaction, other selection criteria should also be considered for a comprehensive handling. In this chapter we are interested in merging the complexity measure as a selection criterion according to the information axiom in the objective function of our integer programming of the selection formulation. In addition to the information axiom, the inclusion of complexity in the selection objective function, an optimization index, is further justified because it relates to many design entity attributes, such as quality loss, as presented in Chapter 5.

For each functional requirement, FR_i, its complexity (information content) is a function of the DPs selected to deliver it. Assume that there are k DPs by which

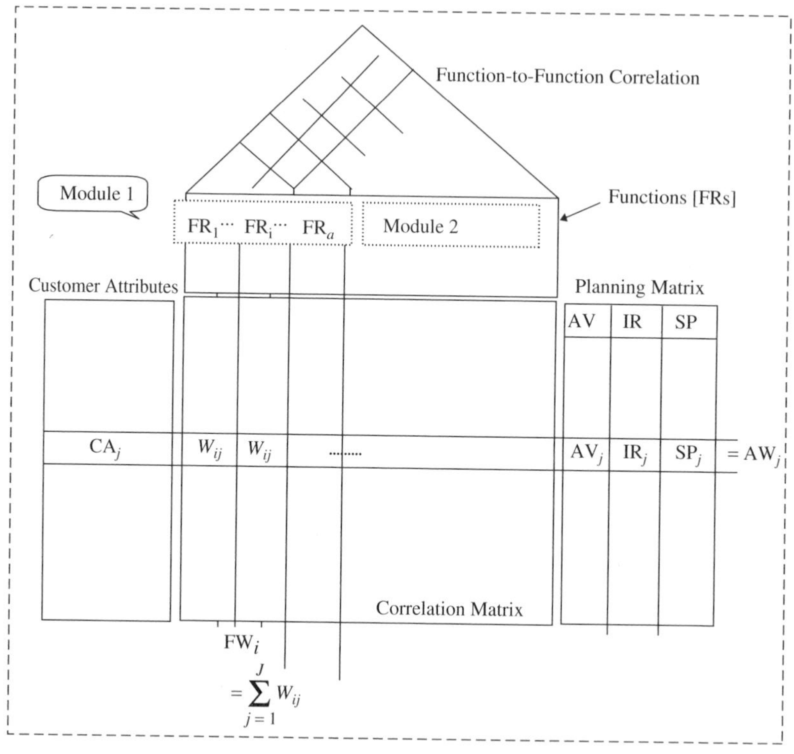

Figure 7.2 QFD house of quality.

FR_i can be delivered (i.e., there are k design instances). The entropy of FR_i in module k ($\text{FR}_i \rightarrow \text{module}_k$), with $\text{DP}_k \in F_i$, $k = 1, 2, \ldots, N_i$, can be denoted as H_{ik} using Chapter 5 derivations (e.g., Theorem 4.1).

Mathematically, let j be the index of customer attributes, $j = 1, 2, \ldots, J$; then the index $I_{ik} = (\sum_{j=1}^{J} W_{ij}\text{AW}_j)/H_{ik}$ can be used to evaluate FR_i at the kth design module (a design instance). A larger index value indicates more customer satisfaction with a simpler design, a better design.

The weights in a QFD matrix, that is, W and FW, are variant to particular DPs at different design modules. For example, electrical solution entities are usually highly rated in the "convenience of operation" and "ease of maintenance" attributes compared to mechanical entities. Therefore, W_{ik} and FW_{ik} should be assessed carefully by the design team.

Now, we can formulate the design concept selection problem as maximizing the selection index, I_{ik}, subject to the technology and assembly feasibility constraints discussed in Section 7.2. Specifically, the concept selection problem can be formulated as the following integer programming problem:

$$\text{Maximize} \quad \frac{\sum_{i=1}^{M} \sum_{k=1}^{K} \sum_{j=1}^{J} W_{ik} Y_{ik} \text{AW}_j}{\sum_{i=1}^{M} \sum_{k=1}^{K} H_{ik} Y_{ik}} \tag{7.3}$$

subject to

$$\sum_{k=1}^{K} Y_{ik} T_{ik} = 1 \qquad \forall i, i = 1, 2, \ldots, m, \quad \mathrm{DP}_k \in \mathbf{F} \tag{7.4}$$

$$\sum_{i=1}^{M} \sum_{k=1}^{K} \sum_{j=1}^{J} \mathrm{FW}_{ik} Y_{ik} \mathrm{AW}_j > \left(\sum_{i=1}^{M_d} \sum_{j=1}^{J} \mathrm{FW}_{ik} Y_{ik} \mathrm{AW}_j \right)_{\text{datum}} \tag{7.5}$$

$$\sum_{i=1}^{M} \sum_{k=1}^{K} H_{ik} Y_{ik} < \left(\sum_{i=1}^{M_d} H_i \right)_{\text{datum}} \tag{7.6}$$

$$Y_{ik} = 0 \text{ or } 1 \tag{7.7}$$

where M_d is the number of functional requirements in the datum design and the T_{ik} are the entries of matrix $\mathbf{T}$. In this technology feasibility formulation, the decision variables are the binary variables Y_{ik}, where $Y_{ik} = 1$ indicates that DP_k is selected to deliver FR_i. The objective function is clearly the objective function of the entire design, in which the numerator is the customer satisfaction index sum and the denominator is the design degree of complexity. This objective maximizes customer satisfaction while minimizing design complexity. Constraint (7.4) forces the selection of one solution entity according to a given function. Constraints (7.5) and (7.6) translate the word *best* into its mathematical equivalent. The best design selected is therefore the design that outperforms the datum design from the perspectives of customer satisfaction and design simplicity.

The mathematical program above [equations (7.4) to (7.7)] does not eliminate the possibility of obtaining an overall assembly infeasible solution. The assembly feasibility can be viewed as a *tour* (loop) between the design DPs selected, where each is visited once starting from a DP of reference depending on the sequence conveyed by the design matrix, if any. A complete loop implies a complete assembly. An overall assembly feasible solution is one that has only one tour (loop) such that all subtours are eliminated. This reasoning is adopted from the traveling salesman problem (Salkin and Mathur, 1989). The program in (7.4) to (7.7) can be rectified to account for assembly feasibility when augmented by $\sum_{i=1}^{M-1} \sum_{u=1+i}^{M} Y_{ik} Y_{ul} Z_{kl} \leq M - 1, i \neq u$, where the binary characterization Z_{kl}'s are the entries of matrix $\mathbf{Z}$ (Section 7.2). An assembly-feasible design with M selected DPs is the one that has at most $M - 1$ nonzero Z_{kl}'s. That is, to synthesize a solution, we need to satisfy simultaneously the technology requirement between a pair of functional requirements through proper selection of DP_k for FR_i and DP_l for FR_u and the assembly requirement between DP_k and DP_l (i.e., $Z_{kl} = 1$). This feasibility assurance process is expanded to all possible pairs of functional requirements. Use of this constraint prevents the selection of subloops of physical entities that are assembly feasible only at the subsystem level.

This formulation can be enhanced further to include a cost performance index, CI. In this case, the formulation can be written as

$$\text{Maximize} \frac{\sum_{i=1}^{M}\sum_{k=1}^{K}\sum_{j=1}^{J} \text{CI}_{ik} W_{ik} Y_{ik} \text{AW}_j}{\sum_{i=1}^{M}\sum_{k=1}^{K} H_{ik} Y_{ik}} \tag{7.8}$$

subject to

$$\sum_{k=1}^{K} Y_{ik} T_{ik} = 1 \qquad \forall i, i = 1, 2, \ldots, M; \quad \text{DP}_k \in \mathbf{F} \tag{7.9}$$

$$\sum_{i=1}^{M}\sum_{k=1}^{K}\sum_{j=1}^{J} \text{CI}_{ik}\text{FW}_{ik} Y_{ik} \text{AW}_j > \left(\sum_{i}^{M_d}\sum_{j=1}^{J} \text{CI}_{ik}\text{FW}_{ik} Y_{ik} \text{AW}_j\right)_{\text{datum}} \tag{7.10}$$

$$\sum_{i=1}^{M}\sum_{k=1}^{K} H_{ik} Y_{ik} < \left(\sum_{i=1}^{M_d} H_i\right)_{\text{datum}} \tag{7.11}$$

$$\sum_{i=1}^{M-1}\sum_{u=1+i}^{M} Y_{ik} Y_{ul} Z_{kl} \leq M - 1 \qquad i \neq u; \quad i = 1, 2, \ldots, M - 1;$$

$$u = 2, 3, \ldots, M \tag{7.12}$$

$$Y_{ik} = 0 \text{ or } 1 \tag{7.13}$$

where CI_{ik} is the cost performance index of the function i at module (solution or design instance) k. The overall solution entity selected for the formulation proposed will achieve higher performance of design requirements from a multi-criterion perspective: cost, customer satisfaction, and complexity.

The elimination or reduction of design coupling may result in added complexity whether by substituting for a coupling DP or by expanding the design mapping. The use of additional DPs to eliminate or reduce coupling may increase the overall design complexity because the cardinality, M, will increase. As formulated here, the design entity's overall complexity takes M as an argument and is a function of the underlying probability distributions of the design parameters or process variables. The reader is encouraged to revisit Chapter 5. The use of probability distributions to assess complexity H_{ik} implies the case of the incremental design classification as in Figure 6.3, that is, experienced situations with applicable and valid data that allow calculation of H_{ik}. Incremental design is a design that is within *minor* variation of the current design in terms of design mapping and physical structure. In many design situations, in particular those classified as creative design projects, the design team does not have the data luxury enabling the calculation of H_{ik}. The type of information in the creative situation is qualitative and fuzzy in a form of engineering judgment warranting another formulation, however, with a fuzzy flavor to assess complexity and other arguments in the integer program presented above.

Fuzzy axiomatic quality concept selection is an extension of the formulation presented above and deals with design situations where there is not sufficient information to warrant the use of deterministic optimization. The formulation assumes the existence of a design alternatives pool with enough expertise to score a ranking against the criteria selected. The fuzzy formulation builds on the rationale and derivation presented in Sections 7.2 and 7.3 for the selection of integer programming. An index pieced from complexity, cost, and customer satisfaction will be fuzzified as an objective function. Again, this rationale is rooted in the concepts of QFD and axiomatic design methodologies. The formulation is expanded to include technical and assembly feasibility as constraints. The formulation uses some concepts of fuzzy set theory to quantify complexity as a formulation ingredient. The concept selection fuzzy formulation is presented next.

7.4 CONCEPT SELECTION FUZZY MODELING

In this section a formulation is presented of the selection problem as an integer programming problem with two principal selection criteria: customer satisfaction and design complexity. Again, the choice of design complexity as a selection criterion stems from the information axiom (Suh, 1990), while the selection problem formulation is built around measures borrowed from QFD.

The type of information in the creative situation is qualitative or fuzzy in the form of engineering judgment. The existence of fuzzy information can be utilized to infer probability distribution from fuzzy distributions using the concepts of *possibility distribution, possibility–probability principle, and maximum entropy principle*. The possibility distribution is key in the formulation below.

7.4.1 Fuzzy Concepts

Linguistic inexactness (imprecision) is the most common feature of many real-life situations. Dutta (1985) classifies imprecision according to its source: measurement, stochastic, ambiguous definitions, incomplete knowledge, and so on. In decision making, for example, the usefulness of mathematical algorithms is in having clearly defined objective criteria and constraints. They are only as good as the information they are given. Information has to be crisp (precise) to yield precise decisions (Zimmermann, 1985).

Certainty formulations require structure with precise parameters. However, most real-life situations are characterized linguistically with degrees of imprecision. Precision implies no ambiguity by assuming that variables, parameters, and structure represent deterministic situations, as we did in Section 7.3. The imprecision issue is further complicated in the creative design classification. In the early stages of the development cycle, a design is a collection of scattered conceptual thoughts and rough drawings. The difficulty in design problem formulation often lies in establishing precise mappings, constraints, and functional requirements which are uncertain, do not fall between what we consider as definite and precise. Even when the design matures to a physical structure via the mapping process, it may still need further tuning and optimization. It is almost the case that we cannot make *deterministic* assertion with respect to certain phenomena because we cannot measure, do not know, cannot calculate all factors involved (Stark and Woods, 1986). We attribute variance between products passing the same processes to randomness by discounting the system to its average behavior. To do that, we use probability theory to handle randomness. As such, design models cannot be described as unequivocal. No comprehensive design models can be written even for incremental design situations. Unfortunately, existing knowledge is normally centered around the crisp incremental (adaptive) classifications. Under these circumstances, a design problem complexity can be lessened using empirical knowledge producing dominating formal models. In customer-oriented design, customers have wants and needs that are hard to interpret. They are expressed, linguistically, using terms that have no precise definition. A statement is not always right or wrong, as people are not always classified as intellectual or not, and a linear programming (LP) problem is not always feasible or infeasible. Yet to classify an LP problem as for most classical decision making, one description or the other must be chosen. This is in accordance with the *law of the excluded middle* (Klir and Folger, 1988). This dichotomous property is the basis of *classical set theory*. By the same analogy, measures, indiexes, and metrics may be viewed as continuous measures of some *possibility distributions*, analogous to crisp random variable probability distributions.

An example that may be used to facilitate the fuzzy concepts is as follows. Assume that there are four design proposal (solution entities): say, the set $S = \{S_1, S_2, S_3, S_4\}$. We would like to select a solution entity at random from S. The probability distribution in this case is $p(\{S_1\}) = p(\{S_2\}) = p(\{S_3\}) = p(\{S_4\}) = \frac{1}{4}$. If we were asked to select randomly a *successful* design, we cannot use the probability distribution above because of the fuzziness in the word *successful*. The answer is in defining a design solution, say G, as a variable that takes in values in the set S according to a distribution constructed around "G is successful."

A fuzzy set accepts objects of a certain degree, called the *membership function* (Zadeh, 1965). The fuzzy set $\tilde{A}$ is represented as $\tilde{A} = \{(\text{FR}, \mu_A(\text{FR}))/\text{FR} \in \text{FRs}\}$, with $m_A(\text{FR})$ understood to represent a mapping of membership of FR, m_A: FRs $\rightarrow$ [0,1]: FR $\rightarrow m_A(\text{FR})$. It is understood that in the crisp case, $\forall \text{FR} \in A, \mu_A(\text{FR}) = 1$ and zero otherwise. Every mapping of this nature with some conceptual realization (in alignment with intuitive semantics of imprecise description of FR) is a fuzzy set. For example, FRs can be the universe of fuzzy functional requirements, such as stylish, cheap, convenient, and so on.

7.4.2 Possibility–Probability Consistency Principle

The fuzzy information about the elements of a finite set can be represented by a possibility distribution. Possibility theory was introduced by Zadeh (1978) as an interpretation of a fuzzy set. The concept was developed further by both Dubois and Prade (1988). Possibility is concerned with linguistic uncertainty that is assumed to be *possibilistic* rather than probabilistic. For example, the proposition "X_1 is $\tilde{A}$" is a possibility proposition where X_1 is a variable taking the values x_1 and $\widetilde{A}$ is a fuzzy set with $m_A(x_1)$. Possibility distribution is considered somehow a modeling to fuzzy restriction. Zadeh (1975, 1978) proposed the following definitions:

Definition 7.1 Let $\widetilde{A}$ be a fuzzy set in the universe X with membership $m_A(x)$, interpreted as the compatibility of $x \in X$ with concept label $\widetilde{A}$. Let X_1 be a variable with values in X and $\widetilde{A}$ acting as a fuzzy restriction, $R(X_1)$, associated with X. Then the proposition "X_1 is $\widetilde{A}$," which translates into $R(X_1) = \widetilde{A}$, associates a possibility distribution, π_x, in which X_1 is postulated to be equal to $R(X_1)$. The possibility distribution is $\pi_x = m_A(x)$.

The relation between probability and possibility has been the focus of work by Zadeh (1978) and Dubois and Prade (1982). The possibility–probability consistency principle is the foundation of such a relationship. Based on this principle, Leung (1980) suggested deriving the probability (p_x) of success based on fuzzy information (π_x) using the consistency principle as evidence in the framework of the *maximum entropy principle*. The important advantage of this formulation lies in transforming the fuzzy information into a deterministic measure for creative design situations.

7.4.3 Maximum Entropy Formulation

There would be much controversy if the design team assigns, rather than assesses, the probability of success in the concept phase to quantify complexity. From the perspective of our earlier discussion, it would appear that the problem is simply deciding how to encode available information. However, the problem is not that simple. It is indeed difficult to answer fundamental questions about design knowledge. Often, we can be explicit about what we know about a specific question. However, this knowledge can be incomplete and must be encoded in a possibility distribution before we can make use of inferential methods. The concept of entropy, with the average entropy given by $H = -\sum \text{Prob}_i \log(\text{Prob}_i)$, was introduced in Section 7.3. The concept of entropy and its extended notions are used to handle the issue of uncertainty. Jaynes (1957a,b) proposed the principle of maximum entropy, and this principle has been employed in various disciplines, including thermodynamics (Tribus, 1961) and urban modeling (Wilson, 1970). The maximum entropy principle addresses the assignment of prior probabilities based on prior knowledge. Jaynes (1957a,b) showed that the least presumptuous way to assign prior probability is by maximizing the entropy function in (7.14) subject to the normalization constraint, (7.15). In this meaning, Jaynes (1957a) added: "The minimally prejudiced probability distribution is that which maximizes entropy subject to constraints supplied by the given information."

Maximum entropy is most beneficial when the knowledge is characterized in average form. The formulation of maximum entropy can be characterized as follows: A DP is a variable that can have different possible nominal values in the concept stage, but we do not know the value. However, we know the possibilities, and we wish to find the probabilities. We would like to generate a probability distribution which agrees with the averages but is maximally noncommittal with respect to anything else.

In the following formulation we treat the variable DP as a discrete variable that takes its value from the universal set of DPs. In the continuous DP form, we need to substitute sums by integrals and take the index of difference or differential intervals as discrete arguments. With LI discrete values (or indexed intervals), the maximum entropy problem has been expressed mathematically by Leung (1980) as

$$\text{maximize } H = -\sum_{l=1}^{LI} p_l \log p_l \tag{7.14}$$

subject to

$$\sum_{l=1}^{LI} p_l = 1 \tag{7.15}$$

$$\text{“}\pi \text{ is consistent with } p\text{''} \tag{7.16}$$

where p_l is the probability that DP will have the value DP_l, $\text{DP}_l \in (\text{DP}_l - \delta/2, \text{DP}_l + \delta/2)$ and LI is the number of discrete intervals.

The distribution that maximizes (7.14) is considered a minimally prejudiced assignment in that it makes the distribution maximally vague or general. The term *minimally prejudiced* implies that the distribution is general and is maximally influenced by new data. Equation (7.16) indicates that at least one of the assertions is true. In (7.17), π is the possibility distribution (the membership function) of the interval with mean DP_l. Zadeh (1978) suggested the following definition of the *consistency principle* in (7.16):

$$\sum_{l=1}^{LI} p_l \pi_l = \alpha \tag{7.17}$$

where $\alpha \in [0,1]$ and near 1.

Dubois and Prade (1982) proposed their own definition of the consistency principle: A probability distribution (p) and a possibility distribution (π) are consistent if $\forall \text{DP} \subset \text{DPs}$,

$$\pi(\text{DP}_l) \geq p(\text{DP}_l) \qquad \forall \text{DP}_l \in \text{DP} \tag{7.18}$$

Either definition (7.17) or (7.18) can be substituted for (7.16).

Example 7.2 The surface finish of a transmission oil pan is a significant design parameter for a sealing requirement. A design team is considering using a silicon elastomer as a possible replacement for the current solid plastic seal. The use of silicon elastomers has very attractive cost advantages over the current design. The design team has no experience with silicon applications, and they would like to determine the nominal value of the surface finish of the oil pan that will maximize the probability of success and hence reduce complexity. The Material Engineering Department was consulted and provided the following possibility distribution of success at four possible nominal discrete values, $\{DP_1, DP_2, DP_3, DP_4\}$, of the surface finish in micrometers:

$$\begin{Bmatrix} DP_1 \\ DP_2 \\ DP_3 \\ DP_4 \end{Bmatrix} = \begin{Bmatrix} 0.50 \\ 0.60 \\ 0.80 \\ 1.00 \end{Bmatrix}$$

The design organization would like to know the probability of success at the consistency levels $\alpha = 0.85$, 0.90, and 0.95.

Nonlinear optimization software was used to solve the program (7.14)–(7.17). The solution is the probability distribution shown in Table 7.1.

The sum of the probability of success of the combined discrete ranges $\sum_{l=3}^{4} DP_l$, $DP_l \in (DP_l - \delta/2, DP_l + \delta/2)$ is greater than 80% even at low consistency levels. The combined range represents the design range for this design parameter. If the system range is the entire range (i.e., $\sum_{l=1}^{4} DP_l$), the area $Prob_{SR}$ is unity. Thus, the common range area, $Prob_{CR}$, is the same as the design range. By using the traditional definition of complexity in Chapter 4, the following probability of success (Table 7.2) and complexity levels can be determined.

The probability of success for a given functional requirement (e.g., the sealing function) can be calculated when the respective DP is selected (e.g., a compression-based gasket versus a chemical elastomer). Since the sealing functional requirement is a function of the DP = "elastomer," which is in turn a random variable, the FR probability of success, as well as the complexity of its

TABLE 7.1 Probability Distribution Obtained from Fuzzy Data p

α	Dist.	DP_1	DP_2	DP_3	DP_4
	π	0.500	0.600	0.800	1.000
0.85	p	0.092	0.130	0.260	0.518
0.90	p	0.047	0.079	0.226	0.649
0.95	p	0.013	0.030	0.156	0.800

TABLE 7.2 Complexity Levels of the Probabilities

α	P_{cr}	P_{sr}	p	H (nats)
0.85	0.778	1.000	0.778	0.251
0.90	0.875	1.000	0.875	0.134
0.95	0.956	1.000	0.956	0.045

degree, can be found via the transfer function vehicle (if known) as presented in Chapter 5. Other elements of this example can be found in El-Haik et al. (1997). Hence, H_{ik} can be calculated and substituted into the formulation presented in Section 7.3 as follows:

1. Determine the discrete set values for all the DPs at module k identified by the technical, manufacturing (production), and assembly feasibility analysis.
2. Determine the membership function of the fuzzy set "successful" around these set values.
3. Solve the discrete mathematical formulation [(7.14)–(7.17)] to obtain the probabilities of success.
4. Substitute the probabilities in (7.2) to obtain H_{ik}.
5. Repeat steps 1 through 4 for every module of the structure (i.e., all feasible physical solution entities).
6. Substitute H_{ik} in the integer programming formulation that was presented in program (7.3)–(7.7) or program (7.8)–(7.13).
7. Solve to select the best concept (a solution entity or physical structure).

In addition to the probability–consistency principle, there have been many attempts to combine probabilistic and fuzzy measures in a discrete framework. Zadeh (1968) first introduced the entropy of a fuzzy set with respect to a discrete probabilistic as the weighted Shannon entropy. Other frameworks to combine probabilistic and fuzzy measures were suggested by Xie and Bedrosian (1984) and Pal and Pal (1992). We found that these measures are difficult to justify in our context.

7.5 AXIOMATIC QUALITY FUZZY CONCEPT SELECTION FORMULATION

So far in this chapter, the development focuses on obtaining an estimate for design complexity from fuzzy data. This simplifies the effort to find a solution to a fuzzy version of the IP formulation of the selection as presented in Section 7.3. In this formulation, information content is quantified using deterministic quantities rather than fuzzy quantities. In this section we use an alternative approach that "fuzzifies" the integer programming itself as a totality. The fuzzification of the integer program can be carried out by fuzzifying any combination of the variables CI, W, or AW. In the derivation below, the variables W and AW were used as fuzzy concepts. Extension of the development to include CI does not contribute to the formulation clarity and it was dropped. Nevertheless, the reader can follow the derivation here to include CI as a fuzzy concept when desired.

The modeling of W and AW as fuzzy numbers allow more realistic and robust representation of the imprecision and the linguistic inexactness experienced in the selection process. In this case, the functional weight rating can be viewed as a fuzzy linguistic variable (denoted as $\tilde{W}$) for QFD's *correlation* (relationship) *factor*, and the attribute weight rating can be viewed as a fuzzy linguistic variable for QFD's *importance factor*. Both factors take linguistic values in a set of rating with elements modeled as fuzzy numbers. A *fuzzy number* is a convex normalized piecewise continuous fuzzy set on the real line. For computational efficiency, Dubois and Prade (1979) suggested a fuzzy number representation that depends on the identification of two reference functions: L for left and R for right and the spreads α and β, respectively. A fuzzy number $\tilde{A}$ in LR representation can be written as $\tilde{A} = (t, \alpha, \beta)_{\mathrm{LR}}$ and is defined by

$$\mu_{\tilde{A}}(x) = \begin{cases} L\dfrac{t-x}{\alpha} & x \le t \\ R\dfrac{x-t}{\beta} & x \ge t \end{cases} \qquad x \in R, \text{ the set of real numbers} \tag{7.19}$$

For example, the attribute weight variable (AW) can be fuzzified when it assumes labels in the set {low, medium, high}. Each value in this label set can be modeled as a fuzzy number that is described by the parametric form in (7.19). Note that $\mathrm{AW}_j = \mathrm{AV}_j \cdot \mathrm{IR}_j \cdot \mathrm{SP}_j$ can only be fuzzified by fuzzifying at least one of its variable arguments in the multiplication form.

In the LR representation, if $\tilde{B} = (s, \gamma, \tau)_{\mathrm{LR}}$, then $\tilde{A} + \tilde{B} = (t + s, \alpha + \gamma, \beta + \tau)_{\mathrm{LR}}$ and $\tilde{A} - \tilde{B} = (t - s, \alpha + \tau, \beta + \gamma)_{\mathrm{LR}}$. For multiplication, we have the following rules:

$$\tilde{A} \cdot \tilde{B} \approx \begin{cases} (ts, s\alpha + t\gamma, s\beta + t\tau)_{\mathrm{LR}} & \text{when } \tilde{A} \text{ and } \tilde{B} \text{ are both positive numbers} \\ (ts, -s\beta - t\tau, s\alpha - t\gamma)_{\mathrm{LR}} & \text{when } \tilde{A} \text{ and } \tilde{B} \text{ are both negative numbers} \\ (ts, s\alpha - t\tau, s\beta - t\gamma)_{\mathrm{LR}} & \text{when } \tilde{A} \text{ is positive and } \tilde{B} \text{ is negative} \end{cases}$$

7.5.1 Case Study: Global Commercial Process

This case study is adapted from Yang and El-Haik (2003) with some alterations to illustrate the QFD fuzzy calculations by a design team.

The inconsistent, global process for selling to, setting up, and servicing current and future accounts justifies the business case of this project. Current sales and customer service information management systems do not enable measurement of accuracy and timeliness on a global basis. Also, enterprise-wide customer care is a requirement—failure to improve the process threatens growth and retention of the portfolio.

The project objective is to design a global commercial process with six-sigma functional requirement performances. More specifically, the project goals are to reduce the prospecting cycle time from 16 to 5 business days, reduce the discovery cycle time from 34 to 10 business days, reduce the deal cycle time from 81 to 45 business days (all sales metrics net of customer wait time), reduce the setup cycle time from 51 to 12 business days, and increase the percentage of service requests closed by commitment date from 54% (1.6σ) to 99.97% (5.0σ).

Stage 1 QFD of this example will be used following the QFD practice discussed in Chapter 6.

1. Identify the customer attributes and critical-to-satisfaction and their relationship. The design team identifies customers and establishes customer wants, needs, delights, and usage profiles. In addition, corporate, regulatory, and social requirements should also be identified. The value of this step is to greatly improve the understanding and appreciation team members have for customer, corporate, regulatory, and social requirements. At this stage the team should be expanded to include market research. A market research professional might help the team assume leadership during startup activities and perhaps later, help them to remain active participants as the team gains knowledge about customer engagement methods. The black belt,[1] the team leader, should put plans in place to collaborate with identified organizations and/or employee relations to define tasks and plans in support of the project and to train team members in customer processes: forward-thinking methods such as brainstorming, visioning, and conceptualizing. The following CAs or "what's" are used:

- Direction of improvement
- Products available
- Professional staff
- Flexible processes
- Knowledgeable staff
- Easy-to-use products
- Speedy processes
- Cost-effective products
- Accuracy

2. Identify the CTSs ("how's") and relationship matrix. The purpose of this step is to define a "good" product/process in terms of customer expectations, benchmark projections, institutional knowledge, and interface requirements, and to translate this information into CTSs. These will then be used to plan an effective and efficient design project.

One of the major reasons for customer dissatisfaction and warranty costs is that the design specifications do not adequately reflect customer use of the product or process. A poorly planned design commonly does not allocate activities/resources in areas of importance to customers and waste engineering resources by spending too much time in activities that provide marginal value. Because missed customer requirements are not targeted or checked in the design process, procedures to handle field complaints for these items are likely to be incomplete. Spending time overdesigning and overtesting items not important to customers is waste. Similarly, not spending development time in areas important to customers is not only a missed opportunity, but significant warranty costs are sure to follow.

[1]See Section 6.8 for axiomatic quality process deployment.

In axiomatic quality, time is spent up front understanding customer wants, needs, and delights, together with corporate and regulatory requirements. This understanding is then translated into CTSs, which then drive product and process design. The following CTSs ("how's") are used:

- Importance to the customer
- Meet time expectations
- Know my business and offers
- Save money/enhance productivity
- Do it right the first time
- Consultative
- Know our products and processes
- Talk to one person
- Answer questions
- Courteous
- Adequate follow-up

A mapping begins by considering the high-level requirements for the product or process. These are the true CTSs, which define what the customer would like if the product or process were ideal. This consideration of a product or process from a customer perspective must address the requirements from higher-level systems, internal customers (other processes), external customers, and regulatory legislation. Customer "what's" are not easily operational in the world of the black belt. For this reason it is necessary to relate true quality characteristics to CTSs—design characteristics that may readily be measured and when targeted properly will substitute or assure performance to the "what's." Such a diagram, which relates true quality characteristics to substitute quality characteristics, is called a *relationship matrix*.

The mapping of customer characteristics to CTS characteristics is extremely valuable when done by the design team. A team typically begins by differing in opinion and sharing stories/experiences when the logic is only a few levels deep. An experiment may even be conducted to better understand the relationships. When completed, the entire team understands how product and process characteristics that are detailed on drawings relate to functions that are important to customers.

The full QFD stages 1, 2, and 3 are given in Chapter 7 of Yang and El-Haik (2003). Our analysis below applies to stage 1. The stage 1 crisp QFD house of quality is repeated in Figure 7.3. The fuzzy calculation will be carried it by fuzzifying the importance rating calculations of the CTSs. For example, $\sum_{j=1}^{J} W_{ij}\text{CA}_j$, $j = 1, \ldots, J$, is a measure of a design-perceived satisfaction index for CTS_i (where $J = 8$, the number of customer attributes in QFD). The crisp calculations are given in the row "Importance of Product Attributes," the importance rating room of HOQ, in Figure 7.3. Importance ratings are a relative comparison of the importance of each CA or "what" to the quality of the

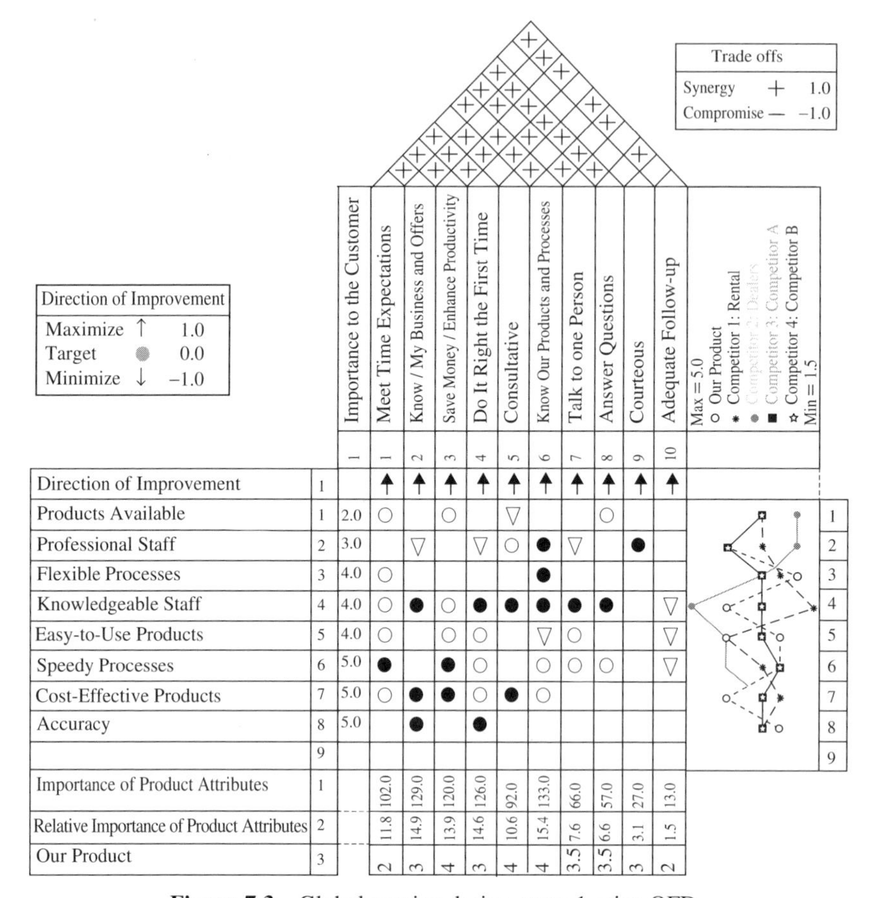

Figure 7.3 Global service design stage 1 crisp QFD.

design. The 9–3–1-relationship matrix strength rating is used. Theses values are multiplied by the customer importance rating obtained from customer engagement activities (e.g., surveys), resulting in a numerical value. The CTS importance rating are summed by adding all values of all relationships. For example, the first CTS of the Figure 7.3 importance rating is calculated as

$$(2.0 \times 3.0) + (4.0 \times 3.0) + (4.0 \times 3.0) + (4.0 \times 3.0) + (5.0 \times 9.0) + (5.0 \times 3.0) = 102$$

Other CTS importance ratings can be calculated accordingly.

3. Calculate the fuzzy importance rating. We first fuzzify the 9–3–1-relationship strength that will be used in the relationship matrix using fuzzy numbers with the labels {weak, moderate, strong} as depicted in Table 7.3. These will be the fuzzy numbers with LR representation to be substituted for $\tilde{W}_{ik} = (t_{ik}, \alpha_{ik}, \beta_{ik})_{\mathrm{LR}}$. For example, the "weak" label number can be written as $\tilde{W}_{\mathrm{weak}} = (1.0, 1.0, 1.0)_{\mathrm{LR}}$ using equation (7.19) in triangular form.

TABLE 7.3 Strength of Fuzzy Numbers

Strong	●	$(1.0, 1.0, 1.0)_{LR}$
Moderate	○	$(3.0, 2.0, 2.0)_{LR}$
Weak	∇	$(9.0, 3.0, 1.0)_{LR}$

TABLE 7.4 Relationship Matrix: Strength of Fuzzy Numbers

Customer Attributes	s	α	β
Products available	2	0.1	0.1
Professional staff	3	0.1	0.1
Flexible processes	4	0.1	0.1
Knowledgeable Management	4	0.1	0.1
Easy-to-use products	4	0.1	0.1
Speedy processes	5	0.1	0.1
Cost-effective products	5	0.1	0.1
Accuracy	5	0.1	0.1

In this work, the ranges of values for quantifying the relationship were predetermined by intuition. The voice of the customer was then sought through interview and was expressed in the figure. To simplify the data entry process, the uncertainty value (i.e., the width of the fuzzy number) was sought first. Subsequently, the mean ratings of the customer needs were obtained through interview. The ranges of initial ratings $[\tilde{CA}_j = (s_j, \gamma_j, \tau_j)_{LR}]$ were then derived from the mean ratings and the predetermined uncertainty value. In this work, the uncertainty value was initially fixed at ± 0.1, as shown in Table 7.4.

Using the fuzzy mathematics approach, customer ratings were obtained after taking into consideration the interactions between different customer needs. The ratings for design requirements were then derived from the customer ratings obtained and the relationships between customer needs and design requirements. Subsequently, the ratings for CTSs were refined using the relationships in Figure 7.3 using the strength fuzzy numbers in Table 7.3. The results generated were compared with those derived from the crisp approach (Figure 7.3) and are tabulated in Table 7.5. The left and right spreads were calculated first. For $\tilde{CA}_j = (s_j, \gamma_j, \tau_j)_{LR}$, the mean $(s_j)_{LR}$ was calculated midway between the left $(\gamma_j)_{LR}$ and right $(\tau_j)_{LR}$ spreads. The fuzzy number multiplication role was used as follows: $\tilde{CA} \cdot \tilde{W} \approx (ts, s\alpha + t\gamma, s\beta + t\tau)_{LR}$ when $A\tilde{W}$ and $\tilde{W}$ are both positive numbers in this example.

As shown in the table, the results obtained by both crisp and fuzzy approaches exhibited an identical trend. It could be inferred that CTS = "know our products and processes," CTS = "know my business and others," CTS = "do it right the first time," and CTS = "save money/enhance productivity" are still the critical design satisfaction requirements for the QFD stage. However, the crisp approach

TABLE 7.5 CTS Fuzzy Rating of Importance

Customer Attributes	*s*	?	?
Meet time expectation	97	55.4	45.4
Know my business and others	119.85	38.1	19.8
Save money/enhance productivity	110	52.7	32.7
Do it right the first Time	117	60.8	42.8
Consultative	92	19.2	19.2
Know our products and processes	122	60.4	38.4
Talk to one person	62	34.6	26.6
Answer questions	53	27.5	19.5
Courteous	15	9.9	3.9
Adequate follow-up	13	13.3	13.3

tended to produce ratings that were close to the upper limits of the ranges registered in Figure 7.3. This might not be desirable since the ratings would possibly affect the selection of critical CTS requirements. The ratings generated by the fuzzy approach, however, were expressed in terms of ranges of values. This would provide an overall picture about the design requirement concerned and could ensure that the decision made in the subsequent selection process would not be biased. As an example, the design requirement, CTS = "consultative," has a rating of [72.8, 111.2]. Qualitatively, this suggests that it is moderately important but far from being a critical requirement. However, a crisp rating of 92 generated by the crisp approach may imply differently.

4. Select the fuzzy concept. The fuzzy integer programming for the concept selection problem is formulated next. Let $\widetilde{W}_{ik} = (t_{ik}, \alpha_{ik}, \beta_{ik})_{\mathrm{LR}}$ and $\mathrm{A}\widetilde{\mathrm{W}}_j = (s_j, \gamma_j, \tau_j)_{\mathrm{LR}}$. Let the set of CAs be partitioned into four subsets. Let $\mathrm{CA}^{\rangle+,+\langle}$, $\mathrm{CA}^{\rangle+,+\langle} \subset \mathrm{CA}$ be the subset, with cardinality $J^{\rangle+,+\langle}$, where both $\widetilde{W}_{ik}$ and $\mathrm{A}\widetilde{\mathrm{W}}_j$ are positive fuzzy numbers; $\mathrm{CA}^{\rangle-,-\langle}$, $\mathrm{CA}^{\rangle-,-\langle} \subset \mathrm{CA}$ be the subset, with cardinality $J^{\rangle-,-\langle}$, where both $\widetilde{W}_{ik}$ and $\mathrm{A}\widetilde{\mathrm{W}}_j$ are negative fuzzy numbers; $\mathrm{CA}^{\rangle+,-\langle}$, $\mathrm{CA}^{\rangle+,-\langle} \subset \mathrm{CA}$ be the subset, with cardinality $J^{\rangle+,-\langle}$, where $\widetilde{W}_{ik}$ is a positive fuzzy number and $\mathrm{A}\widetilde{\mathrm{W}}_j$ is a negative fuzzy number; $\mathrm{CA}^{\rangle-,+\langle}$, $\mathrm{CA}^{\rangle-,+\langle} \subset \mathrm{CA}$ be the subset, with cardinality $J^{\rangle-,+\langle}$, where $\widetilde{W}_{ik}$ is a negative fuzzy number and $\mathrm{A}\widetilde{\mathrm{W}}_j$ is a positive fuzzy number. Then $\mathrm{CA} = \mathrm{CA}^{\rangle+,+\langle} \cup \mathrm{CA}^{\rangle-,-\langle} \cup \mathrm{CA}^{\rangle+,-\langle} \cup \mathrm{CA}^{\rangle-,+\langle}$ and $J = J^{\rangle+,+\langle} + J^{\rangle-,-\langle} + J^{\rangle+,-\langle} + J^{\rangle-,+\langle}$. The fuzzy integer programming formulation can be written as

$$\text{Maximize } \widetilde{O} = \sum_{i=1}^{M}\sum_{k=1}^{K}\sum_{j=1}^{J}(\widetilde{W}_{ik})_{\mathrm{LR}}(\mathrm{A}\widetilde{\mathrm{W}}_j)_{\mathrm{LR}}\left(\frac{Y_{ik}}{\sum_{i=1}^{m} H_{ik}Y_{ik}}, 0, 0\right)_{\mathrm{LR}} \qquad (7.20)$$

subject to

$$\sum_{k=1}^{K} Y_{ik}T_{ik} = 1 \qquad \forall i, i = 1, 2, \ldots, m, \quad \mathrm{DP}_k \in \mathbf{F} \qquad (7.21)$$

$$\sum_{i=1}^{M}\sum_{k=1}^{K}\sum_{j=1}^{J}(\widetilde{W}_{ik})_{LR}(\mathrm{A}\widetilde{\mathrm{W}}_j)_{\mathrm{LR}}(Y_{ik},0,0)_{\mathrm{LR}}$$

$$> \left[\sum_{i=1}^{M_d}\sum_{j=1}^{J}(\widetilde{W}_{ik})_{\mathrm{LR}}(\mathrm{A}\widetilde{\mathrm{W}}_j)_{\mathrm{LR}}(Y_{ik},0,0)_{\mathrm{LR}}\right]_{\mathrm{datum}} \tag{7.22}$$

$$\sum_{i=1}^{M}\sum_{k=1}^{K}H_{ik}Y_{ik} < \left(\sum_{i=1}^{M_d}H_i\right)_{\mathrm{datum}} \tag{7.23}$$

$$\sum_{i=1}^{M-1}\sum_{u=1+i}^{M}Y_{ik}Y_{ul}Z_{kl} \le M-1 \qquad i \ne u;\quad i=1,2,\ldots,M-1;$$

$$u=2,3,\ldots,M \tag{7.24}$$

$$Y_{ik} = 0 \text{ or } 1 \tag{7.25}$$

By applying LR mathematics, the sum of the LR fuzzy number representation in each subset can be calculated as follows: For $j \in \mathrm{CA}^{\rangle+,+\langle}$, we have

$$\left[\sum_{i=1}^{M}\sum_{k=1}^{K}\sum_{j=1}^{j\in J^{\rangle+,+\langle}} t_{ik}s_jY_{ik}, \sum_{i=1}^{M}\sum_{k=1}^{K}\sum_{j=1}^{j\in J^{\rangle+,+\langle}}(t_{ik}\gamma_j + s_j\alpha_{ik})Y_{ik},\right.$$
$$\left.\sum_{i=1}^{M}\sum_{k=1}^{K}\sum_{j=1}^{j\in J^{\rangle+,+\langle}}(t_{ik}\tau_j + s_j\beta_{ik})Y_{ik}\right]_{\mathrm{LR}} \tag{7.26}$$

For $j \in \mathrm{CA}^{\rangle-,-\langle}$, we have

$$\left[\sum_{i=1}^{M}\sum_{k=1}^{K}\sum_{j=1}^{j\in J^{\rangle-,-\langle}} t_{ik}s_jY_{ik}, -\sum_{i=1}^{M}\sum_{k=1}^{K}\sum_{j=1}^{j\in J^{\rangle-,-\langle}}(t_{ik}\tau_j + s_j\beta_{ik})Y_{ik},\right.$$
$$\left.\sum_{i=1}^{M}\sum_{k=1}^{K}\sum_{j=1}^{j\in J^{\rangle-,-\langle}}(-t_{ik}\gamma_j + s_j\alpha_{ik})Y_{ik}\right]_{\mathrm{LR}} \tag{7.27}$$

For $j \in \mathrm{CA}^{\rangle+,-\langle}$, we have

$$\left[\sum_{i=1}^{M}\sum_{k=1}^{K}\sum_{j=1}^{j\in J^{\rangle+,-\langle}} t_{ik}s_jY_{ik}, \sum_{i=1}^{M}\sum_{k=1}^{K}\sum_{j=1}^{j\in J^{\rangle+,-\langle}}(t_{ik}\tau_j - s_j\alpha_{ik})Y_{ik},\right.$$
$$\left.\sum_{i=1}^{M}\sum_{k=1}^{K}\sum_{j=1}^{j\in J^{\rangle+,-\langle}}(t_{ik}\gamma_j - s_j\beta_{ik})Y_{ik}\right]_{\mathrm{LR}} \tag{7.28}$$

For $j \in \mathrm{CA}^{\rangle -,+\langle}$, we have

$$\left[\sum_{i=1}^{M}\sum_{k=1}^{K}\sum_{j=1}^{j\in J^{\rangle -,+\langle}} t_{ik}s_jY_{ik}, \sum_{i=1}^{M}\sum_{k=1}^{K}\sum_{j=1}^{j\in J^{\rangle -,+\langle}} (-t_{ik}\tau_j + s_j\alpha_{ik})Y_{ik}, \sum_{i=1}^{M}\sum_{k=1}^{K}\sum_{j=1}^{j\in J^{\rangle -,+\langle}} (-t_{ik}\gamma_j + s_j\beta_{ik})Y_{ik}\right]_{\mathrm{LR}} \tag{7.29}$$

The overall objective function $\widetilde{O}$ in the LR format is given as

$$\widetilde{O} = \left(\frac{\sum_{i=1}^{M}\sum_{k=1}^{K}\sum_{j=1}^{J} t_{ik}s_jY_{ik}}{\sum_{i=1}^{M}\sum_{k=1}^{K} H_{ik}Y_{ik}}, \frac{\begin{array}{c}\sum_{i=1}^{M}\sum_{k=1}^{K} Y_{ik}\Big[\sum_{j=1}^{j\in J^{\rangle +,+\langle}}(t_{ik}\gamma_j + s_j\alpha_{ik}) \\ -\sum_{j=1}^{j\in J^{\rangle -,-\langle}}(s_j\beta_{ik} + t_{ik}\tau_j)(s_j\alpha_{ik} - t_{ik}\tau_j) \\ +\sum_{j=1}^{j\in J^{\rangle +,-\langle}}\sum_{j=1}^{j\in J^{\rangle -,+\langle}}(t_{ik}\tau_j - s_j\alpha_{ik})\Big]\end{array}}{\sum_{i=1}^{M}\sum_{k=1}^{K} H_{ik}Y_{ik}}, \frac{\begin{array}{c}\sum_{i=1}^{M}\sum_{k=1}^{K} Y_{ik}\Big[\sum_{j=1}^{j\in J^{\rangle +,+\langle}}(t_{ik}\tau_j + s_j\beta_{ik}) + \sum_{j=1}^{j\in J^{\rangle -,-\langle}} \\ \times(s_j\alpha_{ik} - t_{ik}\gamma_j)(s_j\beta_{ik} - t_{ik}\gamma_j) + \sum_{j=1}^{j\in J^{\rangle +,-\langle}}\sum_{j=1}^{j\in J^{\rangle -,+\langle}}(t_{ik}\gamma_j - s_j\beta_{ik})\Big]\end{array}}{\sum_{i=1}^{M}\sum_{k=1}^{K} H_{ik}Y_{ik}} \right)_{\mathrm{LR}} \tag{7.30}$$

An optimum and feasible physical configuration is the one that maximizes the mean of $\widetilde{O}$ while minimizing imprecision, that is, minimizing the left and right spreads in its LR representation. Using this reasoning, the problem can be formulated as a deterministic nonlinear {0,1} integer programming problem. That is, the problem is transferred from a fuzzy nonlinear {0,1} integer program to a deterministic nonlinear {0,1} integer program with constraints derived from fuzzy quantities. In the deterministic domain, the objective function can take a quotient form where the numerator is the mean of $\widetilde{O}$ while the denominator is the sum (or the product) of the left and right spreads of (7.30). The spread sum form of the objective was adopted and is given in (7.32). In addition, the fuzzy constraint in (7.22) should be converted to its deterministic form. Let the datum design performance [the right-hand side of (7.22)] be given as

$$\left(\sum_{j=1}^{J} \widetilde{W}_{ik}\mathrm{A}\widetilde{\mathrm{W}}_j\right)_{\mathrm{datum}} = (w, \varepsilon, \omega)_{\mathrm{LR}} \tag{7.31}$$

Then by employing the subtraction rule of fuzzy numbers of LR representation and by using $(0,0,0)_{\mathrm{LR}}$ as a neutral element for the addition operation, we get constraint (7.33)–(7.35). The overall program can be assembled by appending constraints (7.36)–(7.39).

$$
\text{maximize}\frac{\left(\sum_{i=1}^{M}\sum_{k=1}^{K}\sum_{j=1}^{J} t_{ik}s_j Y_{ik}\right)\left(\sum_{i=1}^{M}\sum_{k=1}^{K} H_{ik}Y_{ik}\right)}{\left(\sum_{i=1}^{M}\sum_{k=1}^{K} Y_{ik}\left(\begin{array}{l}\sum_{j-1}^{j\in J\rangle+,+\langle}(t_{ik}\gamma_j + S_j\alpha_{ik})\\ -\sum_{j=1}^{j\in J\rangle-,-\langle}(S_j\beta_{ik} + t_{ik}\tau_j)\\ +\sum_{j=1}^{j\in J\rangle+,-\langle}(t_{ik}\tau_j - S_j\alpha_{ik})\\ +\sum_{j=1}^{j\in J\rangle-,+\langle}(S_j\alpha_{ik} - t_{ik}\tau_j)\end{array}\right)\right) + \left(\sum_{i=1}^{M}\sum_{k=1}^{K} Y_{ik}\left(\begin{array}{l}\sum_{j=1}^{j\in J\rangle+,+\langle}(t_{ik}\tau_j + S_j\beta_{ik})\\ +\sum_{j=1}^{j\in J\rangle-,-\langle}(S_j\alpha_{ik} - t_{ik}\gamma_j)\\ +\sum_{j=1}^{j\in J\rangle+,-\langle}(t_{ik}\gamma_j - S_j\beta_{ik})\\ +\sum_{j=1}^{j\in J\rangle-,+\langle}(S_j\beta_{ik} - t_{ik}\gamma_j)\end{array}\right)\right)} \tag{7.32}
$$

subject to

$$
\frac{\sum_{i=1}^{M}\sum_{k=1}^{K}\sum_{j=1}^{J} t_{ik}s_j Y_{ik}}{\sum_{i=1}^{M}\sum_{k=1}^{K} H_{ik}Y_{ik}} - w > 0 \tag{7.33}
$$

$$
\frac{\sum_{i=1}^{M}\sum_{k=1}^{K} Y_{ik}\left(\begin{array}{l}\sum_{j=1}^{j\in J\rangle+,+\langle}(t_{ik}\gamma_j + s_j\alpha_{ik}) - \sum_{j=1}^{j\in J\rangle-,-\langle}(s_j\beta_{ik} + t_{ik}\tau_j)\\ \sum_{j=1}^{j\in J\rangle-,+\langle}(s_j\alpha_{ik} - t_{ik}\tau_j) + \sum_{j=1}^{j\in J\rangle+,-\langle}(t_{ik}\tau_j - s_j\alpha_{ik})\end{array}\right)}{\sum_{i=1}^{M}\sum_{k=1}^{K} H_{ik}Y_{ik}} + \omega > 0 \tag{7.34}
$$

$$
\frac{\sum_{i=1}^{M}\sum_{k=1}^{K} Y_{ik}\left(\begin{array}{l}\sum_{j=1}^{j\in J\rangle+,+\langle}(t_{ik}\tau_j + s_j\beta_{ik}) + \sum_{j=1}^{j\in J\rangle-,-\langle}(s_j\alpha_{ik} - t_{ik}\gamma_j)\\ \sum_{j=1}^{j\in J\rangle-,+\langle}(s_j\beta_{ik} - t_{ik}\gamma_j) + \sum_{j=1}^{j\in J\rangle+,-\langle}(t_{ik}\gamma_j - s_j\beta_{ik})\end{array}\right)}{\sum_{i=1}^{M}\sum_{k=1}^{K} H_{ik}Y_{ik}} + \varepsilon > 0 \tag{7.35}
$$

$$
\sum_{k=1}^{K} Y_{ik}T_{ik} = 1 \qquad \forall i, \quad i = 1, 2, \ldots, m, \quad \mathrm{DP}_k \in \mathbf{F} \tag{7.36}
$$

$$
\sum_{i=1}^{M}\sum_{k=1}^{K} H_{ik}Y_{ik} < \left(\sum_{i=1}^{M_d} H_i\right)_{\text{datum}} \tag{7.37}
$$

$$
\sum_{i=1}^{M-1}\sum_{u=1+i}^{M} Y_{ik}Y_{ul}Z_{kl} \le M - 1, \qquad i \ne u;
$$

$$
i = 1, 2, \ldots, M-1; \quad u = 2, 3, \ldots, M \tag{7.38}
$$

$$
Y_{ik} = 0 \text{ or } 1 \tag{7.39}
$$

The decision variables are the binary variables Y_{ik}, which indicates the physical module (indexed k) that delivers the functional requirement FR_i while maximizing customer satisfaction and minimizing uncertainty in the linguistic formulation process as well as design complexity. Minimizing uncertainty increases the design team overall confidence in the modules selected and guards the selection process from being biased toward solutions of questionable confidence. This transferred deterministic formulation of the fuzzy selection problem allows analysis to be conducted at the micro level [i.e., the attributes (parameters) of the fuzzy number]. A macro-level formulation can be obtained when a fuzzy number is replaced by a crisp score (e.g., its centroid or weighted average) (Chen and Hwang, 1992). The crisp score is a function of the left and right sides of the membership function.

Solution to the program in (7.32)–(7.39) can be obtained by branch-and-bound enumeration method. In our case, there are M binary variable, the Y_{ik}'s, which result in exactly $\prod_{i=1}^{M} 2^i = 2^m$ different integer vector solutions. However, as M gets larger, it may be extremely difficult, computationally, to enumerate all integer solutions explicitly. However, with suitable constraint criteria selected, the exhaustive enumeration can be reduced by eliminating sets of the vector solutions that do not fit the criteria or result in improved solutions. These sets are enumerated implicitly. The *branch-and-bound method* requires a solution as a starting requirement, which can be obtained by relaxing the integrality constraints in (7.39) to $0 \leq Y_{ik} \leq 1, \forall i, i = 1, \ldots, M, k = 1, \ldots, K$. In this case the integer program is converted to a nonlinear continuous problem since the decision variables can assume any value in [0,1] inclusive. The objective function becomes a quotient of two linear functions and is subject to linear constraints. A program of this type is called a *fractional program* or a *programming problem with linear fractional functionals*. The different cases of solution treatment can be found in (Murty, 1983). Solutions are obtained after converting the fractional program to a linear program with suitable transformation of variables. Once the continuous solution is obtained, the branch-and-bound enumeration method can be employed.

7.6 SUMMARY

The concept selection problem can be solved using the IP formulation proposed here. The selection criteria include the complexity, customer satisfaction, and cost. Design complexity is measured by information content using Chapter 5 entropy derivation, which in turn takes the probability of success as arguments. In incremental design situations, these probabilities can be quantified and substituted in the deterministic integer programming [(7.8)–(7.13)]. The formulation presented here produces an optimum: that is, the entity selected that best maximizes customer satisfaction and simplicity while minimizing cost and within technical and manufacturing feasible physical structures.

CHAPTER 8

CONCEPTUAL DESIGN FOR CAPABILITY PHASE

8.1 INTRODUCTION

Engineering design can be defined as sets of processes and activities that transform customers' wants into design solutions that are of value to the society. With the current business environment, it is imperative that the design houses and manufacturing companies conceive and produce *healthy* products within a single development cycle. To achieve this goal, design entities (modules) should suffer no or minimal vulnerabilities if the "firefighting" cycles are to be avoided. The design vulnerabilities can be categorized as vulnerabilities leading to lack of conceptual robustness and operational vulnerabilities leading to lack of robustness at the operational level postlaunch. Both types often lead to quality issues in the designed entity. Conceptual robustness enables operational robustness, and not vice versa. The first category is the subject of this chapter.

The objective of the axiomatic quality process is to develop a robust analytical approach that provides solution methods to the two major categories of vulnerabilities mentioned above. The process has three phases, as depicted in Figure 6.1:

1. Customer attributes-to-functional requirements mapping
2. Conceptual design for capability (CDFC) phase (prerequisite to the optimization phase)

Axiomatic Quality: Integrating Axiomatic Design with Six-Sigma, Reliability, and Quality Engineering, by Basem Said El-Haik
ISBN 0-471-68273-X

3. Optimization phase (the last phase of the axiomatic quality process, concerned with the operational vulnerabilities, discussed in Chapter 9).

In this chapter we address the first phase, and focusing on providing a solution framework for the conceptual vulnerabilities to enable the conception of entities with six-sigma quality potential. Most of the steps of the CDFC phases are presented in Chapter 6. This chapter complements the CDFC framework by zooming in key CDFC steps that were discussed briefly in Chapter 6.

8.2 PROBLEMS THAT CAN BE SOLVED BY AXIOMATIC QUALITY

The conception of an uncoupled (independent) design does not automatically guarantee that high quality levels such as six-sigma capability can be obtained. A coupling-free design has a better potential to establish such capability. The task of operational vulnerability optimization up to a six-sigma capability is easier, to a large degree, in uncoupled and decoupled designs versus a coupled design.

Useful information usually lags design activities in a development cycle. It is not until the prototype phase that the design team will have useful and credible actionable feedback, which is based on some aspects of actual and tested performance (Figure 1.8). This usually happens following the midcycle of the product development. As such, there is not much room for making hard changes in the design entity if unpleasant issues do arise. When such events occur, companies usually resort to improving the explicit design operational vulnerability rather than concentrating on enhancing the implicit conceptual vulnerabilities. For example, modeling will help an aggressive operational vulnerability optimization when weak correlation to the actual use environment can be avoided. Such practices do not guarantee a high capability in the system, especially when week concepts were conceived and pushed into production. The challenge, in a given design assignment, is the nonavailability of useful information to lead design activity up front, where most influential decisions are to be made. The objective of axiomatic quality is to design it *right* the first time that a particular design project is encountered. In this dilemma, a sound strategy should provide design principles that directionally lead to the conception of good design alternatives to enable a sound optimization phase in the absence of credible data.

A mapping that is hosted in one design module (e.g., component, subsystem, or system) can be depicted in a robust design diagram called a *P-diagram* (Figure 8.1). Let us assume further that this mapping is at a level of impact in the context of Figure 6.9. The useful module outputs are designated as the array **{FR}**, which is in turn affected by three kinds of variables[1]: the signals represented by the array **{M}**, the design parameters represented by the array **{DP}**, and the noise factors represented by array $\{\mathbf{NF}\}_{b\times 1}$. Variation in **{FR}** and its drift from a desired target performance are usually caused by the noise factors. Ideally,

[1]By borrowing from robust design dynamic classification discussed in Chapter 5.

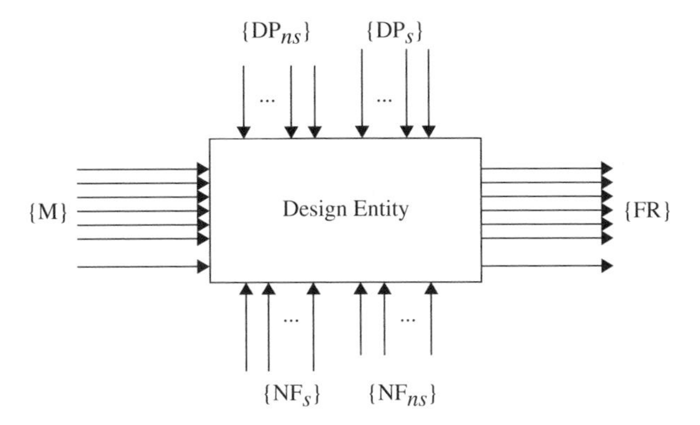

Figure 8.1 P-diagram.

the norms of $\{\mathbf{M}\}$ and $\{\mathbf{FR}\}$ arrays are equal when they are expressed in terms of energy in dynamic systems by assuming the effects of the noise factors are null. In addition to leveraging transfer function nonlinearity and heteroscedasticity as robustness mechanisms, robustness may be achieved by reducing the difference array norm $|\Delta| = |\mathbf{FR}| - |\mathbf{M}|$ through utilizing the interaction $\{\mathbf{DP}\} \times \{\mathbf{NF}\}$. Let the array $\{\mathbf{DP}\}$ be split into significant and nonsignificant factors, denoted as $\{\mathbf{DP}_s, \mathbf{0}\}$, and nonsignificant factors, $\{\mathbf{0}, \mathbf{DP}_{ns}\}$, relative to the array $\{\mathbf{FR}\}$, respectively. That is, $\{\mathbf{DP}\}_{p\times 1} = \{\mathbf{DP}_s, \mathbf{0}\}_{p\times 1} + \{\mathbf{0}, \mathbf{DP}_{ns}\}_{p\times 1}$. Also, let the array $\{\mathbf{NF}\}$ be spilt into significant and nonsignificant factors, denoted as $\{\mathbf{NF}_s, \mathbf{0}\}_{b\times 1}$ and $\{\mathbf{0}, \mathbf{NF}_{ns}\}_{b\times 1}$, respectively. That is, $\{\mathbf{NF}\}_{b\times 1} = \{\mathbf{NF}_s, \mathbf{0}\}_{b\times 1} + \{\mathbf{0}, \mathbf{NF}_{ns}\}_{b\times 1}$. Off course, the level of significance should be decided by any knowledge the team already has from the perspective of the functional requirement.

Usually, the nonsignificant factors are trivially many in number, whereas the significant factors are critically few. There are four possibilities of an axiomatic quality project from the standpoint of design parameter (control) versus noise factor classifications in the context of this section. They are listed in Table 8.1. The effects of nonsignificant factors, whether design parameters or noise factors, are usually week and sparse in a manner that bears more credibility to the Pareto principle. As such, their existence does not add to the complexity of the design or its solution efforts.

Only when the set of significant factors, the $\{\mathbf{DP}_s\}$ array, exists is there a potential for the axiomatic quality method to produce the operational robustness capability desired in the concerned functional requirements. The axiomatic quality process has a potential for use and applicability in this case by employing a rich toolbox of conceptual and statistical tools.

8.3 CONCEPTUAL DESIGN FOR THE CAPABILITY PHASE

The core steps of the CDFC phase were presented in Chapter 6 in the context of the axiomatic quality process depicted in Figure 6.3. In this section we zoom

TABLE 8.1 Possibilities of an Axiomatic Quality Design Problem

	$\{NF_s\}$ Exists	$\{NF_s\}$ Does Not Exist
$\{DP_s\}$ exists	The CDFC and optimization phases have an application potential. A six-sigma design may be possible. This case fits the organized complexity design classification discussed in Section 4.3 (see Weaver, 1948).	The CDFC phase has the potential to uncouple or decouple the design (step A.3 of axiomatic quality in Figure 6.3) where needed in the hierarchy. The operational vulnerability optimization phase may not be needed. A deterministic optimization formulation can be used when step A.3 fails.
$\{DP_s\}$ does not exist	The CDFC phase has a potential. The operational vulnerability optimization phase will not be successful.	Such design problems usually do not exist or are not needed.

into several key steps that need further clarification due to its deep implication design practices.

The design DNA or hierarchy can be defined using the zigzagging method presented in Chapter 3. The zigzagging method of axiomatic design is a conceptual modeling technique that reveals design conceptual vulnerability in the FRs, the array **{FR}**. Coupling of the FRs is a design vulnerability that negatively affects controllability and adjustability of the design entity and can be defined as the degree of lack of independence between the FRs (see Chapter 3). Coupling propagates across the design mappings hierarchy and limits the potential for six-sigma vulnerability optimization.

The integration of quality methods with the axiomatic method yields a robust axiomatic quality process with many advantages. For example, the employment of abstraction at high levels of the design structure facilitates decision making toward vulnerability-free concepts, while the use of mathematical formulation at low levels of the structure facilitates operational vulnerability optimization. Axiomatic design applies principles to structure design synthesis and select entities that are conceptually robust. Reduction of operational vulnerability of an uncoupled or decoupled design is much easier than that of a coupled design.

8.3.1 Implication of Coupling in the CDFC Phase

The term *module* or *physical structure* in the context of this book should not be limited to product design. The term can be extended to manufacturing (i.e., the processes by which the design entity is embodied). The conceptual design for

[2]See Yang and El-Haik (2003) and Theorem 2.9.

capability (CDFC), the first phase of axiomatic quality, is concerned with design synthesis such that a healthy vulnerability-free concept can be selected. *The CDFC phase ensures that the potential for a high quality level in the design entity is established by enabling conceptual robustness. This assurance materializes in both the physical and process mappings by optimization (Chapter 9).*

Design mappings can be expressed mathematically as $\{\mathbf{FR}\}_{m\times1} = [\mathbf{A}]_{m\times p}$ $\{\mathbf{DP}\}_{p\times1}$, $\{\mathbf{DP}\}_{p\times1} = [\mathbf{B}]_{p\times n}\{\mathbf{PV}\}_{n\times1}$, and $\{\mathbf{FR}\}_{m\times1} = [\mathbf{C}]\{\mathbf{PV}\}_{n\times1}$, where **[A]** is the physical mapping matrix, **[B]** the process mapping matrix, and $[\mathbf{C}]_{m\times n} = [\mathbf{A}][\mathbf{B}]$ the overall matrix prescribing the design. In either mapping we seek to satisfy the independence axiom. Therefore, the product matrix **C** should ideally be diagonal. The **A** and **B** matrices can be categorized from coupling perspective using equations (3.1)–(3.3). Accordingly, the various possibilities that can be taken by matrix [**C**] are given in Figure 8.2. Accordingly, the following points can be made:

- A decoupled design may be an upper or a lower triangular matrix, depending on the formulation.
- For the overall design entity (product and process) to be totally uncoupled, both matrixes should be uncoupled.

Not only are uncoupled designs desirable from the controllability and robustness standpoints, it also desirable because of the potential for a high probability of

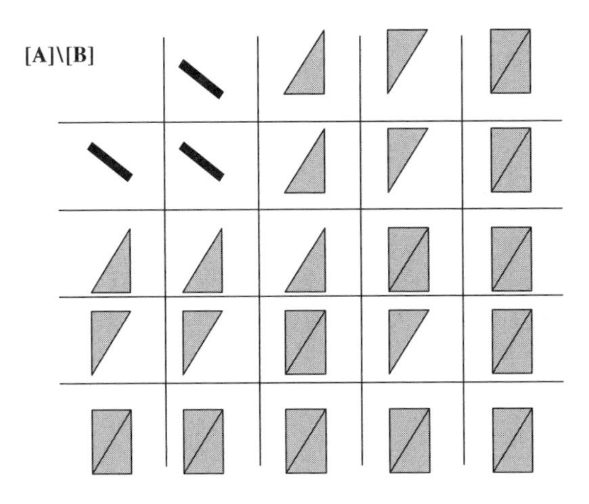

Figure 8.2 CDFC possibilities of matrix [C].

TABLE 8.2 Design Probabilities

Design Classification	Probability
Uncoupled	1/16
Decoupled	6/16
Coupled	9/16

producibility [i.e., a reduced defect per million opportunity (DPMO)].[3] Decoupled design is the next target alternative when uncoupled design cannot be achieved. However, the revealed sequence of adjustment should be followed in executing the synthesis process of creative and incremental decoupled design situations. *Design for producibility in the context of axiomatic quality is defined as having an overall uncoupled or decoupled design by conducting the process mapping and physical mapping concurrently.*

As depicted in Figure 8.2, we have the following CDFC scenarios:

- An overall *uncoupled design* is achieved only when both mappings are uncoupled.
- An overall *decoupled design* is achieved when:
  - Both mappings are decoupled and have similar triangular orientations.
  - Either mapping is uncoupled while the other is decoupled.
- An overall *coupled design* is achieved when:
  - At least one mapping is coupled.
  - Both mappings are decoupled with different triangular orientations.

With everything equal and left to chance, the odds are given by the probability distribution in Table 8.2. Obviously, the odds are not on the design team's side.

In addition, Table 8.2 gives an indication as to where it is *easier* to implement a change for problem solving, before or after release, without causing new problems or amplifying existing problematic symptoms of the FRs. A design change is easier to implement and control in the uncoupled and decoupled design classifications than in a coupled design. As mentioned in Chapter 6, design change may be soft or hard. Soft changes imply adjusting the targets (nominal values) within the tolerances specified, changing the tolerance ranges, or both. Hard changes imply eliminating or adding DPs or PVs in a mapping.

Whether soft changes are solution effective depends on the mappings and the nominal and tolerance settings. Hard changes require alterations of the PV or DP

[3]DPMO is a six-sigma concept. It defines a mix of defects and opportunities for DPs in manufacturing. A number of defects and a number of opportunities are defined for each process step:

$$\text{DPMO}_{\text{step}} = \frac{\text{number of defects}}{\text{number of opportunities}} \times 10^6$$

array or both, a new incremental design project. Hard changes are usually followed by a soft changes phase for tuning and adjustment. In either case, the cost of controlling solution implementation is a major cost element, as discussed in Chapter 9.

Altering (hard changes) or adjusting (soft changes) the DPs can be used to uncouple or decouple a system. Unfortunately, hard changes are not always feasible, due to incapable or outdated technology, the cost of change limitation, the organizational culture, or other inhibitors. A company may choose to make an educated decision on keeping the coupled design entities but with a reduced (minimized) degree of coupling among the FRs. While the company should recognize that this decision should be a short-term strategy, its adoption as a long-term strategy may result in a big loss in several dimensions. Conceptual robustness targeting coupling resolution means eliminating conceptual vulnerabilities (i.e., fire prevention), in contrast to coupling management, which implies practicing an endless cycle of firefighting operation modes.

The hard changes that target design decoupling may be difficult and costly to implement after launch, a scenario that could have been avoided when the product is based on design axioms. Usually, companies resort to soft changes first to improve their current product portfolio. The objective of such practices is problem solving of immediate and pressing customer concerns.

8.3.2 Step A.3: Uncouple or Decouple the Design Mappings

The axiomatic quality process CDFC phase is repeated in Figure 8.3 for reference. The objective in this section is to explore the uncoupling or decoupling activity. In the axiomatic quality process, the decoupling phase should be conducted when coupled designs are conceived in the zigzagging process across all design hierarchical levels prior to physical structure synthesis. It is imperative that the conceptual vulnerability be characterized clearly in all mappings prior to proceeding to decoupling activity (step A.3, Figure 6.3). Prior to beginning such an activity, the design team needs to characterize all design mappings clearly by rearranging or reordering of the design matrices in one of the (3.1)–(3.3) equation format. An example of such reordering is exhibited in Figure 8.4. It is obvious that the design matrix in case 2 is a decoupled design, a picture that was vague in case 1 prior to reordering.

There are many ways to decouple a design, depending on the situation:

1. Make the size of array **{FR}** equal the size of array **{DP}** (i.e., $m = p$) According to Theorem 2.2 (Section 2.5), when a design is coupled because the number of FRs is greater (less) than the number of design parameters, it may be decoupled by smartly adding (fixing) parameters so that the number of FRs equals the number of design parameters. The hint to which desirable design parameter to add (fix) depends on the coupling situation already present post the reordering step discussed earlier. Consider case 1 in Figure 8.5, which shows a subset of the design matrix containing $m \times m$ elements which constitutes a triangular matrix.[4]

[4]The entry X in the design matrices is a shorthand notation for nonzero sensitivities.

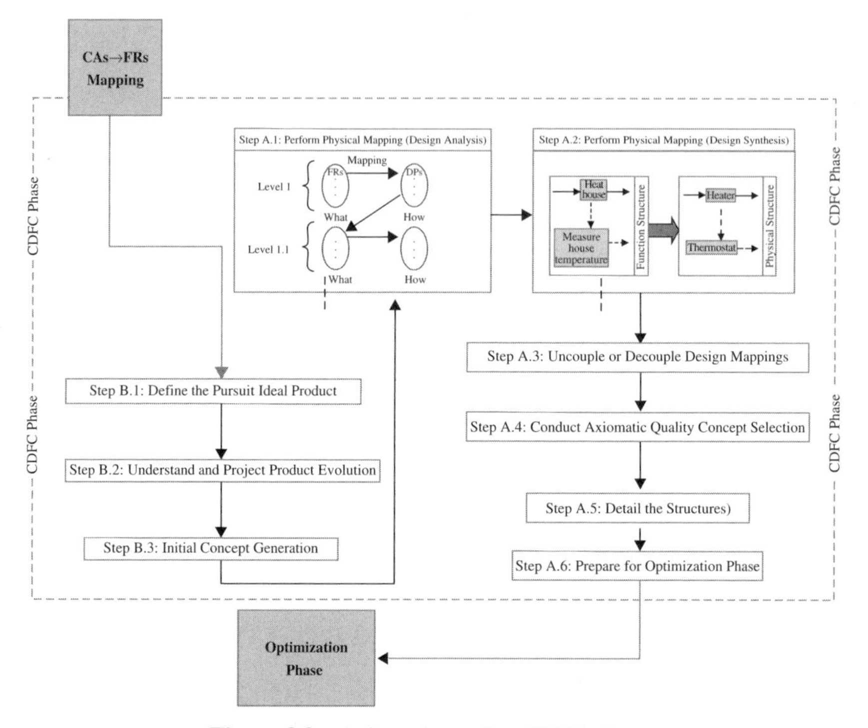

Figure 8.3 Axiomatic quality CDFC phase.

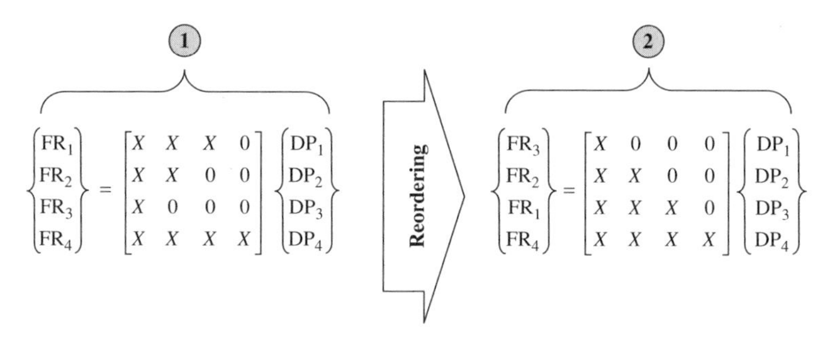

$$\begin{Bmatrix} FR_1 \\ FR_2 \\ FR_3 \\ FR_4 \end{Bmatrix} = \begin{bmatrix} X & X & X & 0 \\ X & X & 0 & 0 \\ X & 0 & 0 & 0 \\ X & X & X & X \end{bmatrix} \begin{Bmatrix} DP_1 \\ DP_2 \\ DP_3 \\ DP_4 \end{Bmatrix} \quad \begin{Bmatrix} FR_3 \\ FR_2 \\ FR_1 \\ FR_4 \end{Bmatrix} = \begin{bmatrix} X & 0 & 0 & 0 \\ X & X & 0 & 0 \\ X & X & X & 0 \\ X & X & X & X \end{bmatrix} \begin{Bmatrix} DP_1 \\ DP_2 \\ DP_3 \\ DP_4 \end{Bmatrix}$$

Figure 8.4 Design mapping reordering.

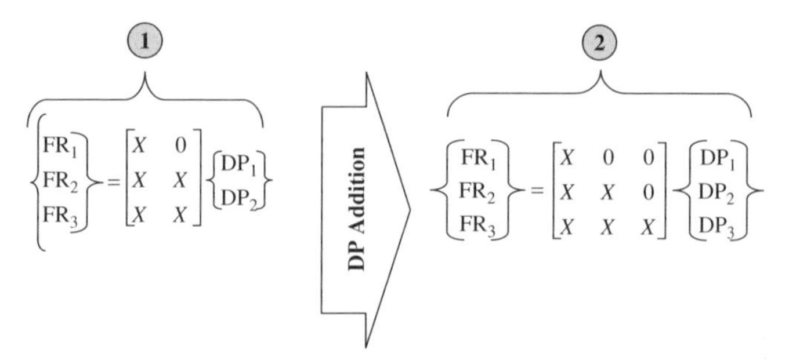

$$\begin{Bmatrix} FR_1 \\ FR_2 \\ FR_3 \end{Bmatrix} = \begin{bmatrix} X & 0 \\ X & X \\ X & X \end{bmatrix} \begin{Bmatrix} DP_1 \\ DP_2 \end{Bmatrix} \quad \begin{Bmatrix} FR_1 \\ FR_2 \\ FR_3 \end{Bmatrix} = \begin{bmatrix} X & 0 & 0 \\ X & X & 0 \\ X & X & X \end{bmatrix} \begin{Bmatrix} DP_1 \\ DP_2 \\ DP_3 \end{Bmatrix}$$

Figure 8.5 Decoupling by adding extra design parameters.

Another DP, in this case DP_3, needs to be added in this case that will only map to FR_3 without affecting FR_1 or FR_2, producing case 2 of the matrix. For a redundant design ($p > m$), the design team should fix the extra DPs to keep the mapping entries that achieve an uncoupled or decoupled design in the FRs, generally using *isogram* plots and/or equations (3.4) and (3.5).

The FR–DP isogram usually plots in design mappings where m is less than or equal to 3. Assessment using independence measures is also a method to assure the decision of this uncoupling or decoupling step.

2. Decouple by utilizing the nonlinear design sensitivity of some FR. In this case the design team is seeking parameters that have a minimal effect on FRs other than the targeted FR. This can be done by analyzing the magnitude of the off-diagonal elements in the design matrix on reangularity and semangularity measures by varying the DPs over the design range. Maximizing both measures should be a goal.

Methods 1 and 2 seek decoupling or uncoupling by adding, replacing, or changing the sensitivity of design parameters. These methods may greatly benefit from other axiomatic design theorems and corollaries in Suh (1990, 2001). In addition, a great solution synergy can be gained using TRIZ (Section 8.5) contradiction elimination principles to reduce or eliminate coupling vulnerability.

Decoupling methods 1 and 2 have many opportunities when engaged at the start of stage 1 of Figure 1.8. The degree of freedom in applying them is limited in redesign situations with binding physical and financial constraints. Redesign incremental scenarios that are classified as coupled call for another method that is based on tolerance optimization to reduce operational vulnerability (see Sections 9.4 and 9.5).

3. Apply the axiomatic design theorems and corollaries (nontolerance based). How is a design decoupled or uncoupled? Considering the independence axiom, we want a 1-to-1 relationship (i.e., a bijection mapping) between FRs and DPs. Ideally, we would want a square diagonal matrix. The design corollaries[5] give clues for possible uncoupling treatment, however axiomatic in nature. The relevant set in this regard is listed below.

Corollary 2.1: Decoupling of a Coupled Design Decouple or separate parts or aspects of a solution if FRs are coupled or become interdependent in the design proposed.

Theorem 2.1: Coupling Due to an Insufficient Number of DPs When the number of DPs is less than the number of FRs, either a coupled design results or the FRs cannot be satisfied.

Theorem 2.2: Decoupling of a Coupled Design When a design is coupled due to the greater number of FRs than DPs (i.e., $m > p$), it may be decoupled by

[5]For a list of axiomatic design theories and corollaries used in this book, see Section 2.5. A complete list can be found in Suh (2001).

the addition of new DPs so as to make the number of FRs and DPs equal each other if a subset of the design matrix containing $p \times p$ elements constitutes a triangular matrix.

Theorem 2.3: Redundant Design When there are more DPs than FRs, the design is either a redundant design or a coupled design.

Theorem 2.6: Path Independence of Uncoupled Design The information content of an uncoupled design is independent of the sequence by which the DPs are changed to satisfy the given set of FRs. (See Section 1.3 for more details.)

Theorem 2.7: Path Dependency of Coupled and Decoupled Designs The information contents of coupled and decoupled designs depend on the sequence by which the DPs are changed to satisfy the given set of FRs. (See Section 1.3 for more details.)

Theorem 2.11: Invariance Reangularity, R, and semangularity, S, for a design matrix [**DM**] are invariant under alternative orderings of the FR and DP variables as long as orderings preserve the association of each FR with its corresponding DP.

4. Decouple by tolerance optimization. The tolerances of the FRs have a strong role to play in decoupling a design mapping. This method has more leverage with nonlinear design mappings. The FRs are always specified with some tolerances, $T_i \pm \Delta \mathrm{FR}_i$, $i = 1, \ldots, m$, where $\Delta \mathrm{FR}_i$ is the half tolerance of FR_i and m is the number of FRs in the array $\{\mathbf{FR}\}$ of a given design mapping. Assume that we have a 2×2-coupled design with

$$\begin{Bmatrix} \mathrm{FR}_1 \\ \mathrm{FR}_2 \end{Bmatrix} = \begin{bmatrix} A_{11} & A_{12} \\ A_{21} & A_{22} \end{bmatrix} \begin{Bmatrix} \mathrm{DP}_1 \\ \mathrm{DP}_2 \end{Bmatrix}$$

In method 4, the question is: Can A_{12} or A_{21}, the off-diagonal mapping entries, be neglected by leveraging the nonlinearity in the DPs, for example? In effect, this activity will make either or both off-diagonal sensitivities approach zero (i.e., $A_{12} \rightarrow 0$ and/or $A_{21} \rightarrow 0$). That is, can the design be improved to a uncoupled or decoupled design? If the answer is "no," method 3 may be applied. The transferred variation of FR_1 in the 2×2 coupled mapping above is given by

$$\partial \mathrm{FR}_1 = \frac{\partial \mathrm{FR}_1}{\partial \mathrm{DP}_1} \partial \mathrm{DP}_1 + \frac{\partial \mathrm{FR}_1}{\partial \mathrm{DP}_2} \partial \mathrm{DP}_2$$

Based on customer specification, we need to maintain $\partial \mathrm{FR}_1 \leq \Delta \mathrm{FR}_1$ (i.e., the change in the FR_1 due to changes in the design parameters is less than the tolerance specified by the customer). To achieve a decoupled design, we need to make A_{12} negligibly small, which translates into making $\Delta \mathrm{FR}_1 \geq (\partial \mathrm{FR}_1 / \partial \mathrm{DP}_2) \partial \mathrm{DP}_2$

(i.e., neglecting the off-diagonal element). This is the essence of Theorem 8 in Suh (1990, p. 122) quoted here: "A design is an uncoupled design when the design team–specified tolerance is greater than

$$\sum_{\substack{j \neq i \\ j=1}}^{p} \frac{\partial \mathrm{FR}_i}{\partial \mathrm{DP}_j} \Delta \mathrm{DP}_j$$

so that the nondiagonal elements of the design matrix can be neglected from design consideration."

In summary, the decoupling or uncoupling actions are:

1. Start from high-level FRs obtained from stage 2 QFD.
2. Define high-level DPs.
3. Use the zigzagging process to map FRs to DPs to get the design matrices and physical structure.
4. Reorder and categorize design matrices at all levels as coupled, decoupled, or uncoupled. Employ TRIZ for rich reordering, replacement, or DP substitution in the design mappings.
5. Maintain independence of FRs at all levels of physical structure by employing the methods presented in this section.

In Chapter 9, a tolerance-decoupling concept will be carried further by optimizing the tolerances of the design parameters or process variables such that the FRs are released at the six-sigma quality level by minimizing the sum of quality losses and the tolerance control cost of the design parameters.

The problem is formulated as a non-linear optimization problem using tolerances of the design parameters as decision variables to achieve robustness at Six-Sigma levels of all mappings simultaneously. The formulation also constrains coupling vulnerability in an effort to satisfy the Independence Axiom where CDFC was not applied. First, we will establish a relationship between the sigma level and robustness. Second, we will use the single FR case as an introduction for FRs (array {**FR**}) optimization formulation. In either scenario, we use mathematical programming and borrow from robust design concepts and nonlinear optimization (Luenberger, 1989). The solution provides an analytical framework to the axiomatic quality optimization phase.

8.4 CASE STUDY: TRANSMISSION VANE OIL PUMP CDFC

A hydraulic pump is a mechanism by which an external power source (i.e., the engine) is used to apply force to a hydraulic medium. Usually, the front pump drive is attached to the converter hub in an automatic transmission as depicted

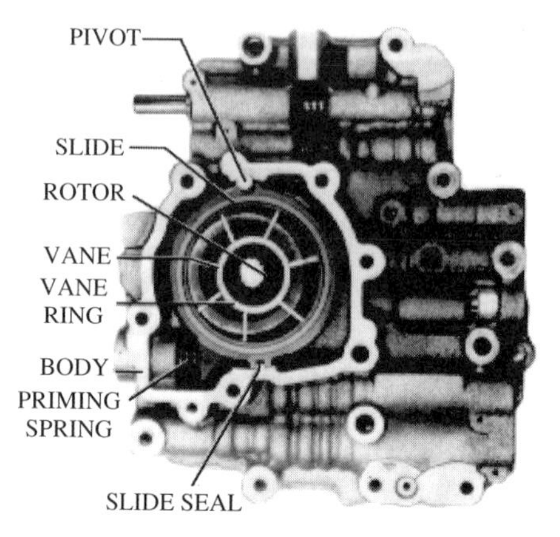

Figure 8.6 Automatic transmission pump. (From Brejcha, 1982.)

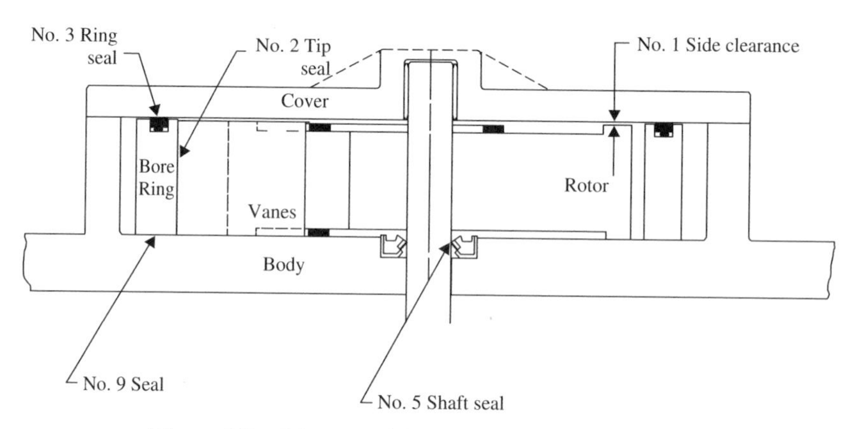

Figure 8.7 Oil pump side view. (From Brejcha, 1982.)

in Figure 8.6. Figure 8.7 represents a side view and Figure 8.8 represents a top view of the vane pump.

A hydraulic pump creates work when it transmits force and motion in the form of flow and pressure. In other words, the pump is the heart of the automatic transmission. Most currently used pumps are rotary type, with the following mechanism of operation: The hydraulic medium (also, fluid) is trapped in chambers that are expanding and collapsing cyclically. Expanding is needed at the pump inlet to draw fluid into the pump with collapsing at the outlet to force fluid into the system under pressure. The variable vane pump is a rotary pump with variable capacity. The output will vary according to the requirements of the transmission to conserve power. The advantages are many, chief among them being the ability to deliver a large capacity when the demand is high, especially at low speeds, and the minimal effort to drive at high speeds.

The mechanism of operation is as follows: When the priming spring moves the slide to the full-extended position, the slide and rotor are eccentric. As the rotor

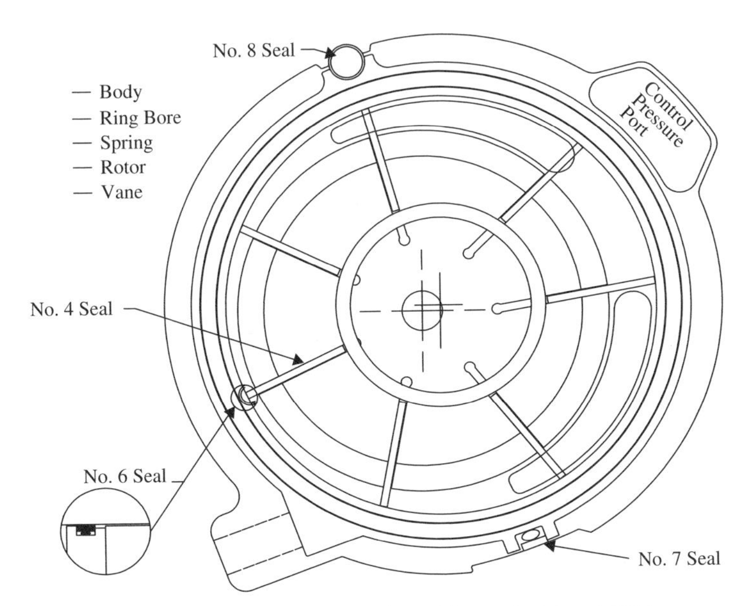

Figure 8.8 Oil pump top view. (From Brejcha, 1982.)

and vanes rotates within the slide, the expanding and contracting areas form suction (expanding) and pressure (collapsing) chambers. The hydraulic media trapped between the vanes at the suction side is moved to the pressure side. A large quantity of fluid is moved from the pressure side back to the suction side as the slide moves toward the center (Figure 8.8). A neutral zone (no volume change) is created when concentricity is attained between the slide and the rotor.

The function of the priming spring is to keep the slide in the fully expanded position such that full output can be commanded when the engine starts. Movement of the slide against the spring occurs when the pump pressure regulator valve reaches its predetermined value. At the design regulating point, the pressure regulator valve opens a port feed to the pump slide and results in a slide movement against the priming spring to cut back on volume delivery and maintain regulated pressure.

8.4.1 Pump Zigzagging Process

The FRs, array **{FR}**, and design parameters, array **{DP}**, must be decomposed into a hierarchy using the zigzagging process until a full structure in terms of design mappings is obtained. The design team must zigzag between the domains to create such a structure. In the physical mapping, we first have to define the high-level FRs. In the pump case, there is one high-level requirement, $FR_1 =$ convert external power to hydraulic power. This requirement is delivered by five design parameters: $FR_1 =$ displacement mechanism, $DP_2 =$ power source, $DP_3 =$ inlet system, $DP_4 =$ outlet system, $DP_5 =$ hydraulic media, and $DP_6 =$ external power coupling system. The mapping is depicted in (8.1), where X denotes a nonzero functional relationship. In a P-diagram, DP_2 is the signal and DP_1, DP_3, DP_4, DP_5, and DP_6 are control factors.

FR_1 = convert external power to hydraulic power

DP_1 = displacement mechanism

DP_2 = power source—engine crank via torque converter and pumping shaft

DP_3 = inlet system

DP_4 = outlet system

DP_5 = hydraulic media

DP_6 = external power coupling system—shaft seal and bearing

The scope of this case study is technically limited to high-level subsystems of the vane pump. Other mappings are presented in Appendix 8A.

Level 1: High-Level Design Matrix

$$\{FR_1\} = [\,X \quad X \quad X \quad X \quad X \quad X\,] \begin{Bmatrix} DP_1 \\ DP_2 \\ DP_3 \\ DP_4 \\ DP_5 \\ DP_6 \end{Bmatrix} \tag{8.1}$$

The level 1 mapping in (8.1) represents a *zag* step. Next, each of the design parameters will be *zagged* to the FR domain. The design parameter DP_2 and DP_5 will not be decomposed further, as they are determined by other transmission requirements outside the scope of the pump. They can be treated as noise factors. Equation (8.2) is a mapping of DP_1 = displacement mechanism. We have four FRs, array **{FR}** with $m = 4$, and eight design parameters, array **{DP}** with $p = 8$, a redundant design since $p > m$. Appendix 8A lists the balance of mappings of the pump example that are not discussed here. The mappings in Appendix 8A have a mix of the different design categories in light of the independence axiom.

Level 2: Mappings

Design Matrix of DP_1 = Displacement Mechanism

FR_{11} = charge chamber

FR_{12} = discharge chamber at uniform rate

FR_{13} = prohibit slip between flow to pass from outlet to inlet

FR_{14} = provide displacement change based on external hydraulic signal

DP_{11} = expanding chamber

DP_{12} = collapsing chamber

DP_{13} = sealing device—geometry boundary between inlet and outlet

DP_{14} = movable bore ring

DP_{15} = bias spring

DP_{16} = control pressure

DP_{17} = rigid cover

DP_{18} = rigid body

$$\begin{Bmatrix} FR_{11} \\ FR_{12} \\ FR_{13} \\ FR_{14} \end{Bmatrix} \begin{bmatrix} X & 0 & X & X & 0 & 0 & X & X \\ 0 & X & X & 0 & 0 & X & X & X \\ 0 & 0 & X & 0 & 0 & 0 & X & X \\ 0 & 0 & X & X & X & X & X & X \end{bmatrix} \begin{Bmatrix} DP_{11} \\ DP_{12} \\ DP_{13} \\ DP_{14} \\ DP_{15} \\ DP_{16} \\ DP_{17} \\ DP_{18} \end{Bmatrix} \tag{8.2}$$

Design Matrix of DP_3 = Inlet System

FR_{31} = provide flow of fluid media to pumping chamber

DP_{31} = inlet geometry

DP_{32} = degree of roughness

DP_{33} = fluid properties

DP_{34} = local losses, direction, and velocity change

$$\{FR_{31}\} = [X \quad X \quad X \quad X] \begin{Bmatrix} DP_{31} \\ DP_{32} \\ DP_{33} \\ DP_{34} \end{Bmatrix} \tag{8.3}$$

Design Matrix of DP_4 = Outlet System

FR_{41} = conduct fluid to hydraulic systems feeds

FR_{42} = collect flow from pumping chambers

DP_{41} = discharge port geometry

DP_{42} = rigid cover

DP_{43} = seal at discharge port (pump interface sealing)

DP_{44} = Seal at discharge port (square-cut seal system)

$$\begin{Bmatrix} FR_{41} \\ FR_{42} \end{Bmatrix} = \begin{bmatrix} X & X & X & X \\ X & 0 & 0 & 0 \end{bmatrix} \begin{Bmatrix} DP_{41} \\ DP_{42} \\ DP_{43} \\ DP_{44} \end{Bmatrix} \tag{8.4}$$

Design Matrix of DP_6 = External Power Coupling System: Shaft, Seal, and Bearing

FR_{61} = transmit shaft power to displacement mechanism

FR_{62} = locate rotor

FR_{63} = transfer fluid to and from torque converter

FR_{64} = seal the displacement machinery

FR_{65} = support hydraulic loads

DP_{61} = geometry: size of the shaft

DP_{62} = material properties of shaft strength

DP_{63} = tolerances of geometry between shaft and housing

DP_{64} = concentric geometry

DP_{65} = hydraulically sound shaft material

DP_{66} = surface finish

DP_{67} = support bearing

DP_{68} = seal

$$\begin{Bmatrix} FR_{61} \\ FR_{62} \\ FR_{63} \\ FR_{64} \\ FR_{65} \end{Bmatrix} = \begin{bmatrix} X & X & 0 & 0 & X & 0 & 0 & 0 \\ 0 & X & X & 0 & X & 0 & X & 0 \\ X & 0 & X & X & X & X & 0 & X \\ X & X & X & 0 & X & 0 & 0 & X \\ X & X & X & 0 & 0 & 0 & X & 0 \end{bmatrix} \begin{Bmatrix} DP_{61} \\ DP_{62} \\ DP_{63} \\ DP_{64} \\ DP_{65} \\ DP_{66} \\ DP_{67} \\ DP_{68} \end{Bmatrix} \tag{8.5}$$

Level 3 Mappings

Design Matrix of DP_{11} = Expanding Chamber

FR_{111} = provide pressure differential between atmospheric pressure and inside chamber pressure

FR_{112} = displace the fluid from inlet to outlet

DP_{111} = rate of volume increase

DP_{112} = position of vanes to cam ring

DP_{113} = chamber seal from all sides (eight different seals)

DP_{114} = power source

$$\begin{Bmatrix} FR_{111} \\ FR_{112} \end{Bmatrix} = \begin{bmatrix} X & X & X & X \\ X & X & 0 & X \end{bmatrix} \begin{Bmatrix} DP_{111} \\ DP_{112} \\ DP_{113} \\ DP_{114} \end{Bmatrix} \tag{8.6}$$

Design Matrix of DP_{12} = Collapsing Chamber

FR_{121} = create flow at controllable rate

DP_{121} = rate of volume decrease

DP_{122} = position of vanes to cam ring

DP_{113} = chamber seal from all sides (eight different seals)

DP_{114} = power source

$$\{FR_{121}\} = [X \quad X \quad X \quad X] \begin{Bmatrix} DP_{121} \\ DP_{122} \\ DP_{113} \\ DP_{114} \end{Bmatrix} \tag{8.7}$$

Design Matrix of DP_{131-1} = Ring–Rotor Clearance Chamber Seal

FR_{131-11} = minimize leak (loss of flow) from high pressure to low pressure

FR_{132-12} = lubricate running surface of chamber
to minimize leak (loss of flow)

DP_{131-11} = close clearance between vane and rotor

$$\begin{Bmatrix} FR_{131-11} \\ FR_{132-12} \end{Bmatrix} = \begin{bmatrix} X \\ X \end{bmatrix} \{DP_{131-11}\} \tag{8.8}$$

Design Matrix of DP_{131-2} = Vane Tip Sealing Between Vane and Bore Ring

FR_{131-21} = prevent leakage

DP_{132-21} = outward radial centrifugal force

DP_{132-22} = maintain position of vane to bore ring

$$\{FR_{132-21}\} = [X \quad X] \begin{Bmatrix} DP_{132-21} \\ DP_{132-22} \end{Bmatrix} \tag{8.9}$$

Design Matrix of DP_{131-3} = Ring Cover Seal

$$FR_{131-31} = \text{prevent leakage between ring and cover}$$

$$DP_{131-31} = \text{clamping force}$$

$$DP_{131-32} = \text{clamping force location}$$

$$\{FR_{131-31}\} = [\,X \quad X\,]\begin{Bmatrix} DP_{131-31} \\ DP_{131-32} \end{Bmatrix} \tag{8.10}$$

Design Matrix of DP_{131-4} = Vane to Rotor Slot Seal

$$FR_{131-41} = \text{minimize leakage between vane and rotor}$$

$$DP_{131-41} = \text{vane surface flatness}$$

$$DP_{131-42} = \text{surface flatness of slot}$$

$$\{FR_{131-41}\} = [\,X \quad X\,]\begin{Bmatrix} DP_{131-41} \\ DP_{131-42} \end{Bmatrix} \tag{8.11}$$

Design Matrix of DP_{131-5} = Lip Shaft Seal

$$FR_{131-51} = \text{prevent leakage around shaft (ID sealing)}$$

$$FR_{131-52} = \text{prevent leakage between body and seal OD (OD sealing)}$$

$$DP_{131-51} = \text{clamping force on shaft}$$

$$DP_{131-52} = \text{outward reaction forces on bore}$$

$$\begin{Bmatrix} FR_{131-51} \\ FR_{131-52} \end{Bmatrix} = \begin{bmatrix} X & 0 \\ 0 & X \end{bmatrix} \begin{Bmatrix} DP_{131-51} \\ DP_{131-52} \end{Bmatrix} \tag{8.12}$$

Design Matrix of DP_{131-6} = Seal Between Vane and Ring Bore

$$FR_{131-61} = \text{minimize leakage between chambers}$$

$$DP_{131-61} = \text{dimensional control of bore ring height to bucket height}$$

$$\{FR_{131-61}\} = [X]\{DP_{131-61}\} \tag{8.13}$$

Design Matrix of DP_{131-7} = Control Chamber Seal

$$FR_{131-71} = \text{minimize leakage between control chamber and pump inlet}$$

$$FR_{131-72} = \text{position bore ring relative to pivot pin}$$

$$DP_{131-71} = \text{redial compression force on contact surface (rubber)}$$

$$DP_{131-72} = \text{maintain contact between body and slide seal}$$

$$\begin{Bmatrix} FR_{131-71} \\ FR_{131-72} \end{Bmatrix} = \begin{bmatrix} X & X \\ X & X \end{bmatrix} \begin{Bmatrix} DP_{131-71} \\ DP_{131-72} \end{Bmatrix} \tag{8.14}$$

Design Matrix of DP_{131-8} = Cylindrical Pivot Seal

FR_{131-81} = minimize leakage between control chamber and pump inlet

DP_{131-81} = zero radial clearance between pivot pin and bore ring

DP_{131-82} = controlled axial clearance

$$\{FR_{131-81}\} = \begin{bmatrix} X & X \end{bmatrix} \begin{Bmatrix} DP_{131-81} \\ DP_{131-82} \end{Bmatrix} \tag{8.15}$$

Design Matrix of DP_{131-9} = Seal Between Ring Bore and Body

FR_{131-91} = minimize leakage between ring bore and body

DP_{131-91} = compression force from rubber (DP_{131-3})

DP_{131-92} = flatness of the body

DP_{131-93} = surface finish of body

$$\{FR_{131-91}\} = \begin{bmatrix} X & X & X \end{bmatrix} \begin{Bmatrix} DP_{131-91} \\ DP_{131-92} \\ DP_{131-93} \end{Bmatrix} \tag{8.16}$$

At level 3, sealing devices (DPs) 2, 3, 4, 8, and 9 are all redundant designs; devices 5 and 6 are uncoupled designs; devices 1 and 7 are coupled designs. Once the zigzagging process is completed, the decoupling phase starts according to the axiomatic quality process depicted in Figures 8.4 and 6.3.

8.4.2 Decoupling Phase

Sealing device 1, rotor clearance to chamber mapping in (8.8), will be used as a decoupling example. In this device we have $m = 2$ and $p = 1$, clearly a coupled design. They are FR_{131-11} = minimize leak (loss of flow) from high pressure to low pressure, FR_{132-12} = lubricate running surface of chamber to minimize leak (loss of flow), and DP_{131-11} = close clearance.

This device is selected in this example because sealing devices within the pump are not robust, resulting in low pump efficiency. Without the axiomatic quality process, the pump manufacturer will resort to improving the robustness of the seal through an empirical experiment, an operational vulnerability improvement phase. This is depicted in Figure 8.9. This may not be sufficient because of the conceptual vulnerability of the seal being a coupled design. Without resolving the coupling, the best that can be done is a trade-off between FR_{131-11} = minimize leak from high pressure to low pressure and FR_{131-12} = lubricate running surfaces of the chamber, since both are delivered by one design parameter DP_{131-11} = tolerance (clearance) between the vane and the rotor. The coupling occurs because the seal device 1 system is charged with delivering two FRs and one design parameters; that is, the number of FRs ($m = 2$) is greater than the number of design parameters ($p = 1$). Clearly, another design parameter,

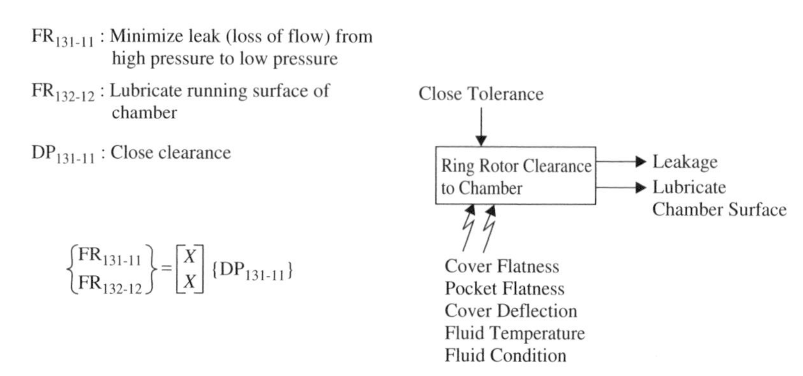

Figure 8.9 P-diagram of sealing device 1 prior to CDFC phase.

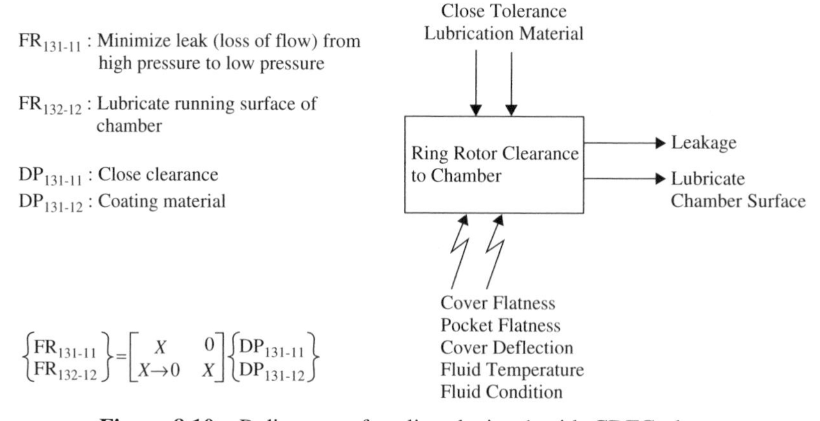

Figure 8.10 P-diagram of sealing device 1 with CDFC phase.

say DP_{131-12}, needs to be introduced to resolve the coupling. This parameter should also be smartly introduced to produce uncoupled or at least decoupled designs. This parameter should deliver one of the FRs without adversely affecting the other FR. These characteristics fit the coating to be declared as DP_{131-12} in sealing 1 mapping. Coating materials such as Lube Rite or LMC are proposed. Coating will help the lubrication by reducing the surface tensions of the hydraulic media, keeping the surfaces wet and lubricated all the time. Coating does not affect FR_{131-11}, allowing the tolerance to be tightened to reduce the leakage. The resulting mapping and the P-diagram are given in Figure 8.10.

8.4.3 Step A.5: Detail the Design

Engineering and physical disciplines have been used for many generations to affect design vulnerabilities. However, in today's' comprehensive design process they are no longer enough. The trend now in many industries requires a disciplined engineering process that ties together the multitude of tools available. Axiomatic quality serves this trend very well. It provides perspectives that usually overlooked by other methods.

Quality engineering is a disciplined approach that seeks to find the best expression of a product. That is, the lowest-cost solution to meet product specifications is based on customer needs. Quality engineering focuses on parameter optimization. This is done by reducing the variation of the key FRs, caused by noise factors, and ensuring that those FRs can easily be adjusted onto the nominal value. Minimizing variation or making the system less sensitive to variation enables possible cost gains by eliminating the need to control DP quality.

Noise factors are defined, in general, as anything that causes an FR to deviate from its target value. A complete enumeration of noise factors includes:

- *Manufacturing variability (unit-to-unit noise).* This is a result of the inability to manufacture two parts exactly alike. Manufacturing processes and machines are two major sources.
- *Customer usage noise.* Customers exhibit different patterns of use of a given product, and hence different duty cycles are generated.
- *Deterioration (internal) noise.* This represents product aging.
- *Environment (external) noise.* This comes from the outside of the product (e.g., temperature and humidity).
- *Coupling and interaction noises.* This is noise that happens because of physical mapping decisions or interaction among DPs.

In the context of axiomatic quality detailing step, step A.5 of Figure 8.4, the noise factor effect on a given FR can be assessed, at this initial step, based on engineering and historical knowledge prior to optimization. This activity will also help in developing a sound noise strategy to obtain transfer functions within the context of optimization phase. We suggest the noise assessment scheme depicted in Figure 8.11. It exhibits five hashed blocks (tags) in the FR cell (Figure 8.12). When tagged by any hashed block, an FR implies that design engineering should take the respective noise factor into consideration in the optimization phase to minimize operational design vulnerabilities. The activity of tagging blocks to FRs requires some evidence testifying to the effect of that noise factor on the respective FR. The density of hashing inside the block (tag) perimeter expresses the strength of the casual relationship between a noise factor and an FR. All FRs of a given mapping should be tagged with such a noise assessment scheme. This activity should continue across all design mapping hierarchical levels. The end result provides a visual exploration of what type of noise factors are present in their design and which should be considered in any future parameter and tolerance design optimization DOE. Parameter design as an operational vulnerability optimization technique is discussed in Section 9.3. For example, consider sealing device 1: rotor clearance to the chamber seal DP in Figure 8.12. We note that the manufacturing variability, deterioration, and external noise factors are believed to be effective based on historical evidence from previous testing. It is wise to include such noise factors in testing.

When coupled with quality engineering, the zigzagging process allows the identification of noise effect areas where further improvement can be sought.

Sequence Code	Variability Source
1	Manufacturing Variability
2	Customer Usage
3	Degradation over Time
4	Environment
5	Coupling with Other Systems

Legend:

NA The variability source is not applicable to the FR.

No evidence can be identified for the variability source.

An evidence can be identified, but with no solid proof (e.g., data, analytical study, etc.).

An evidence can be identified with solid proof (e.g., data, analytical study, etc.).

Rules:

1. Evidence can be analytical or hardware.
2. Evidence need to be noted according to Ford system.
3. An evidence should identify quantitatively how an FR is affected by respective variability source.
4. Variability sources are sequenced from left to right.

Figure 8.11 Axiomatic quality noise factor assessment scheme.

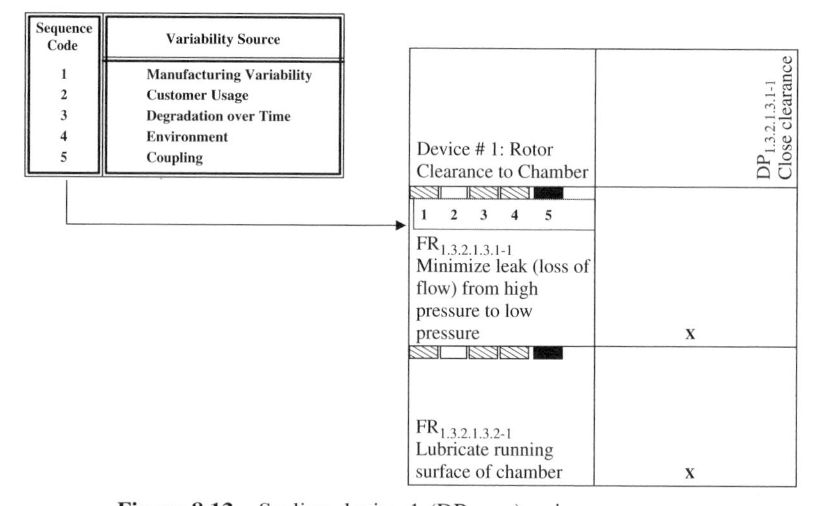

Figure 8.12 Sealing device 1 (DP_{131-1}) noise assessment.

This in turn permits better allocation of the engineering and testing resources. In this example it is obvious that more work needs to be done to gain more robustness in the usage noise. The team is started on the process of developing the necessary testing to further improve operational robustness in this category.

8.5 THEORY OF INVENTIVE PROBLEM SOLVING

In this section an existing design improvement methodology developed in Russia is introduced. It is a very promising design improvement technique used in situations where design vulnerability inhibits or restricts design improvement practices. TRIZ has ample opportunities for application of axiomatic quality process steps: A.3, A.5, B.1, B.2, and B.3. The TRIZ premise is founded on the idea that the design problems encountered contain key elements that have already been solved in different fields or industries. TRIZ is the answer to those inventive design projects for which strong engineering expertise, in particular operational vulnerability improvement practices, is not sufficient. Such projects contain a technical *contradiction* where adjusting one parameter or a requirement of a system degrades another. The applicability of TRIZ as a design improvement tool is discussed briefly in Chapter 6.

TRIZ was formulated from over 3.1 million worldwide patents analyzed, with 21% deemed innovative. These were studied further, leading to three key discoveries:

1. Patterns of technical evolution were repeated across industries and sciences.
2. Problems and solutions were repeated across industries and sciences.
3. Innovations used scientific effects outside the field where they were developed.

In TRIZ terminology, removal of dependence is defined as resolving contradictions in design. Altshuller (1990) developed 40 principles that comprised the first TRIZ tool developed. Altshuller has identified 39 design characteristics (DRs), such as weight, strength, and speed, which are frequently involved in technical contradictions. The majority of the DRs may be considered DPs in an axiomatic quality context. Separation principles are utilized to resolve these physical contradictions.

TRIZ is an algorithmic tool that provides a powerful tool for solving conceptual design problems. It is based on the laws of evolution of engineering systems and consists of three subsystems: the algorithms for inventive problem solving, the standard solutions to inventive problems, and the database of physical, chemical, and geometrical effects (Fey et al., 1994).

TRIZ suggests a systematized approach to reconsider (reformulate) the design project (secondary failures) and formulate a set of principles (pathways) to resolve contradictions in secondary failures. The TRIZ methodology is based on the premise that contradictions and couplings in products can be eliminated by changing (redesigning), basically altering the $\{\mathbf{DP}\}$ array. Based on the successful innovation premise, it is easier to resolve system contradictions (the presence of secondary failures) by reformulating the problem (Zoltin et al., 1996). In other words, reformulate the problem and redesign the system to remove couplings.

In TRIZ, design is recognized as a multiple objective problem. The goal of design is to deliver every desired functional requirement, whereas the expenditures should be as minimal as possible. Here the expenditures are all kinds of energy, material, and information resources necessary to perform functionality in the entire product life cycle. The ratio between useful functionality performed by the product and the expenditure, termed *ideality*, may be defined conceptually as $\sum \text{benefits}/(\sum \text{costs} + \sum \text{harm})$. Clearly, the goal of design is to maximize the ideality. Ideality describes the solution to a design project, independent of the mechanism and constraints of baseline design. It defines an ideal product, which delivers benefit without cost or harm: for example, requires no maintenance, occupies no space, has no weight, requires no labor, takes no time. Ideality provides a technology forecasting of the design project to establish a sense of direction. It temporarily gets design teams to think "outside the box" by removing perceived or real barriers to offer alternative innovative solutions. Starting with this perfection thinking encourages breakthrough design, inhibits the moves to less ideal solutions, and leads to discussions that will clearly establish boundaries of project.

However, in the actual design process of a product, there are several problems that may reduce the ideality of a product. The problems that reduce the ideality of designs are:

1. *Engineering conflicts.* Since a design module has multiple functional requirements, an engineering conflict occurs if when we try to improve one functional requirement, it will inadmissibly harm other functional requirements. For example, when we try to improve the strength of a component, we may increase its weight. Engineering conflict will affect the product's ability to deliver the desired functionality.
2. *Harmful or unintended functions.* We design products only for useful functions, but many actual design entities will deliver harmful and unintended responses as well, such as noise and heat. Harmful and unintended outputs may affect the delivery of useful functions, increase the cost, and cause failures (Sushkov et al., 1995).
3. *Failure costs.* Excessive costs may arise due to improper designs.

To resolve the three problems mentioned above, TRIZ suggests a systematic approach to reformulate (defined as evolutionary concept generation) the design problem and use a set of principles (pathways) and techniques to resolve these problems. Specifically, these techniques resolve engineering conflicts and harmful outputs. These techniques are derived from the extensive studies of past patents in Soviet Union.

Clearly, the concept of engineering conflict in TRIZ draws a parallel to the concept of coupling in axiomatic design. In the effort to improve functional robustness, TRIZ can be utilized to develop a vulnerability free design methodology that removes coupling to increase system function robustness and reduces design complexity due to increased reliability.

8.5.1 TRIZ in the Axiomatic Quality Process

TRIZ provides a set of tools and strategies in terms of contradiction that can best be leveraged to reduce conceptual vulnerabilities. Specifically, contradiction, a coupling vulnerability synonymous, provides ample opportunities for further conceptual robustness, as depicted in Figure 8.13. This in effect will improve the ease of conducting an operational vulnerability phase when uncoupled or decoupled

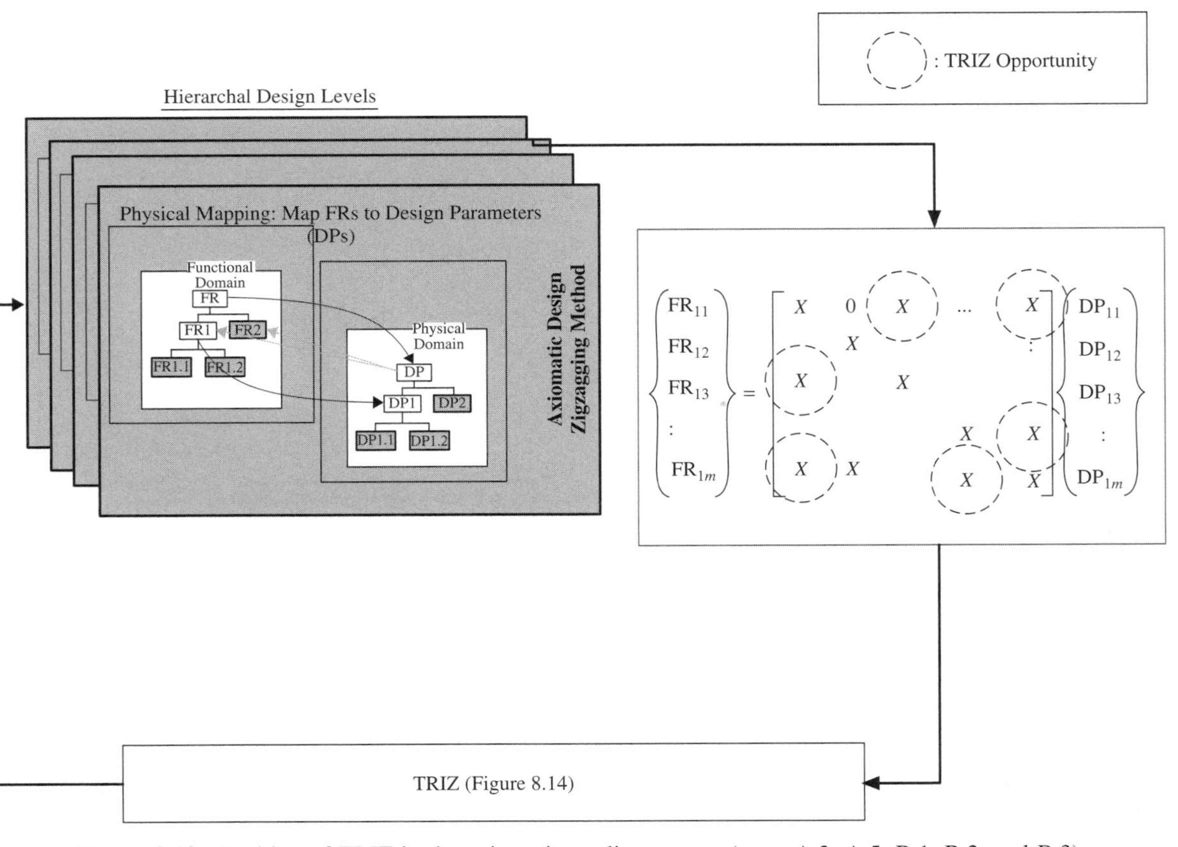

Figure 8.13 Position of TRIZ in the axiomatic quality process (steps A.3, A.5, B.1, B.2, and B.3).

design is accomplished. Independence helps producing additive transfer functions, a desirable property for operational robustness optimization.

Some of the tools used in TRIZ methodology are given below, with an implementation sequence given in Figure 8.14.

1. There are 40 principles first developed by Altshuller. Their purpose was to guide the TRIZ practitioner in developing useful concepts for inventive solutions. For example, one possible principle to resolve the conflict between adequate strength and excessive weight could be to optimize the shape. Separation principles are a special type of contradiction in having one requirement and its opposite in the same DP (e.g., hot and cold).
2. The TRIZ standard solutions represent frequently used solutions or specific problems. Altshuller identified 76 standard solutions and grouped them into five classes. For example, given a specific weight/strength conflict problem, using a shell structure might be a standard solution based on the principle of optimizing the shape.
3. Frequently, real-world problems do not appear as contradictions, and thus it is not always obvious how or where to apply TRIZ tools. ARIZ is a step-by-step method whereby given an unclear technical problem, the inherent contradictions are revealed, reformulated, and resolved.
4. One software product that has been effective in processing TRIZ as a module is TechOptimizer. This software is an intelligent problem solver. The

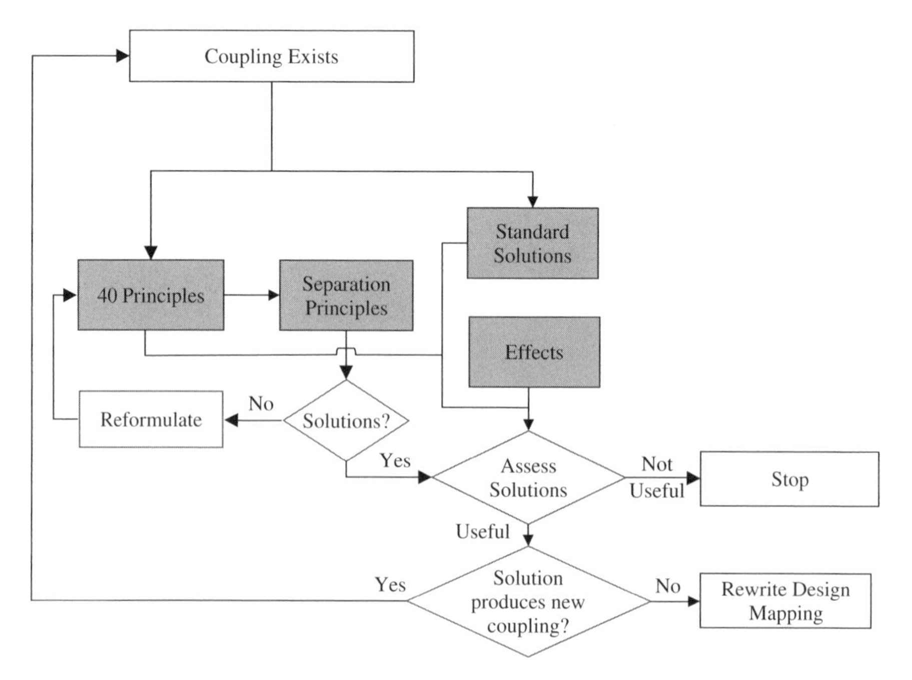

Figure 8.14 TRIZ implementation steps within the axiomatic quality process.

direct results of using TechOptimizer are efficient and patentable engineering solutions.

8.6 SUMMARY

Coupling is a design vulnerability indicating lack of design controllability and adjustability at both the development stages and in the use environment. Coupled designs are prone to lower reliability and robustness levels. Study of several inventions indicate that in the majority of the cases, coupled designs are usually short-lived and replaced with new inventions in the spirit of continuous design improvement. Coupling is the major conceptual vulnerability addressed in this chapter.

This chapter was developed to complement the CDFC phase discussed in Chapter 6 by exploring the axiomatic quality process steps that were mentioned briefly. In Section 8.2 we provided a background about the design projects that can be solved by the axiomatic quality process. In Section 8.3 we presented the CDFC phase, identifying its uniqueness, capabilities, and features that were not explored in Chapter 6, and discussed coupling vulnerability, its implications, and methods to uncouple or decouple a design mapping. In Section 8.4 we presented an example. In Section 8.5 we discussed TRIZ methodology and its position in the axiomatic quality process.

APPENDIX 8A: DESIGN MATRIXES

Design Matrix of DP_{14} = Movable Bore Ring

FR_{141} = provide room to confide volume

FR_{142} = provide reach surface for seal device 7

FR_{143} = provide surface for reaction spring force

FR_{144} = provide surface for pivoting

FR_{145} = provide housing for reaction force of sealing device 3

DP_{141} = irregular outside geometry

DP_{142} = regular geometry

DP_{143} = strength

$$\begin{Bmatrix} FR_{141} \\ FR_{142} \\ FR_{143} \\ FR_{144} \\ FR_{145} \end{Bmatrix} = \begin{bmatrix} 0 & X & X \\ X & 0 & X \\ X & 0 & X \\ X & 0 & X \\ X & 0 & X \end{bmatrix} \begin{Bmatrix} DP_{141} \\ DP_{142} \\ DP_{143} \end{Bmatrix} \tag{8A.1}$$

Design Matrix of DP_{15} = Bias Spring

$$\{FR_{151}\} = [X]\{DP_{151}\}$$

FR_{151} = provide reaction force to control signal (achieve eccentricity)

FR_{151} = spring rate

Design Matrix of DP_{16} = Control Pressure

FR_{161} = matches pump flow to transmission flow requirements

DP_{161} = adjusting concentricity of bore to ring to rotor

$$\{FR_{161}\} = [X]\{DP_{161}\}$$

Design Matrix of DP_{17} = Rigid Cover

FR_{171} = confine volume—seal hydraulic fluid

FR_{172} = locate rotor for assembly

FR_{173} = control damping orifice

FR_{174} = provide running surface for vanes (wear)

DP_{171} = flat running surface

DP_{172} = hydraulically sound surface

DP_{173} = locating ring sleeves

DP_{174} = wear resistance surface

DP_{175} = pilot hole and machine datum

DP_{176} = orifice feature geometry

$$\begin{Bmatrix} FR_{171} \\ FR_{172} \\ FR_{173} \\ FR_{174} \end{Bmatrix} = \begin{bmatrix} X & X & 0 & 0 & 0 & X \\ 0 & 0 & X & 0 & X & 0 \\ X & X & 0 & 0 & X & X \\ X & 0 & 0 & X & 0 & 0 \end{bmatrix} \begin{Bmatrix} DP_{171} \\ DP_{172} \\ DP_{173} \\ DP_{174} \\ DP_{175} \\ DP_{176} \end{Bmatrix} \tag{8A.2}$$

Design Matrix of DP_{18} = Rigid Body

FR_{181} = confine volume—seal hydraulic fluid

FR_{182} = provide inlet and outlet connection

FR_{183} = provide reaction surface to bias spring

FR_{184} = locate bore ring

FR_{185} = provide running surface for vanes

FR_{186} = provide running surface for sealing devices 7, 8, and 9

FR_{187} = provide vent for collapsing vanes (inner pump)

DP_{181} = flat running surface

DP_{182} = hydraulically sound surface

DP_{183} = parts geometry

DP_{184} = spring bucket plane is perpendicular to body surface

DP_{185} = bore ring center is between sealing devices 7 and 8

DP_{186} = relieve of bore

$$\begin{Bmatrix} FR_{181} \\ FR_{182} \\ FR_{183} \\ FR_{184} \\ FR_{185} \\ FR_{186} \\ FR_{187} \end{Bmatrix} = \begin{bmatrix} X & X & X & 0 & 0 & 0 \\ 0 & X & X & 0 & 0 & 0 \\ X & 0 & 0 & X & 0 & 0 \\ X & 0 & 0 & X & X & 0 \\ X & 0 & X & 0 & 0 & 0 \\ X & 0 & 0 & X & X & 0 \\ 0 & 0 & 0 & 0 & 0 & X \end{bmatrix} \begin{Bmatrix} DP_{181} \\ DP_{182} \\ DP_{183} \\ DP_{184} \\ DP_{185} \\ DP_{186} \end{Bmatrix} \tag{8A.3}$$

Design Matrix of DP_{42} = Rigid Cover

FR_{421} = minimize deflection of cover

FR_{422} = provide a wear surface for cover

FR_{423} = provide damping function

FR_{424} = provide for attachment of cover to body

FR_{425} = seal hydraulic fluid inside pump

DP_{421} = cover material properties

DP_{422} = cover geometry

DP_{423} = phosphate coating

DP_{424} = finishing operations

$$\begin{Bmatrix} FR_{421} \\ FR_{422} \\ FR_{423} \\ FR_{424} \\ FR_{425} \end{Bmatrix} = \begin{bmatrix} X & X & 0 & 0 \\ X & 0 & X & X \\ 0 & X & 0 & X \\ 0 & X & 0 & 0 \\ X & 0 & 0 & X \end{bmatrix} \begin{Bmatrix} DP_{421} \\ DP_{422} \\ DP_{423} \\ DP_{424} \end{Bmatrix} \tag{8A.4}$$

Design Matrix of DP_{44} = Seal Discharge Port (Sqr Cut Seal System)

FR_{441} = prevent loss of hydraulic fluid

FR_{442} = provide for wear resistance

DP_{441} = seal material properties

DP_{442} = surface finish

DP_{443} = axial loading

$$\begin{Bmatrix} \text{FR}_{441} \\ \text{FR}_{442} \end{Bmatrix} = \begin{bmatrix} X & X & X \\ X & X & X \end{bmatrix} \begin{Bmatrix} \text{DP}_{441} \\ \text{DP}_{442} \\ \text{DP}_{443} \end{Bmatrix} \tag{8A.5}$$

Design Matrix of DP_{44} = Seal Discharge Port (Pump Interface Sealing)

FR_{441} = prevent loss of hydraulic fluid

DP_{441} = seal material properties

DP_{442} = surface finish

DP_{443} = clamping loading

$$\{\text{FR}_{441}\} = \begin{bmatrix} X & X & X \end{bmatrix} \begin{Bmatrix} \text{DP}_{441} \\ \text{DP}_{442} \\ \text{DP}_{443} \end{Bmatrix} \tag{8A.6}$$

CHAPTER 9

AXIOMATIC QUALITY OPTIMIZATION PHASE

9.1 INTRODUCTION

In this chapter we discuss the optimization phase from an axiomatic quality perspective. The goal is to present several operational vulnerability optimization techniques for different handling according to Figure 9.1, depending on the availability of the transfer functions of a given mapping. Following the CDFC phase, operational vulnerability techniques takes FR mean settings (to target the performance desired by the customer) and their variability reduction as objectives. This is conducted to achieve results similar to that depicted in Figure 1.5 with imposed six-sigma targets for FRs. Taguchi's parameter design, Taguchi's tolerance design,[1] six-sigma, and the axiomatic quality optimization techniques presented in Sections 9.5 and 9.6 are premier techniques to address the variability minimization objective within the context of operational vulnerability optimization. Taguchi methods are experimental in nature. The availability of credible transfer functions will be used as the deciding factor as to whether to use experimental methods (Taguchi methods) or analytical methods (axiomatic quality optimization formulation). Besides, if these objectives can be accomplished while

[1]Tolerance design was introduced briefly in Chapter 5. It is beyond the scope of this book. For more complete handling the reader is encouraged to consult other references [see, e.g., Taguchi (1993) and Creveling (1997)].

Axiomatic Quality: Integrating Axiomatic Design with Six-Sigma, Reliability, and Quality Engineering, by Basem Said El-Haik
ISBN 0-471-68273-X

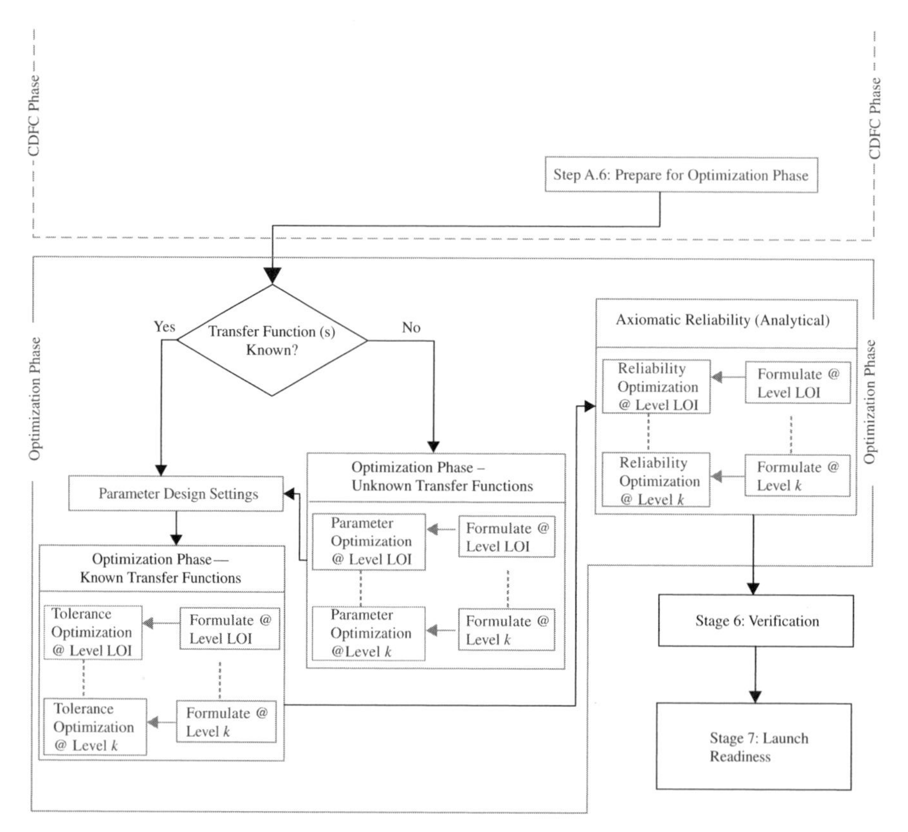

Figure 9.1 Axiomatic quality optimization phase.

further reducing conceptual vulnerabilities that were not resolved in the CDFC phase, an enhanced conceptually robust system can be obtained.

In practical terms, the deployment of design axioms is not a trivial activity for existing systems that fall within the coupled category. Constraints such as budget and technology are usually binding in addition to the ignorance of design axioms and associated concepts. Under these circumstances, conceptual vulnerabilities are established in the criteria related to the unsatisfied axioms. A design that does not satisfy the independence axiom will have some degree of coupling vulnerability. A conceptually weak system has limited success opportunity in its operation environment. Variability optimization of weak conceptual entities may produce marginally effective results. Prior to this optimization phase, conceptual vulnerability should be eliminated as discussed in the CDFC phase (Chapters 6 and 9). Therefore, the axiomatic quality process to design conceptually and operationally robust systems is imperative if the firefighting mode of operations is to be eliminated.

This chapter is developed as follows: In Section 9.2 we present axiomatic quality strategy FR mean settings (shift) techniques, depending on the availability of transfer functions. It features an analytical technique in algorithmic format to optimize for conceptual robustness at the level of impact (LOI) or even lower levels. The LOI is the design mapping level that can be detailed mathematically by transfer functions. At this level, the objective is to find the settings of DPs or PVs that minimize conceptual and operational vulnerabilities, both coupling and complexity. This formulation is more difficult for nonlinear designs. The optimum DP (or PV) settings are substituted through the transfer functions, at the LOI, producing a suggested setting for the FRs. Remember that the FRs with their customer tolerances were obtained from QFD stage 2. The FR suggested settings obtained from solving the formulation presented in Section 9.2 needs to be contrasted with those obtained in the QFD stage 2 to decide on final specification limits.

Section 9.3 is the empirical equivalent of Section 9.2. It features a robust parameter design overview. At its core, Taguchi parameter design (Chapter 5) can be used to achieve operational robustness. In Sections 9.2 and 9.3 we handle, in two different ways, the first objective of vulnerability optimization phase, the functional requirements mean setting. In Section 9.2 we present an analytical formulation based on a conceptual and vulnerability optimization routine, and in Section 9.3 we present an experimental setting to accomplish the same objective using the parameter design method. Only key aspects of parameter design are reviewed.

In Section 9.4 we present an analytical tolerance optimization formulation. It relates axiomatic quality to robust design on top of what is presented in Chapter 5 by quantifying robustness measures such as the signal-to-noise ratio and quality loss function at a six-sigma quality level. The effect of degradation on such measures is also discussed. Section 9.5 features a mathematical formulation to establish a six-sigma quality level in a uni-FR. Section 9.6 is a generalization of Section 9.5, where the nonlinear optimization model for a system with an array of FRs is formulated and solved.

9.2 AXIOMATIC QUALITY OPERATIONAL VULNERABILITY OPTIMIZATION

In this section a suggested optimization routine which will simultaneously optimize for conceptual robustness (independence maximization) and operational robustness (complexity or variability minimization) is presented. The objective is to find the optimum DP or PV settings that can be used to set the FRs where possible in the hierarchy (at or below the LOIs) and where permitted upon contrasting with FRs and their tolerances obtained in QFD stage 2.

The routine is composed from two nested loops. The first loop is concerned with maximizing independence measures with several proposals for a selected objective function, denoted as O, of the first loop. Table 9.1 presents different forms of the objective function O. In the first two proposals of Table 9.1, O is a function of the two independence measures listed in (3.4) and (3.5), reangularity,

TABLE 9.1 Mathematical Forms of the Objective Function and Its Optimization Direction

No.	Objective Function, O
1	Minimize $2 - (R + S)$
2	Maximize RS
3	Maximize $\sum_{i=1}^{m} \Delta\text{FR}_i - \sum_{i=1}^{m} \sum_{\substack{j=1 \\ j \neq i}}^{p} \frac{\partial\text{FR}_i}{\partial\text{DP}_j} \Delta\text{DP}_j$

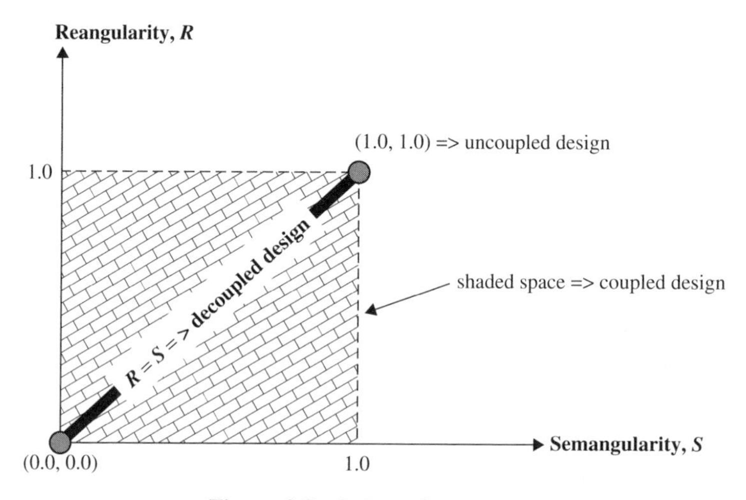

Figure 9.2 Independence space.

R, and semangularity, S, respectively. The first form is a linear objective function in R and S that can be written as $O = (1 - R) + (1 - S)$, where $0 \leq R \leq 1$ and $S \geq 0$. The idea behind this formulation is to minimize the difference from the optimum case of unity for both measures (i.e., the design is uncoupled when $R = S = 1$). The independence space is depicted in Figure 9.2. Notice that for a decoupled design class, $R = S$, but not unity. For a coupled design class, equality is lost. As R approaches unity ($R \to 1$) and S approaches unity ($S \to 1$), the design approaches the uncoupled design classification.

The second formulation of O in Table 9.1 is nonlinear. In this case, the objective function O has R and S as the sensitivities when differentiated (i.e., $\partial O/\partial R = S$ and $\partial O/\partial S = R$). In terms of the DPs, the objective function O may be simplified by Taylor series expansion in the tolerance space of the respective DPs.

For the first two formulations of Table 9.1, the tolerance constraint $\{\mathbf{T} \pm \Delta\mathbf{FR}\} = [\mathbf{A}]\,\{\boldsymbol{\mu} \pm \Delta\mathbf{DP}\}$ is used. The third formulation utilizes Theorem 8 (Suh,

1990). The theorem states that an uncoupled design may be achieved if for all FRs, the nondiagonal element contribution to the tolerance can be neglected as compared with the tolerance of the diagonal DP, say DP_l, $l \in \{1, \ldots, p\}$. The condition that assures that this is the FR tolerance (in the design range, DR), ΔFR, should be greater than or equal to $\sum_{j=1, j\neq l}^{p}(\partial FR_i/\partial DP_j)\Delta DP_j$. This formulation manipulates tolerance to maximize design independence and will be used for illustration in this section. The numerical values of R and S will increase as the design converges to the uncoupled design class.

9.2.1 Vulnerability Optimization Routine

In the linear design case, the first loop is used to optimize independence and the second loop is used to minimize complexity. Reduction in complexity can be translated into reduction in variability based on Chapter 4 development. This routine targets both types of design vulnerability. The constraint set includes the tolerance constraints and lower limits representing a tolerable independence measure value. The decision variables in the outer loop, the independence loop, are the design matrix coefficients (the entries of matrix **[A]**). For example, in a given resistive electric circuit, if voltage is a functional requirement, an FR, and the current is its DP, the design matrix coefficient is the resistance. After finding the design matrix entries that maximize independence, the inner loop is entered for complexity minimization. The output of this loop is an approximate value, a surrogate indicating the infinitesimal intervals within which the optimum design parameters **{DP}** values lie. This is in effect a discrete version of Boltzmann entropy used for simplification by allowing a linear programming formulation. The entries of matrix **[A]** can be approximated by constant values within the space of interests for nonlinear design situations. It also serves as a zooming mechanism into the decision variable space, the DPs. In this setup, we first find the discrete subset, the intervals δ, that contains the optimum DP settings from complexity minimization standpoint, and then we zoom inside δ intervals to find the numerical optimum values up to a predetermined precision limit. Integer programming is utilized in the inner loop, the complexity reduction loop, while nonlinear optimization is used at the outset loop, the independence maximization loop.

Let $\mathbf{A}_{m \times p}$ be a design matrix with m FRs (indexed by $i, i = 1, \ldots, m$) and p continuous DPs (indexed by $j, j = 1, p$). Each DP will be fragmented into several discrete intervals. Assume the uniform discrete scheme for the sake of simplification. Let δ_{jd}, D_j be a uniform dth discrete interval length and the number of discrete intervals of DP_j where $d = 1, \ldots, D_j$ such that $\delta_{jd} = d\delta_j$. Let $\{\mathbf{FR}_{m \times 1}\}$ be the array of functional requirements. Let Y_{ij} be a binary variable that is defined as $Y_{ij} = 1$ if $FR_i \rightarrow DP_j$ and 0 otherwise. The variable Y_{ij} is a binary variable that codes the existing physical mapping between FR_i and DP_j. Each DP is defined in the respective support set: $DP_j \in [\mu_j \pm \Delta DP_j]$. In the initial loop, the DPs may assume baseline design settings to solve for $[A_{ij}]$:$i = 1, m$ and $j = 1, p$. At any other iteration, the value of the DPs is delivered by the inner loop to the outer loop. In the first loop, the tolerance constraint $\{\boldsymbol{T} + \Delta\text{FR}\} \leq [\mathbf{A}]\ \{\mu + \Delta\mathbf{DP}\}$ and $\{\boldsymbol{T} - \Delta\text{FR}\} \geq [\mathbf{A}]\{\mu - \Delta\mathbf{DP}\}$ should be met. The tolerance of the **{FR}** array is the design range as determined from QFD stage 2 in Chapter 6. The tolerance of a given DP array is determined by the system ranges and represents normal (not controlled) machining conditions. Credible equivalent baseline (datum) design data can be used. Additional constraints may be needed to maximize the independence such as $S \geq \omega$ and $R \geq \varpi$, where ω and ϖ are very conservative values (approximately unity).

The decision variables in the outer loop are the sensitivity coefficients; in the second loop they are the binary variables X_{jd}. The variable X_{jd} is a binary variable that denotes the discrete interval d, which contains the optimum value of DP_j. The solution of the first loop will determine the coupling contribution to complexity, the logarithm of the determinant of matrix $\mathbf{A}'$ (see Theorem 4.1). In addition, cost and other physical constraints may be added to complement the constraint set of the inner loop. For example, the coefficients k_{ijd} are the cost of accomplishing DP_j in the interval indexed d for FR_i used in (9.1). The termination strategy includes one or more of the following criteria: a multiple of the computational underflow limits in the logarithmic calculations, the precision required on the DP and FR quantification, or the increment reduction gain, ε, between successive iterations h of the total design complexity, h_D. For example, we may terminate the optimization if $(h_u - h_{u-1})/h_{u-1} = \Delta h_D/h_{u-1} \leq \varepsilon$, where u is the inner loop iteration index. In the outer loop, iterations are indexed by v.

The optimization routine is as follows:

$$\text{Initialize} \left\{ \begin{array}{l} \text{DP}_j = \mu_j - \Delta\text{DP}_j(a_j = 0) \text{ at } v = 1 \\ D_j\text{: number of discretized uniform intervals } \delta_j \\ \delta_j = \dfrac{2\Delta\text{DP}_j}{D_j} \text{ at } u = 1 \end{array} \right\} \quad \forall j,\ j = 1, \ldots, p$$

Loop $v = 1, \ldots, V$

$$\text{Maximize } O = \sum_{i=1}^{m} \Delta\text{FR}_i - \sum_{i=1}^{m} \sum_{\substack{j=1 \\ j \neq i}}^{p} Y_{ij} A_{ij} (\delta\text{DP}_{jd})_u \tag{9.1}$$

subject to

$$\begin{bmatrix} A_{11} & A_{12} & \cdot & A_{1p} \\ A_{21} & A_{22} & \cdot & \cdot \\ & & \cdot & \cdot \\ \cdot & \cdot & \cdot & \cdot \\ & \cdot & \cdot & \\ A_{m1} & \cdot & \cdot & A_{mp} \end{bmatrix} \left\{ \begin{array}{c} \text{DP}^*_{1d} \\ \vdots \\ \text{DP}^*_{pd} \end{array} \right\}_u \geq \left\{ \begin{array}{c} T_1 - \Delta\text{FR}_1 \\ \vdots \\ T_m - \Delta\text{FR}_m \end{array} \right\} \tag{9.2}$$

$$\begin{bmatrix} A_{11} & A_{12} & \cdot & A_{1p} \\ A_{21} & A_{22} & \cdot & \cdot \\ & & \cdot & \cdot \\ \cdot & \cdot & \cdot & \cdot \\ & \cdot & \cdot & \\ A_{m1} & \cdot & \cdot & A_{mp} \end{bmatrix} \begin{Bmatrix} \mathrm{DP}^*_{1d} \\ \vdots \\ \mathrm{DP}^*_{pd} \end{Bmatrix}_u \leq \begin{Bmatrix} T_1 + \Delta\mathrm{FR}_1 \\ \vdots \\ T_m + \Delta\mathrm{FR}_m \end{Bmatrix} \tag{9.3}$$

$$\prod_{\substack{j=1,p-1 \\ k=1+i,p}} \sqrt{\frac{[1 - (\sum_{k=1}^{p} A_{kj} A_{kj})^2}{(\sum_{k=1}^{p} A_{kj}^2)(\sum_{k=1}^{p} A_{kj}^2)]}} \geq \varpi \tag{9.4}$$

$$\prod_{j=1}^{p} \left(\frac{|A_{jj}|}{\sqrt{\sum_{k=1}^{p} A_{kj}^2}} \right) \geq \omega \tag{9.5}$$

Loop $u = 1, U$

$$\text{Minimize } h_D = -\sum_{i=1}^{m} \sum_{j=1}^{p} Y_{ij} \left[\sum_{d=1}^{D_j} X_{jd} f(\mathrm{DP}^*_{jd}) \delta_{jd} \ln f(\mathrm{DP}^*_{jd}) - \left(\sum_{d=1}^{D_j} X_{jd} f(\mathrm{DP}^*_{jd}) \delta_{jd} \right) (\ln \delta_{jd}) \right] + \ln \left| \begin{bmatrix} A_{11} & A_{12} & \cdot & A_{1p} \\ & & \cdot & \\ A_{21} & A_{22} & \cdot & \cdot \\ & & \cdot & \cdot \\ \cdot & \cdot & \cdot & \cdot \\ & \cdot & \cdot & \\ A_{m1} & \cdot & \cdot & A_{mp} \end{bmatrix}_v^{\prime} \right| \tag{9.6}$$

subject to

$$\begin{bmatrix} A_{11} & A_{12} & \cdot & A_{1p} \\ & & \cdot & \\ A_{21} & A_{22} & \cdot & \cdot \\ & & \cdot & \cdot \\ \cdot & \cdot & \cdot & \cdot \\ & & \cdot & \\ A_{m1} & \cdot & \cdot & A_{mp} \end{bmatrix} \begin{Bmatrix} \sum_{d=1}^{D_j} X_{1d} \mathrm{DP}^*_{1d} \\ \vdots \\ \sum_{d=1}^{D_j} X_{pd} \mathrm{DP}^*_{pd} \end{Bmatrix} \geq \begin{Bmatrix} T_1 - \Delta\mathrm{FR}_1 \\ \vdots \\ T_m - \Delta\mathrm{FR}_m \end{Bmatrix} \tag{9.7}$$

$$\begin{bmatrix} A_{11} & A_{12} & \cdot & A_{1p} \\ & & \cdot & \\ A_{21} & A_{22} & \cdot & \cdot \\ & & \cdot & \cdot \\ \cdot & \cdot & \cdot & \cdot \\ & & \cdot & \\ A_{m1} & \cdot & \cdot & A_{mp} \end{bmatrix} \begin{Bmatrix} \sum_{d=1}^{D_j} X_{1d}\mathrm{DP}^*_{1d} \\ \vdots \\ \sum_{d=1}^{D_j} X_{pd}\mathrm{DP}^*_{pd} \end{Bmatrix} \leq \begin{Bmatrix} T_1 + \Delta\mathrm{FR}_1 \\ \vdots \\ T_m + \Delta\mathrm{FR}_m \end{Bmatrix} \tag{9.8}$$

$$\mathrm{DP}^*_{jd} = f^{-1}\left(\delta_{jd}^{-1}\int_{\mathrm{DP}_{j(d-1)}}^{\mathrm{DP}_{jd}} f(\mathrm{DP}_j)d\mathrm{DP}_j\right)$$

$$= f^{-1}\left(\frac{p_d}{\delta_d}\right), \qquad \mathrm{DP}_{jd} = \mu + \left(\frac{2d}{D_j} - 1\right)\Delta\mathrm{DP}_j$$

$$\forall j, j = 1, \ldots, p; \quad \forall d, d = 1, \ldots, D_j \tag{9.9}$$

$$\sum_{d=1}^{D_j} \delta_{jd} f(\mathrm{DP}^*_{jd}) = 1 \qquad \forall j, j = 1, \ldots, p \tag{9.10}$$

$$\sum_{d=1}^{D_j} X_{jd} = 1 \qquad \forall j, j = 1, \ldots, p \tag{9.11}$$

$$X_{jd} = \{0, 1\} \qquad \forall j, j = 1, \ldots, p; \quad \forall d, d = 1, \ldots, D_j \tag{9.12}$$

$$\mathrm{DP}_j \in [\mu_j \pm \Delta\mathrm{DP}_j] \qquad \forall j, j = 1, \ldots, p \qquad (9.13)^2$$

Additional constraints:

Cost: $\sum_{j=1}^{p} Y_{ij} \sum_{d=1}^{D_j} X_{jd} k_{ijl} \mathrm{DP}^*_{jd} \leq C_{\mathrm{FR}_i} \qquad \forall i, i = 1, \ldots, m,$
where C is cost in dollars

Packaging weight: $\mathrm{DP}^*_{jd} \leq \mathrm{DP}_j^{\mathrm{up}}$ where $\mathrm{DP}_j^{\mathrm{up}}$ is an upper limit (9.14)

Check exit criteria. If

$$\frac{h_u - h_{u-1}}{h_{u-1}} = \frac{\Delta h_D}{h_{u-1}} \leq \varepsilon, \qquad \text{exit} \tag{9.15}$$

[2] Consideration for unrestricted in sign DPs may be added if needed.

Otherwise:

Loop updating

$$u \leftarrow u - 1$$

$$(\delta_{jd})_u \leftarrow \left(\frac{\delta_{jd}}{D_j}\right)_{u-1} \qquad \forall j, j = 1, \ldots, p \tag{9.16}$$

$$(\text{DP}^*_{jd})_u \leftarrow (\text{DP}^*_{jd})_{u-1} \qquad \forall j, j = 1, \ldots, p$$

$$v \leftarrow v - 1$$

Go to (9.1).

Constraints (9.2) and (9.3) represent the tolerance constraints or design range constraints of the FRs. Constraint (9.4) is the lower limit (ϖ) on R, the reangularity measure, while constraint (9.5) is the lower limit (ω) on S, the semangularity measure. The objective function, h_D, of the complexity minimization loop (the inner loop) is given in (9.6). Note that the complexity due to coupling, $|[\mathbf{A}']|$, is treated as a constant in the second loop. Constraints (9.7) and (9.8) are the tolerance constraint in the complexity loop. In this loop we need to find the interval δ_d for all the DPs where the optimum resides. The selection of the interval is indicated by the binary variable X_{jd}. The value of the DP used between iterations is the interval midpoint DP^*_{jd}. This midpoint is obtained from constraint (9.9) using a closed-form equation for the uniform distribution. Other distributions warrant another closed form or table lookup. The looping is continued until a selected termination criterion limit is satisfied. The use of constraint (9.10) enforces the selection of one interval for every DP in the inner loop. After solving the inner loop, the solution is supplied to the outer loop. The outer loop is then solved for the sensitivities, which in turn will be delivered to the inner loop for the next optimization iteration.

In this formulation, an optimization routine that minimizes both conceptual and operational vulnerabilities is presented. The routine consists of two loops. The first loop (the outer loop) is the independence optimization loop. The decision variables are the coefficients of the design matrix **A**. The second loop (the inner loop) is the complexity optimization loop. The decision variables are the binary variables X_{jd}, indicating the discrete interval indexed d that contains the optimum for DP_j. This optimization routine allows the simultaneous optimization of the axiomatic measures in one framework. The sequence is to achieve independence or at least minimum coupling levels with complexity reduction. Cost is taken as a constraint rather than an objective function.

9.3 PARAMETER DESIGN OPTIMIZATION

Parameter design is an experimental method to obtain transfer functions while desensitizing (operational vulnerability optimization) a design module to various sources of variation called *noise factors*. Parameter design requires arrays of test, robustness performance measures, and optimum selection as well as confirmation tests. At the core of parameter design is the activity of confirming the array of transfer functions corresponding to the design mapping (called *predictive equations* in the context of robust parameter design) by empirical testing. In this section we touch on several key aspects of robust parameter design. The reader is encouraged to consult a textbook on the subject for comprehensive handling.

The design matrices are very informative relative to how to conduct a parameter design or an optimization study of a given mapping at or below the LOI. The sequence of optimization is not arbitrary. The selection for the DPs (constituting the inner array) according to an FR depends on the mapping classification of interest. If the design is uncoupled, each FR can be optimized separately via its respective DP. Hence, we will have m total optimization studies (tests), the number of FRs. If the design is decoupled, the optimization routine has to follow the coupling sequence. The selection of the DPs to be included in a parameter design DOE depends on the needed information as well as cost. The selection of design parameters will be done in a manner that will enable target values to be varied during experiments with no major impact on module cost. The greater the number of potential design parameters that are identified, the greater the opportunity for optimization of function in the presence of noise.

A key philosophy of robust design is that during the design stage, inexpensive parameters can be identified and studied and can be combined in a way that will result in performance that is insensitive to noise. The design team task is to determine the combined best settings (parameter targets) for each of the control parameters, DPs, which have been judged by the design team to have potential to improve the module under study. By varying the parameter target levels in a DOE setup experimentally, a region of nonlinearity can be identified (see, e.g., Figure1.5). This area of nonlinearity is the most robust setting for the parameter under study. In this section we use the dynamic robust design formulation for illustration.

9.3.1 Noise Factors Identification

Noise factors cause the response to deviate from the intended target, which is specified by the input (signal factor) value, if any. Noise factors can be classified into three general categories (see Chapter 8 for a full enumeration of noise factors):

1. *External sources (usage and environment):* include temperature; user use, misuse, and abuse; and loading-related variation
2. *Unit-to-unit sources (manufacturing and supplier variation):* dimensional, assembly-related, or material property requirement variation from target values
3. *Deterioration sources (wear-out):* functional requirements, which degrade from the time the product is new, such as material fatigue or aging and wear, abrasion, and the general effects of use over time

In a robust design study, factors that are uncontrollable in use (or which are not practical to control) are selected to produce a testing condition during the operational vulnerability optimization experiment. The objective is to produce variation in the functional requirements (experimental data set) that would be similar to the effect that would be experienced in actual use of the design module.

Simulating the effects of all the noise factors is not practical and is not necessary. The key requirement in the selection and combination of these noise factors is to select a few important factors at points that cover the spectral range and intensity of real-world noises. Such noises are called *surrogate noises*. The rationale for this simplification approach is that the full spectral continuum of real-world noises should not cause variations much different from a small set of discrete choices positioned across the real-world spectrum.

9.3.2 Parameter Design Optimization DOEs

The purpose of this step is to coordinate all knowledge about the system under development into a comprehensive experimentation and data collection plan. The plan should be designed to maximize research and development efficiency through the application of orthogonal arrays, as well as responses such as quality loss function and signal-to-noise ratios and other statistical data analysis. The design team is encouraged to explore (experimentally) as many design parameters as feasible to investigate the functional performance potential of the design or technology concept that is being applied within the system. Transferrability of the improved functional requirements performance to the customer environment will be maximized because of use of the noise factor experiment strategy during data collection.

Data from the optimization experiment will be used to generate transfer function models, which will improve design robustness. The validity of these models and the resulting conclusions will be influenced by the experimental and statistical assumptions that are made by the team. The dynamic robust design formulation is assumed in what follows (see Chapter 5).

The planning in this step includes deciding on:

- *Noise orthogonal array.* An orthogonal array can be applied for assigning individual noise factors to an experimental test plan. The reason for this approach might be that the experimenter needs to have the ability to predict the effect of specific noise factors on the response. Specific noise factor information is not possible if the noise factors are tested experimentally using the compounded noise format. The approach for noise factors increases the total number of tests required to complete the experiment.
- *Signal range.* The range and number of levels to be studied for the signal factor are determined by the actual use (or intended use) of the system. This factor represents the operating range for the user-defined requirement.
- *Control factor array.* In a robust design study, factors (DPs) at or below the LOI, which are specified freely by the design team, are called *control*

factors. The more control factors studied, the more likely a better solution will be found. Each of these factors can be studied at two or three levels, with each level spaced apart so that a wider experimental region may be covered. Missing an important control factor in an experiment can mean the difference between breakthrough performance and mediocre performance as discussed in the context of Table 8.1. The types of DPs selected will depend on the mapping of the physical structure module that is being optimized. It is typical to have factors that are related to material properties, product or process dimensions, and in the case of manufacturing processes, operating conditions.

If the experiment is exploratory, it is suggested that levels be set at extreme values of the feasible operating tolerance or system range. Two levels will be appropriate for screening purposes, but more information on nonlinear effects will require a three-level strategy. Depending on the FR, there are two available broad forms of signal-to-noise robustness measures. Static forms apply where the FR has a fixed value. Dynamic forms apply where the FR operates over a range of input (signal) values (see Chapter 5 for more details).

9.3.3 Data Collection and Results Analysis

The individual signal-to-noise (SN) ratio or other appropriate robustness measures are calculated using the data from each experimental run. The purpose of determining the SN values is to characterize the ability of DPs to reduce the variability of the FRs over a specified dynamic range. In a dynamic SN experiment, the individual values for ideal function sensitivity (Chapter 5) are calculated using the same data from each experimental run. The purpose of determining the sensitivity values is to characterize the ability of control factors (DPs at or below the LOI) to change the average value of the function across a specified dynamic range. The resulting ideal function (see Chapter 5) sensitivity performance of a system is illustrated by the slope of a *best-fitline* of the functional performance data which is compared to the slope of the ideal function line.

For a uni-FR module parameter design study, we have the following steps:

1. The DP level effects are calculated by averaging SN ratios which correspond to the individual levels as depicted by the orthogonal array diagram.
2. Control factor importance for variability optimization is determined by comparing the gain in SN ratio from level to level for each factor, comparing relative performance gains between each design parameter and then selecting those that produce the largest gains. The level for each DP with the highest SN ratio is selected as the parameter's best target value to optimize the design from an operational vulnerability standpoint. All of these best levels will be selected to produce the optimum, the best parameter target combination.

3. The same analysis and selection process is used to determine DPs that can best be used to adjust the average (mean) of a functional requirement. These may be the same factors that have been chosen on the basis of SN improvement, or they may be factors that do not affect SN optimization. DPs that do not contribute to improvements in the SN ratio or the sensitivity of the transfer function are set to their most economical values.
4. The team needs to run confirmation tests of optimum design combinations and verify assumptions, perform a test with samples configured at the combined optimum design level, and calculate the representative SN ratio and sensitivity performance. This will enable prediction of the combined SN and sensitivity performance for the optimum combination identified by the response tables.
5. The team needs to compare the SN and sensitivity performance values to the values predicted. If the actual performance is within the interval of performance that was predicted, the predictive model validity is confirmed. There is a good chance at this point that the optimum results experienced in the confirmation run will translate to the usage environment. If the confirmation test values fall outside the interval, the team should reassess the original assumptions for this experiment since in all likelihood, other conditions are operating that are not accounted for in the model.

For an array of FRs being optimized simultaneously, some trade-off may be inevitable among the FRs relative to operational vulnerability optimization objectives, mean setting, and variability minimization. Use of a multiobjective optimization treatment such as penalty or ranking (e.g., quality loss function) is strongly encouraged. Such treatment is beyond the scope of this book.

A successful experiment will lead the team to clarify if new technical information has been uncovered that will revolutionize (or greatly improve) the physical structure. The team will want to consider if other DP levels should now form the basis of a revised experimental plan. If the study failed to produce a significant SN gain reflecting, for example, a six-sigma level, the combination of noise factors that were in the original experiment may have overpowered the ability of the control factors to generate improved performance. If improvement cannot be realized and the team has exhausted (and tested) all possible DPs, there may be a reason to conclude that the current concept being optimized will not be able to support the requirements for the system under development. This would justify the consideration and selection of a new incremental (redesign) project or even an altogether new creative design.

9.3.4 Case Study: Axiomatic Quality Parameter Design

An application involving an automobile automatic transmission, a highly coupled and complex electromechanical hydraulic kinematics system, was selected to illustrate key parameter design principles within the framework of the axiomatic quality process. A major subsystem, the planetary assembly of an automobile transmission, requires high-mileage reliability and robustness as demonstrated through field history, life testing, and laboratory fatigue testing. Planetary reliability and robustness are strongly correlated to the life of the engineered system, defined as the interface between the pinion gear bore, needle bearing, and pinion shaft. A planetary gear system is a highly efficient epicyclical kinetic mechanism with two degrees of freedom. A gear train with two degrees of freedom can be used to couple two inputs into one output. For the simple transmission of power from an input to an output that occurs in an automobile automatic transmission, only one degree of freedom is needed. The planetary system (module) is restricted to a single degree of freedom simply by locking individual components to ground. A primary component is designated the *pinion* (also known as a *planet*) *gear* because it is not fixed to the ground and is free to *orbit* the sun gear, a central gear, called the *sun gear* because its center is fixed to ground and it is being orbited by the planet gear. Unlike ordinary gear trains, the system is not grounded and frees up an interconnecting arm to rotate. This arm is referred to as the *carrier*. The pinion gear turns on a shaft fixed in the carrier. The pinion gear is positioned radially on the shaft on a roller bearing and axially between thrust washers. Finally, a *ring* or *annulus gear* can be fixed to ground to eliminate one degree of freedom. Selectively grounding or holding various elements of the planetary may achieve a speed reduction, a speed reversal, and a speed increase, thus providing the key functional requirements of an automatic transmission.

The voice of the customer was processed and translated into engineering terms and functional requirements as a result of a comprehensive automotive system quality function deployment (QFD) house of quality stage 2 (see Chapter 6). A design team is convened to collaborate on the translation of customer attributes (CAs) into FRs in preparation for the required zigzagging process using the zigzagging axiomatic design decomposition process discussed in the CDFC phase.

The cascading required was accomplished through a zigzagging process from the customer level down to the supersystem, system, subsystem, and component levels, as depicted in Figure 9.3. The plethora of component mapping and resulting design matrixes as a result of the zigzagging process converged on one coupled component region. The results allow us to identify the "critical few" design parameters for subsequent optimization within this region. The following automatic transmission planetary gear system was decomposed: (1) annulus gear, (2) planetary carrier, (3) sun gear, (4) pinion gear. The pinion gear needle-bearing component (module) design mapping is the subject of this example and is shown in Figure 9.4.

A P-diagram similar to Figure 8.1 was constructed to identify the ideal function, noise factors, control factors, and the energy transformation concept shown in Figure 9.5. The compound noise strategy, experiment control factors, and final experiment orthogonal array are depicted in Tables 9.2, 9.3, and 9.4, respectively. An example of a main effects plot of mean response at 9000 rpm is shown in Figure 9.6. A sample main effects plot for SN ratios is shown in Figure 9.7.

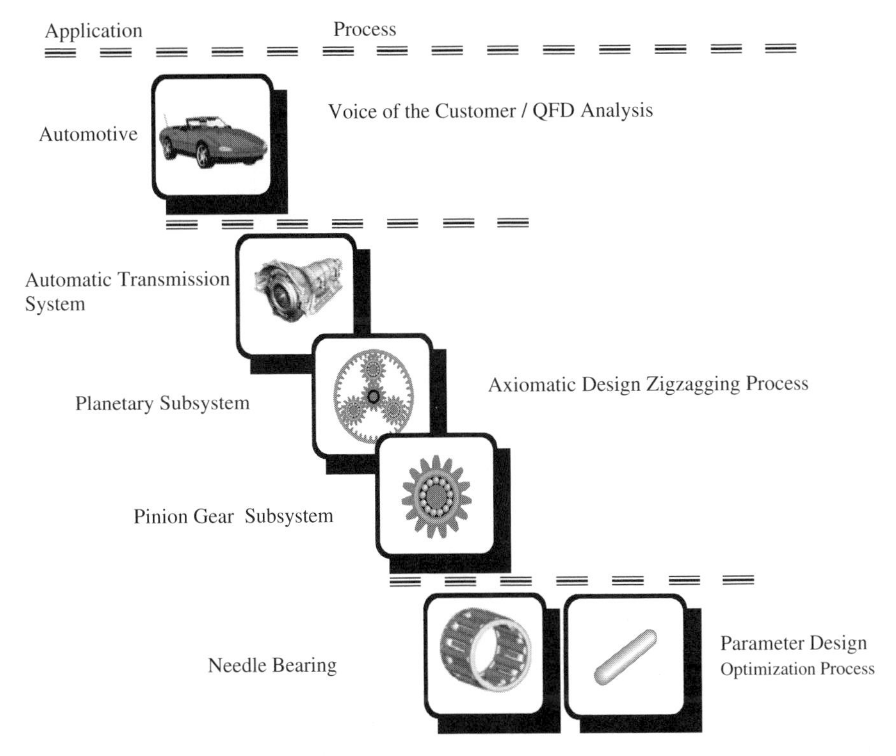

Figure 9.3 Zigzagging method application up to needle bearing.

Functional Requirements of the Needle System:

- FR_1: Transmit Carrier Torque
- FR_2: Transmit Rotation
- FR_3: Create Radial Force
- FR_4: Locate Pinion Gear

Design Parameters of the Needle System:

- DP_1: Diametrical Clearance
- DP_2: Circumferential Clearance
- DP_3: Shaft Surface Characteristics
- DP_4: Pinion Bore Surface Characteristics

$$\begin{Bmatrix} FR_1 \\ FR_2. \\ FR_3 \\ FR_4 \end{Bmatrix} = \begin{bmatrix} 0 & 0 & A_{13}. & A_{14} \\ A_{21} & A_{22} & A_{23} & A_{24} \\ A_{31} & A_{32} & 0 & 0 \\ A_{41} & 0 & 0 & 0 \end{bmatrix} \begin{Bmatrix} DP_1 \\ DP_{2.} \\ DP_3 \\ DP_4 \end{Bmatrix}$$

(Coupled Design)

Figure 9.4 Needle-bearing design mapping.

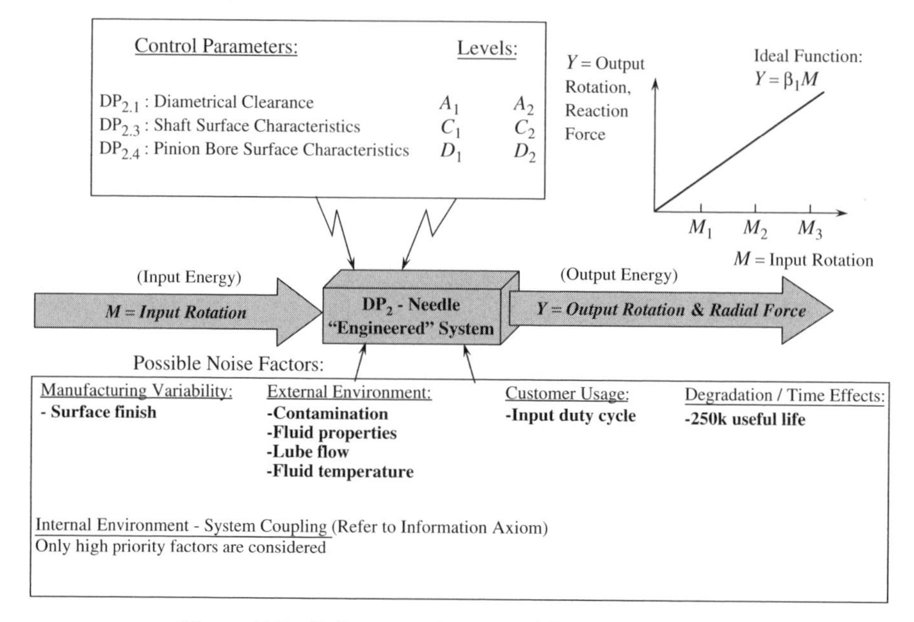

Figure 9.5 P-diagram: planetary pinion needle system.

TABLE 9.2 Compound Noise Strategy

Noise Factor	N_1 Good Level	N_2 Bad Level
Lubrication flow	Improved flow	Current flow
Lubrication properties	New oil	Aged oil
Degrade (usage)	New rollers	Aged rollers

TABLE 9.3 Parameter Design DOE Control Factors

Control Factor	Level 1	Level 2
$DP_{2.1}$: diametrical clearance	A_1: current specification	A_2: reduced clearance
$DP_{2.3}$: shaft surface characteristics	C_1: current specification	C_2: improved finish
$DP_{2.4}$: pinion bore surface characteristics	D_1: current specification	D_2: three-stage hone

TABLE 9.4 DOE Orthogonal Array

	Control Factors			Noise Factors					
	DC	Shaft OD	Gear ID	N_1	N_2	N_1	N_2	N_1	N_2
1	A_1	C_1	D_1	120	105	135	110	150	125
2	A_1	C_2	D_2	120	105	125	115	145	145
3	A_2	C_1	D_2	120	102	135	118	155	208
4	A_2	C_2	D_1	135	100	125	112	155	140
5	A_2	C_2	D_2	130	105	130	115	150	135
6	A_2	C_1	D_1	130	110	135	122	145	150
7	A_1	C_2	D_1	120	100	125	115	150	145
8	A_1	C_1	D_2	120	105	130	120	140	170

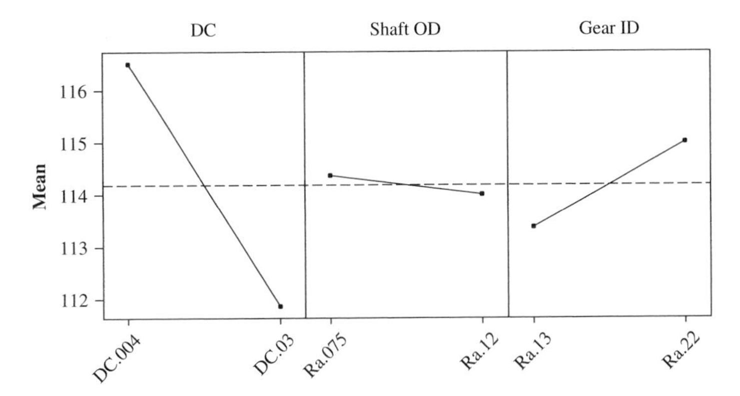

Figure 9.6 Main effects plot for means.

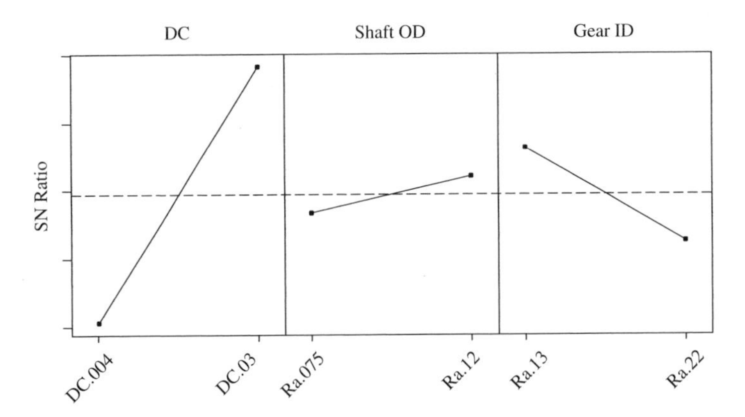

Figure 9.7 Main effects plot for SN ratios.

Overall, the optimal design predicted resulted in an SN ratio of 16.5, which represented a 6-dB[3] improvement over the baseline design. Confirmation results indicated a SN ratio of 16.1, a 5.6-dB improvement over the baseline design. Subsequent life testing and Weibull probability plots confirmed a significant improvement in useful life and high-mileage reliability. The design parameters under study have been proven to affect the life of the planetary system significantly. Through application of this study, reliability and robustness of the planetary system were improved by 28% at 150,000 miles. A transfer function model has now been developed that can be applied to future design iterations and emerging products.

The axiomatic quality methodologies described here represent powerful tools in achieving high-level quality and reliability goals. In particular, axiomatic design (design axioms, cascading process, design matrices, etc.), integrated with parameter design principles, was highly effective in translating customer-based functional requirements into design parameters for operational vulnerability optimization. This case study is a testimony to the power of the axiomatic quality process, providing a practical real-world confirmation of the process.

9.4 AXIOMATIC QUALITY STRATEGY IN THE TOLERANCE OPTIMIZATION PHASE

A design entity can be tagged as defective in a specific FR, say FR_i, when it does not meet customer attribute targets, denoted as T_i, within an acceptable tolerance range. Sample average and standard deviation are used to estimate the FR_i average, μ_i, and the standard deviation, σ_i, respectively. When the FR average equals its target value (i.e., $\mu_i = T_i$), the design entity is centered (i.e., delivered as intended by the customer).

The half-tolerance range of FR_i, ΔFR_i, should be specified by the customer (next process) or the consumer (the buyer). When the customer is external to the design organization, acceptable functional requirements definition can be obtained using CA-to-FR mapping, the first axiomatic quality process, by employing two stages of QFD, customer clinics and surveys. When the customer is internal (i.e., within the company), engineering specifications are easy to obtain. The sigma level of FR_i, η_i, equals $\Delta\text{FR}_i/\sigma_i$. The term *six-sigma quality level* implies that $\eta_i = 6$ for two-sided limits.

The CDFC phase presented in Chapters 6 and 8 endorses the operational robustness philosophy and concepts: in particular, the idea of classification variables as noise f and control factors and their effects on the FRs in the context of Table 8.1. The array of noise factors is denoted as $\{\mathbf{z}\}$. Sources of variation or noise factors should be anticipated and their effects should be assessed. These sources of variation include manufacturing variability (denoted *mv*), customer usage (denoted *cu*), degradation over time (*de*), and environment (denoted as *e*).

[3]This is equivalent to a 50% reduction in variability in output energy (see Section 9.4.1).

Therefore, the overall variance, σ_i^2, can be written as

$$\sigma_i^2 = g(\sigma_{mv}^2, \sigma_{cu}^2, \sigma_{de}^2, \sigma_e^2) \tag{9.17}$$

where g is a real-valued function. Anticipating the effect of noise factors implies incorporating their effect on the overall variance of FR_i. The mathematical form of the function depends on how the noise factors affect FR_i. Study of the noise factor effect can be simulated or tested after finishing the CDFC phase for incremental design projects. This should be done when the design is centered [i.e., $\mathrm{FR}_i|_{\text{at } \mu_{\text{DPs}}} = f_z(z_{mv}, z_{cu}, z_{de}, z_e)$] and the DPs are fixed at their nominal settings (using baseline or prototype data). If the noise effect is additive, and assuming independence (i.e., $\mathrm{FR}_i|_{\text{at } \mu_{\text{DPs}}} = \phi_{mv}z_{mv} + \phi_{cu}z_{cu} + \phi_{de}z_{de} + \phi_e z_e$), we have

$$\sigma_i^2 = \phi_{mv}^2\sigma_{mv}^2 + \phi_{cu}^2\sigma_{cu}^2 + \phi_{de}^2\sigma_{de}^2 + \phi_e^2\sigma_e^2 \tag{9.18}$$

Another generic form that is usually encountered is

$$\mathrm{FR}_i = \phi z_1^{\phi_1} z_2^{\phi_2} z_3^{\phi_3} z_4^{\phi_4} \tag{9.19}$$

and

$$\sigma_i^2 \cong \mu_i^2 \left[\sum_{j=1}^{p} \phi_j^2 \left(\frac{\mu_j}{\sigma_j}\right)^2 + 2\sum_{j=1}^{p-1} \sum_{j'=j+1}^{p} \phi_j \phi_{j'} \rho_{jj'} \frac{\mu_j}{\sigma_j}\frac{\mu_{j'}}{\sigma_{j'}} \right] \tag{9.20}$$

where $j = \{1, 2, 3, 4\}$ corresponds to $\{mv, cu, de, e\}$, ρ is the correlation coefficient, and the ϕ's are constants.

The tolerance optimization phase of the axiomatic quality process prefers σ_i^2 over the practice of using only manufacturing variability σ_{mv}^2. The real form of the functions g and f_z can be derived using physical laws analytically, or empirically using key life tests, DOEs, and/or regression analysis. However, many assumptions need to be examined. First, we will assume that the design is incremental off a baseline so that the company databases can be used. Second, we will assume that these databases are credible (i.e., data were collected with statistical knowledge and error noise has been filtered and reduced). In addition, the databases should always be maintained and updated on a continuous basis to maintain usage in future design projects. A third assumption is that when some of the data are simulated, they should be correlated to reality for both the manufacturing and customer usage environments.

9.4.1 Robustness at Six-Sigma Quality: Signal-to-Noise Ratio and Quality Loss Function

A six-sigma capability in a given target FR can be achieved by either of two means:

(1) by opening the tolerance range (ΔFR) or
(2) by reducing the standard deviation (σ_{FR}).

The first option may be feasible when based on a consensus from the customer. However, this is usually a very remote possibility. The other option is the subject of variability reduction techniques such as robust design and six-sigma. Option 2 above is utilized in the derivations and discussions that follow.

Robustness in Signal-to-Noise Ratio The six-sigma strategy for problem solving, the DMAIC approach, deals with variability and mean performance adjustment in a manner that parallels the treatment entertained in the robustness methodology. In the subject of robustness, Taguchi and Elsayed (1989) and Taguchi et al. (1999) deal with the tolerances, and hence variability, from the perspective of cost and quality when parameter design fails to produce the variability targets desired. From the cost perspective, it is assumed that there exists a target performance array for the FRs, and any deviation from the target value will incur an economic loss usually represented by the QLF (see Section 1.5 for more details). On the other hand, quality is usually expressed using SN, which is given by

$$\mathrm{SN} = 10 \log \frac{\mu_i^2}{\sigma_i^2} \tag{9.21}$$

Improvement in the SN ratio between baseline design and new design (hopefully, optimum), denoted as $\Delta\mathrm{SN}$, is desired when design changes (soft or hard) are implemented to redesign a baseline or to solve a problem. Operational vulnerability optimization includes reducing variability and adjusting the mean to a target value desired for the FRs. The latter is usually a typical engineering problem that can be solved at minimal cost. The former, however, is usually achieved with engineering burden, effort, and cost. In the derivation presented here, we assume that the design is centered for FR_i (i.e., $\mu_i = T_i$). Therefore, the SN ratio and QLF improvements will come from reducing the variance. Let $\sigma_{i,0}$ be the standard deviation of the datum (baseline) design and let σ_i be the standard deviation of the new design (after implementing the changes). For the baseline design, we have $\Delta\mathrm{FR}_i = \eta_{i,0}\sigma_{i,0}$, where $\eta_{i,0}$ is the baseline design's sigma level. Similarly, for the new improved design we have $\Delta\mathrm{FR}_i = \eta_i\sigma_i$, where η_i is the new (optimum or not) design's sigma level. A positive value of $\Delta\mathrm{SN}$ implies that $\sigma_i < \sigma_{i,0}$ and $\eta_i > \eta_{i,0}$. Using these variables, improvement in the SN ratio can be expressed as

$$\begin{aligned}
\Delta\mathrm{SN}_i &= \mathrm{SN}_{i,\mathrm{new}} - S_{i,\mathrm{baseline}} \\
&= 10\log_{10}\frac{\mu_i^2}{\sigma_i^2} - 10\log_{10}\frac{\mu_i^2}{\sigma_{i,0}^2} \\
&= 10\log_{10}\frac{\eta_i^2}{\eta_{i,0}^2} \\
&= 20\log_{10}\frac{\eta_i}{\eta_{i,0}}
\end{aligned} \tag{9.22}$$

Assume that a current quality level equals $\eta_{i,0} = 3$. The six-sigma level means that the ratio $\eta_i/\eta_{i,0}$ equals 2 (i.e., a 50% reduction in variability, which translates to a 6.0206-dB improvement). This is depicted in Figure 9.8.

When improvement is achieved for an array of FRs, say a system with m FRs, (9.22) is written as

$$\begin{aligned}\Delta\text{SN} &= \sum_{i=1}^{m} \Delta\text{SN}_i \\ &= 20 \sum_{i=1}^{m} \log_{10} \frac{\eta_i}{\eta_{i,0}} \\ &= 20 \log_{10} \left(\prod_{i=1}^{m} \frac{\eta_i}{\eta_{i,0}} \right)\end{aligned} \tag{9.23}$$

When $\eta_{i,0} \approx 3, \forall i = 1, \ldots, m$, (9.23) can be expressed as

$$\begin{aligned}\Delta\text{SN} &= \sum_{i=1}^{m} \Delta\text{SN}_i \\ &\cong 20 \log_{10} \left(3^{-m} \prod_{i=1}^{m} \eta_i \right) \\ &\cong 20 \log_{10} \left(\prod_{i=1}^{m} \eta_i \right) - 9.542425m\end{aligned} \tag{9.24}$$

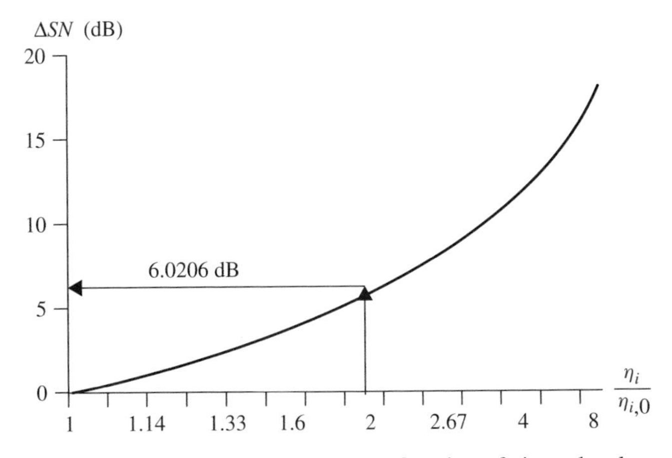

Figure 9.8 SN improvement as a function of sigma level.

The six-sigma capability of a system with m FRs as it relates to SN ratio improvement from the current three-sigma levels can be expressed as

$$\begin{aligned} \Delta \mathrm{SN} &= \sum_{i=1}^{m} \Delta \mathrm{SN}_{ji} \\ &\cong 20 \log_{10} 2^m \\ &\cong 6.0206m \end{aligned} \tag{9.25}$$

The improvement or gain in the SN ratio of a given FR cannot be maintained as constant over time after launch (stage 7 of Figure 1.8). This loss in performance over time, referred to as *degradation*, can be modeled with a real-valued drift function $\theta(t)$ that can be observed over time or may be obtained based on the physical laws of the design entity. The drift function has the following properties:

- It should be increasing monotonically with time.
- $\theta(0) = 0$ as an initial condition.
- There is a maximum for the drift without being noticed by the customer. In six-sigma this maximum is, typically, assumed to be 1.5.

Therefore, (9.22) can be written as

$$\begin{aligned} \Delta \mathrm{SN}_i(t) &= 20 \log_{10} \frac{\eta_i - \theta(t)}{\eta_{i,0}} \\ &= 20 \log_{10} \left(\frac{\eta_i}{\eta_{i,0}} - \frac{\theta(t)}{\eta_{i,0}} \right) \end{aligned} \tag{9.26}$$

For example, $\theta(t) = C(1 - e^{-\theta t})$, where C is a constant, satisfies the first two properties. Note that degradation models have some exponential components with a time constant dependent on the FR and the design entity. Assuming that at $t = 0$, the six-sigma level was established (i.e., $\eta_i/\eta_{i,0} = 2$ with $\eta_{i,0} = 3$). Note that the minimum improvement over time occurs when $\theta(t)$ is maximal. If the maximum value is 1.5, the minimum $\Delta \mathrm{SN}_{\mathrm{minimum}} = 3.5218$ dB. For $C = 1$ and $\theta = 1$, the improvement ΔSN over time is as depicted in Figure 9.9.

Robustness in the Quality Loss Function From a QLF perspective, it is assumed that there exists a target performance for the FRs, denoted as the array $\{\mathbf{T}\}$, and any deviation from the target value will incur an economic loss, usually represented by the QLF. Specifically, let FR_i denote the FR of interest, and let T_i be its target value. The FR transfer function $\mathrm{FR}_i = f(\mathrm{DP}_1, \mathrm{DP}_2, \ldots, \mathrm{DP}_p)$ is used. We will assume that noise factor effects are represented by the variation in DPs via interaction. The loss function, L_i, is given by

$$L_i = k_i[\sigma_i^2 + (\mu_i - T_i)^2] \tag{9.27}$$

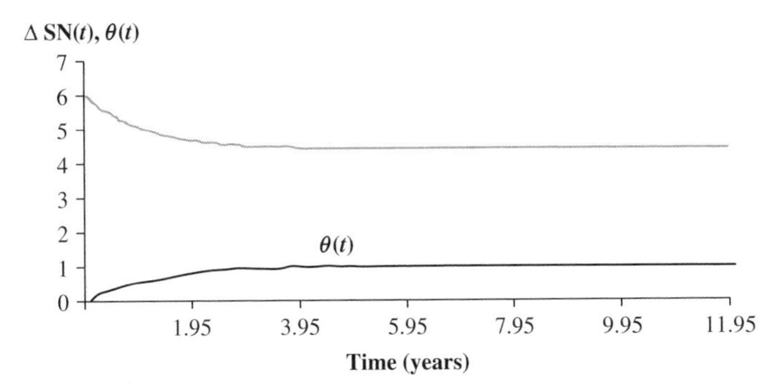

Figure 9.9 Decay in SN improvement over time.

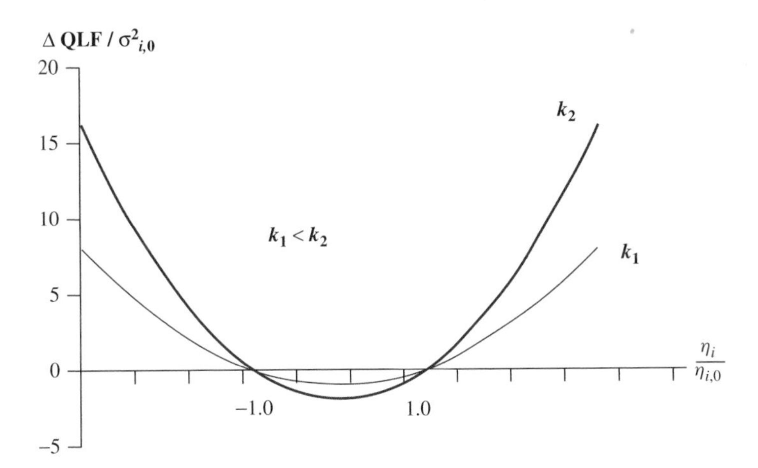

Figure 9.10 QLF improvement as a function of sigma level.

where $\mu_i = E(\text{FR}_i)$ and $\sigma_i^2 = \text{Var}(\text{FR}_i)$. The reader will recognize that (9.27) is a repetition of (5.4).

An improvement in QLF, denoted as ΔQLF, should be negative and implies that $\sigma_i < \sigma_{i,0}$ and $\eta_i > \eta_{i,0}$. Using this notation, the improvement, ΔQLF, can be expressed as

$$\begin{aligned}\Delta\text{QLF}_i &= \text{QLF}_{i,\text{new}} - \text{QLF}_{i,\text{baseline}} \\ &= k_i\left[\left(\frac{\eta_{i,0}}{\eta_i}\right)^2 - 1\right]\sigma_{i,0}^2 \end{aligned} \tag{9.28}$$

A normalized version of (9.28) is shown in Figure 9.10. Note that when the design is centered, the window of opportunity is $\eta_{i,0}/\eta_i \in (-1, 1)$. Outside this window, the baseline system is better or equivalent to the new design, indicating inefficient changes (i.e., $\eta_{i,0}/\eta_i = 1.0$). Therefore, further savings in QLF should come from reducing k_i. This requires reducing all design parameters contributing to repair of the design entity when failures (as defined by the customer) occur. When $\eta_{i,0}/\eta_i = 0.5$ with $\eta_{i,0} = 3$, the six-sigma quality level is established. This corresponds to an SN ratio improvement of 6.0206 dB, a 50% reduction in variability.

As is the case with the SN ratio, discussed earlier, the effect of degradation on ΔQLF can be evaluated using degradation modeling, discussed earlier in this section, by following the same steps and adhering to the same assumptions.

9.5 DESIGN OPERATIONAL VULNERABILITY OPTIMIZATION USING TOLERANCES OF UNI-FR DESIGN MODULES

The derivations in this section are based on the use of design tolerances to achieve operational vulnerability objectives. It assumes that the DP mean settings have been identified prior to tolerance optimization activity. Tolerance research includes many areas that deal with the assignment of tolerances in design parameters, the assessment and control of manufacturing processes, several metrological issues, as well as geometric and cost modeling. In the derivation below, Taguchi's QLF concept is used.

In the context of (9.27), it is clear that the FR_i can generate loss in two ways: deviation of μ_i from T_i, which can be quantified by $(\mu_i - T_i)^2$, and variability in FR_i, which can be quantified by σ_i^2. A reduction in the two components of the loss expected can be achieved by adjusting $E(\text{DPs}) = (\mu_1, \mu_2, \ldots, \mu_p)$, the mean values of the FRs. Usually, μ_i can be adjusted to be equal to T_i by adjusting the array $(\mu_1, \mu_2, \ldots, \mu_p)$ alone. On the other hand, σ_i^2 can be estimated by the error formula (Ku, 1966) and (9.27) approximates to

$$L_i = k_i \sigma_i^2 \cong k_i \sum_{j=1}^{p} \left(\frac{\partial \text{FR}_i}{\partial \text{DP}_j} \right)^2_{\text{DP}_j = \mu_j} \sigma_j^2 \tag{9.29}$$

Therefore, the total quality loss L_i can be minimized by reducing σ_j^2, where $\sigma_j^2 = \text{Var}(\text{DP}_j)$, the variance of DP_j, the jth parameter in the array **{DP}**. The half-tolerance range of DP_j, denoted by ΔDP_j, is equal to $\eta_j \sigma_j$, where η_j is the value of the sigma quality level. That is, the tolerance can be written as $\mu_j \pm \Delta\text{DP}_j$, where μ_j is the nominal (target) setting of DP_j. In a sense, the tolerance vulnerability optimization step in axiomatic quality reduces to the assignment of σ_j. However, a tighter tolerance results in a high manufacturing cost, and a solution is needed that takes into consideration the parameter tolerance settings to fit multiple FRs simultaneously and at minimum cost.

Note that achieving six-sigma capability in the DPs *does not guarantee that this capability will transfer to the* FRs. That is, a design change (good or bad) in the

DPs will propagate to the FR variance and quality loss, via the transfer function, amplified or attenuated by $(\partial \text{FR}_i/\partial \text{DP}_j)^2_{\text{DP}_i=\mu_j}$, $j = 1, \ldots, p$, the square of the sensitivities at the nominal settings. The conclusion is that the overall product term is what is important in (9.29).

Axiomatic quality tolerance techniques address optimization of the total quality cost of an FR by finding the optimum value of σ_j^2, $j = 1, \ldots, p$. Total cost is taken as the objective function and in the formulation derived below has two components:

1. *Tolerance control cost* (incurred by the company). The tolerance control cost is the total of all cost components incurred by the company to institutionalize six-sigma capability in the module before release. This includes the costs associated with machine upgrading, process improvement, labor training, new procedures, monitoring, and others. This cost is usually inversely proportional to the reduction in parameter variances.
2. *Quality loss* (i.e., the FR variability cost incurred by the customer and the society at large). The quality loss cost has two parts, as given by (9.27). The first term represents the cost due to variability and the second, the cost due to the deviation from target. In almost every case, target adjustment capability can be assumed to exist in all systems and is considered a typical engineering problem that can be solved with no or minimal cost. On the other hand, reducing σ_i^2 usually requires the reduction of DP variances, which incurs considerable cost that usually offsets the target adjustment cost. This cost will be denoted as $d_j(\sigma_j)$ in the following derivation, and the tolerances will be given as $\mu_j \pm \eta_j \sigma_j$ and $T_i \pm \eta_i \sigma_i$ of the parameters DP_j and the FR_i, respectively.

Without loss of generality we will assume that μ_i, the FR mean, can be adjusted to T_i. Error propagation formulas can be used to approximate σ_i^2 in tolerance vulnerability optimization formulation problems. For example, equation (9.29) can be used in the following axiomatic quality optimization program:

$$\text{Minimize } \text{TC}_i = \sum_{j=1}^{p} d_j(\sigma_j) + k_i \sigma_i^2 \tag{9.30}$$

subject to

$$\sigma_i^2 \leq \left(\frac{\eta_{i,0}}{\eta_i} \right)^2 \sigma_{i,0}^2 \tag{9.31}$$

The six-sigma quality level is obtained when $\eta_i = 6$. The term $d_j\ (\sigma_j)$ is the tolerance control cost for parameter DP_j, a monotonically decreasing function in σ_j. Several forms were proposed for this cost, such as $d_j(\sigma_j) = b_j + c_j/\sigma_j$, proposed by Chase and Greenwood (1988), and $d_j(\sigma_j) = b_j + c_j/\sigma_j^2$, suggested by Spotts (1973).

9.5.1 Coupling Vulnerability Constraint

According to the independence axiom (Suh, 1990), we would like to adjust a functional requirement primarily through one design parameter, say DP_l, where $l \in J = \{j|j = 1, \ldots, p\}$, which is the greatest contributor to the tolerance of the FR. In other words, we would like the FR tolerance contribution of the off-diagonal DPs to diminish and become negligibly small. In the CDFC phase, the design team should identify the premier parameters in all mappings at all hierarchical levels during the zigzagging process. Mathematically, DP_l can be characterized as the design parameter with $\max_j\{\partial\text{FR}_i/\partial\text{DP}_j\}^2_{\text{DP}_i=\mu_j}\sigma_j^2$, $l \in J = \{j|j = 1, \ldots, p\}$. Sensitivities can be known analytically using CAE or in a DOE setup at a lower hierarchical level at or below the LOI.

A coupled design can be uncoupled or decoupled (see Chapter 8) if the tolerances can be set such that the effect of the nondiagonal elements of the design matrix could be made *small enough* to be neglected. Armed with this thinking, the usefulness of the (9.30)–(9.31) optimization program can be extended to include the enhancement of functional independence. This activity translates to minimizing the coupling vulnerability—this time by operational means. To improve independence, the tolerance of the individual DPs needs to be set such that a premier DP, say DP_l, can be singled out as the most significant. This can be achieved with some degree of ease because coupling cannot be eliminated totally if not done in the CDFC phase. That is, the objective of independence improvement can be achieved to some degree, say α, modeled as a percentage of the half-width tolerance of FR_i. The premier parameter, DP_l, needs to satisfy such a constraint through design sensitivities so that it becomes a significant factor based on tolerances and variability. Therefore, we can supplement the (9.30)–(9.31) optimization program with the following constraint:

$$\eta_l \left(\frac{\partial\text{FR}_i}{\partial\text{DP}_l}\right)_{\text{DP}_l=\mu_l} \sigma_l \geq \alpha \frac{\eta_{i,0}}{\eta_i}\sigma_{i,0} \tag{9.32}$$

The solution to the supplemented model is obtained by applying the Kuhn–Tucker condition of constrained nonlinear programming:

$$\sigma_j = \frac{|\partial d_j(\sigma_j)/\partial\sigma_j|}{2(k_i + \lambda_1)(\partial\text{FR}_i/\partial\text{DP}_j)^2_{\text{DP}_j=\mu_j}} \tag{9.33}$$

$$\sigma_l = \frac{\alpha(\eta_{i,0}/\eta_i)\sigma_{i,0}}{\eta_l|\partial\text{FR}_i/\partial\text{DP}_l|_{\text{DP}_l=\mu_l}} \tag{9.34}$$

$$\lambda_1 = \frac{1}{2}\left\{\frac{\sum_{\substack{j=1\\j\neq l}}^{p} \dfrac{(\partial d_j(\sigma_j)/\partial\sigma_j)^2}{(\partial\text{FR}_i/\partial\text{DP}_j)^2_{\text{DP}_j=\mu_j}}}{\left(\dfrac{\eta_{i,0}}{\eta_i}\right)^2 \sigma_{i,0}^2\left[1 - \left(\dfrac{\alpha(\eta_i/\eta_{i,0})}{\eta_l}\right)^2\right]}\right\}^{1/2} - k_i \tag{9.35}$$

$$\lambda_2 = \frac{\partial d_l(\sigma_l)/\partial \sigma_l}{\eta_l |\partial \mathrm{FR}_i/\partial \mathrm{DP}_l|_{\mathrm{DP}_l=\mu_l}} + \frac{2\alpha\left(\dfrac{\eta_{i,0}}{\eta_i}\right)\left[\sum_{\substack{j=1\\ j\neq l}}^{p} \dfrac{(\partial d_j(\sigma_j)/\partial \sigma_j)^2}{(\partial \mathrm{FR}_i/\partial \mathrm{DP}_j)^2_{\mathrm{DP}_j=\mu_j}}\right]^{1/2}}{\eta_l^2\left[1-\left(\dfrac{\alpha(\eta_i/\eta_{i,0})}{\eta_l}\right)^2\right]^{1/2}} \tag{9.36}$$

with $1/p \leq \alpha < \eta_l/(\eta_{i,0}/\eta_i)$; λ_1 and λ_2 are the Lagrange multipliers of (9.31) and (9.32), respectively.

There is an economic interpretation for the Lagrange multipliers in (9.35) and (9.36). Usually, they are called the *shadow prices* of the constraints, and their optimal values are used in sensitivity analysis. For example, λ_1 is the rate of change of the optimal value of the objective function TC with respect to the right hand side (RHS) of the constraint in (9.35). In other words, the change in the optimal value of the objective function for a unit increase in the RHS constant of constraint is give by λ_1. The same logic applies to λ_2.

The term $\partial d_j(\sigma_j)/\partial \sigma_j$ represents the sensitivity of the tolerance control cost of the ith DP. The following forms have been proposed:

$$\left|\frac{\partial_j(\sigma_j)}{\partial \sigma_j}\right| = \begin{cases} \dfrac{c_j}{\sigma_j^2} & \text{(Chase and Greenwood, 1988)} \\[2ex] \dfrac{c_j}{\sigma_j^3} & \text{(Spotts, 1973)} \end{cases} \tag{9.37}$$

If the solution is infeasible, either the desired six-sigma level η_i in (9.31), or the desired independence level α in (9.32), or both, cannot be achieved. In this case you may change the RHS of the unsatisfied constraint until an optimal and feasible solution can be achieved. In many situations a compromise between the independence desired and the sigma level needs to be accomplished. This practice suggests a classification, a taxonomy, for a quality–independence trade-off. For example, when the maximum α and maximum sigma level can be obtained, an excellent classification can be given, denoted as EE. The other proposed classifications are given in Table 9.5.

Of course, the designer should target an EE design when theoretical independence (uncoupled design) and the six-sigma level cannot be achieved. As $\alpha \to 1$, the system quality (defined here as the independence of the functional requirements) of the design entity will be higher. In addition, as $\eta_i \to 6$, the quality level will be higher.

9.5.2 Meaning of the Solution

The solution above indicates that the optimal tolerance should be defined in such a way that the tolerance for DP_j, $j = 1, \ldots, p$ [a design parameter other than the premier (diagonal) parameter where $j \neq l$] should be proportional to the

TABLE 9.5 Quality–Independence Taxonomy of Design[a]

η_i	α		
	70–100%	50–70%	Less than 50%
Six sigma	EE	EM	EP
Five sigma	ME	MM	MP
Four sigma	PE	PM	PP

[a]E, excellent; M, medium; P, poor.

incremental tolerance reduction cost $\partial d_j(\sigma_j)/\partial\sigma_j$ and inversely proportional to the square of the sensitivity of FR$_i$ at the current nominal settings of the jth DP, as presented in (9.33). The constant of proportionality is $1/[2(k_i+\lambda_1)]$, where λ_1 is the Lagrange multiplier of (9.31), a nonnegative constant. From (9.35), λ_1 is found to be the sum of the tolerance control cost weighted by the reciprocal of the sensitivity squared and adjusted by

$$\tau = \left(\frac{\eta_{i,0}}{\eta_i}\right)^2 \sigma_{i,0}^2 \left[1 - \left(\frac{\alpha(\eta_i/\eta_{i,0})}{\eta_l}\right)^2\right]$$

an unevenness factor due to the nonuniform tolerance allocation that resulted from introduction of the (9.32) constraint. This unevenness may create or eliminate cost, depending on where tolerances need to be tightened or relaxed based on (9.33) and (9.34). However, the opportunity is here to reduce coupling, thus improving the sigma-level quality while reducing cost.

In (9.34), the tolerance of DP$_l$ is equal to and is a fraction of the FR$_i$ tolerance. It is inversely proportional to the sensitivity of the FR with respect to DP$_l$ evaluated at the nominal DP setting. The constant of proportionality equals $\alpha(\eta_{i,0}/\eta_i)/\eta_l$, where η_i is the desired or feasible sigma level and α is the independence level assigned to the diagonal DP (denoted as DP$_l$) as a percentage of the FR$_i$ tolerance. As $\alpha \to 1$, the design approaches the uncoupled category. However, an upper bound on α is equal to the ratio [i.e., $\alpha \le \eta_l/(\eta_{i,0}/\eta_i)$]. The lower limit can be taken as $\alpha \le 1/p$ (i.e., the reciprocal of the number of DPs, indicating a uniform tolerance allocation).

9.6 DESIGN OPERATIONAL VULNERABILITY OPTIMIZATION USING TOLERANCES OF AN FR ARRAY

The uni-FR case is generalized to an array of FRs in this section. Arrays of FRs are obtained in a design mapping hierarchy revealed by the zigzagging process (Chapter 3) and fed into the CDFC phase. The group (or decomposition tree) of all design mappings prescribes the design of interest in a creative or incremental project. As we learned in Chapter 6, not all design mappings will be hosted in the same physical modules, depending on design team logic for a sound physical structure. The axiomatic quality process CDFC phase starts with a high-level array of FRs identified by the design team based on QFD stage 2. From an axiomatic perspective, an **{FR}** array should contain the minimum number of independent functional requirements needed to fulfill the design goals as defined by the customer at a given hierarchical level.

As a requirement for formulation derived here and based on what we learned in the CDFC phase, the number of FRs, m, should be at most equal to the number of DPs, p (i.e., $m \leq p$). This requirement translates to the case of either a redundant design ($m < p$) or an ideal design ($m = p$).

Let i and j be the indexes of the elements of arrays **{FR}** and **{DP}** of given mapping, respectively. The independence requirement implies the existence of a premier design parameter, say DP_l, for every FR in the array **{FR}**. Again we assume that the nominal values of all FRs can be set with known cost [i.e., $E(\mathrm{FR}_i) = T_i$]. In this case, the system quality loss that will be incurred by the customer is given by

$$L = \sum_{i=1}^{m} k_i \sum_{j=1}^{p} \left(\frac{\partial \mathrm{FR}_i}{\partial \mathrm{DP}_i} \right)^2_{\mathrm{DP}_j=\mu_j} \sigma_j^2 \tag{9.38}$$

This leads to the following nonlinear program:

$$\text{Minimize TC} = \sum_{j=1}^{p} d_j(\sigma_j) + \sum_{i=1}^{m} k_i \sum_{j=1}^{p} \left(\frac{\partial \mathrm{FR}_i}{\partial \mathrm{DP}_j} \right)^2_{\mathrm{DP}_j=\mu_j} \sigma_j^2 \tag{9.39}$$

subject to

$$\sum_{j=1}^{p} \left(\frac{\partial \mathrm{FR}_i}{\partial \mathrm{DP}_j} \right)^2_{\mathrm{DP}_j=\mu_j} \sigma_j^2 \leq \left(\frac{\eta_{i,0}}{\eta_i} \right)^2 \sigma_{i,0}^2 \qquad \forall i = 1, \ldots, m \tag{9.40}$$

$$\eta_l \left(\frac{\partial \mathrm{FR}_i}{\partial \mathrm{DP}_j} \right)_{\mathrm{DP}_j=\mu_j} \sigma_l \geq \alpha_i \left(\frac{\eta_{i,0}}{\eta_i} \right) \sigma_{i,0} \qquad \forall i = 1, \ldots, m; \quad l \in W \tag{9.41}$$

Applying the Kuhn–Tucker condition, we have the following theorem (see Appendix 9A for the proof).

Theorem 9.1 Let **{FR}** $= \{\mathrm{FR}_1, \mathrm{FR}_2, \ldots, \mathrm{FR}_m\}$ be the array of FRs and **{T}** $= \{T_1, T_2, \ldots, T_m\}$ be the target vector. The array **{DP}** is a vector of random variables with adjustable mean

$$E\{\mathrm{DPs}\} = \{\mu_1, \mu_2, \ldots, \mu_p\} \quad \text{and} \quad \mathrm{Var}\{\mathrm{DPs}\} = \{\sigma_1^2, \sigma_2^2, \ldots, \sigma_p^2\}$$

Let $\eta_j\sigma_j$ be the half-tolerance width of the jth DP. If $\mathbf{E\{FR\}} = \mathbf{\{T\}}$, the optimal tolerance allocation that achieves independence and six-sigma capability should be defined in equations (9.42)–(9.45), where σ_l^2 is the variance of the diagonal DP, $l \in W$; $W = \{l/l = 1, \ldots, m\}$, and W is the set of indexes of the diagonal DPs. Tolerance allocation starts at the most mapped-to FR in the mapping matrix, starting with the diagonal element.

$$\sigma_j = \frac{|\partial d_j(\sigma_j)/\partial\sigma_j|}{2\sum_{i=1}^{m}(k_i + \xi_i)(\partial \mathrm{FR}_i/\partial \mathrm{DP}_j)^2_{\mathrm{DP}_j=\mu_j}} \qquad \forall j \notin W \tag{9.42}$$

$$\sigma_l = \frac{\alpha(\eta_{i,0}/\eta_i)\sigma_{i,0}}{\eta_l|\partial \mathrm{FR}_i/\partial \mathrm{DP}_l|_{\mathrm{DP}_l=\mu_l}} \qquad \forall i = 1, \ldots, m; \quad l \in W \tag{9.43}$$

$$\xi_i = \frac{1}{2}\left\{\frac{\displaystyle\sum_{i=1}^{m}\sum_{\substack{j=1\\ j\notin W}}^{p}\frac{(\partial \mathrm{FR}_i/\partial \mathrm{DP}_j)^2_{\mathrm{DP}_j=\mu_j}}{\left[\sum_{i=1}^{m}(\partial \mathrm{FR}_i/\partial \mathrm{DP}_j)^2_{\mathrm{DP}_j=\mu_j}\right]^2}\left[\frac{\partial d_j(\sigma_j)}{\partial(\sigma_j)}\right]^2}{\displaystyle\sum_{i=1}^{m}\left(\frac{\eta_{i,0}}{\eta_i}\right)^2\sigma_{i,0}^2\left[1-\left(\frac{\alpha_i(\eta_i/\eta_{i,0})}{\eta_l}\right)^2\right]}\right\}^{1/2} - 1 \tag{9.44}$$

$$\xi_{il} = \frac{|\partial d_l(\sigma_l)/\partial\sigma_l|}{\eta_l|\partial \mathrm{FR}_i/\partial \mathrm{DP}_l|_{\mathrm{DP}_l=\mu_l}} + \frac{2\alpha_i(\eta_{i,0}/\eta_i)^2 k_i\sigma_{i,0}(\partial \mathrm{FR}_i/\partial \mathrm{DP}_l)_{\mathrm{DP}_l=\mu_l}}{\eta_l^2} \qquad \forall i = 1, \ldots, m; \quad l \in W \tag{9.45}$$

where $p_i \leq \alpha_i < \eta_l/(\eta_{i,0}/\eta_i)$ and ξ_i and ξ_{il} are the Lagrange multipliers of constraints (9.40) and (9.41), respectively. The uniform tolerance allocation, p_i, limit on α is given by where Y_{ij} is a binary variable denoting the mapping as given by $Y_{ij} = 1$ if $A_{ij} \neq 0$ and 0 elsewhere.

It is clear that the solution here replicates that of the single-FR-case solution. For example, the sum of squared sensitivities in (9.42) parallels the squared sensitivity in (9.33). Comparisons conducted between matched equation pairs of the uni- and multi-FR solutions reveal the same pattern. In the multi-FR case, every FR needs to be checked against the taxonomy given in Table 9.5. When this is done, the design organization knows where improvements are lacking and how to sustain current strengths.

The purpose of the axiomatic quality tolerance optimization step is to assign tolerances to DPs based on overall tolerable variation in FRs, the relative influence of different sources of variation on the whole, and the cost/benefit trade-offs. In this step, the design team determines the allowable deviations in parameter values, reducing coupling, tightening tolerances, and upgrading materials only where necessary to meet functional requirement optimization targets if not met in the parametric optimization studies to produce the transfer functions. Where possible, tolerances may also be loosened based on the optimization results (if large tolerances were not used in the initial design). This step calls for thoughtful selection of tolerances and material upgrades. Selection is based on the economics of customer satisfaction, the cost of manufacturing and control, and the relative contribution to the whole of sources of functional variation. When this is done, the cost of the product is balanced with the quality of the product, within the context of satisfying customer demands.

By determining which tolerances have the greatest impact on the functional variation, only a few tolerances need to be tightened (at some cost), and often (if large tolerances were not used in the initial design) many can be relaxed (at a savings). The QLF is the basis for these decisions (Chapter 1). The process proposed also identifies key characteristics where functional criteria are met but where further variability reduction will result in corresponding customer benefits.

9.7 SUMMARY

In this chapter we developed several analytical operational vulnerability optimization techniques for the axiomatic quality process. The major contributions are nonlinear optimization formulations, with objective functions pieced from economic consideration of both the customer and the company. Constraints are derived to reduce both the coupling and variability vulnerabilities of the functional requirements to release the system at a six-sigma level of quality. Such formulations address conceptual vulnerabilities that were not resolved in the CDFC phase while achieving operation robustness.

The most important issue in the axiomatic quality process is to recognize that design and manufacturing must become allies in producing six-sigma uncoupled (or decoupled) products to ensure survival in the marketplace. This may sound trite, but the author's experience in practice indicates that this is a major problem in several companies. The axiomatic quality process enhances the poor communication between design and manufacturing by providing means to resolve design vulnerabilities. The optimization phase formulation serves as a link toward a common definition of tolerance settings that is usable by design and production. Its use by both parties will permit communicating the needs of each in terms of quality data currently available. It makes the assumption that it is plausible to the design teams and can be monitored in production.

APPENDIX 9A: PROOF OF THEOREM 9.1

From Kuhn–Tucker conditions of constrained nonlinear programming, we have the following equations:

$$\xi_i \left[\sum_{j=1}^{p} \left(\frac{\partial \mathrm{FR}_i}{\partial \mathrm{DP}_j} \right)^2_{\mathrm{DP}_j=\mu_j} \sigma_j^2 - \left(\frac{\eta_{i,0}}{\eta_i} \right)^2 \sigma_{i,0}^2 \right] = 0 \qquad \forall i = 1, \ldots, m \tag{9A.1}$$

$$\xi_{il} \left[\eta_l \left(\frac{\partial \mathrm{FR}_i}{\partial \mathrm{DP}_j} \right)_{\mathrm{DP}_j=\mu_j} \sigma_l - \alpha_i \left(\frac{\eta_{i,0}}{\eta_i} \right) \sigma_{i,0} \right] = 0 \qquad \forall i = 1, \ldots, m \tag{9A.2}$$

$$\frac{dd_j(\sigma_j)}{d\sigma_j}+2\sum_{i=1}^{m}k_i\left(\frac{\partial \mathrm{FR}_i}{\partial \mathrm{DP}_j}\right)^2_{\mathrm{DP}_j=\mu_j}\sigma_j+2\xi_i\left(\frac{\partial \mathrm{FR}_i}{\partial \mathrm{DP}_j}\right)^2_{\mathrm{DP}_j=\mu_j}\sigma_j=0$$

$$\forall j=1,\ldots,p,\quad j\notin W;\quad \forall i=1,\ldots,m \tag{9A.3}$$

$$\frac{dd_l(\sigma_l)}{d\sigma_l}+2k_i\left(\frac{\partial \mathrm{FR}_i}{\partial \mathrm{DP}_j}\right)^2_{\mathrm{DP}_j=\mu_j}\sigma_l+2\xi_i\left(\frac{\partial \mathrm{FR}_i}{\partial \mathrm{DP}_j}\right)^2_{\mathrm{DP}_j=\mu_j}\sigma_l$$

$$-\xi_{il}\eta_l\left(\frac{\partial \mathrm{FR}_i}{\partial \mathrm{DP}_j}\right)_{\mathrm{DP}_j=\mu_j}=0\qquad \forall i=1,\ldots,m;\quad l\in W \tag{9A.4}$$

where ξ_i and ξ_{il} are the Lagrange multipliers of constraints (9.40) and (9.41), respectively. Let (9.40) be the only binding constraint (i.e., $\xi_{il}>0$); then we have

$$\sigma_j=\frac{|dd_j(\sigma_j)/d\sigma_j|}{2\sum_{i=1}^{m}(k_i)(\partial \mathrm{FR}_i/\partial \mathrm{DP}_j)^2_{\mathrm{DP}_j=\mu_j}}\qquad \forall j\notin W \tag{9A.5}$$

$$\sigma_l=\frac{\alpha(\eta_{i,0}/\eta_i)\sigma_{i,0}}{\eta_l|\partial \mathrm{FR}_i/\partial \mathrm{DP}_l|_{\mathrm{DP}_l=\mu_l}}\qquad \forall i=1,\ldots,m;\quad l\in W \tag{9A.6}$$

Substituting (A.5) in (A.4), we get

$$\xi_{il}=\frac{|dd_l(\sigma_l)/d\sigma_l|}{\eta_l|\partial \mathrm{FR}_i/\partial \mathrm{DP}_l|_{\mathrm{DP}_l=\mu_l}}+\frac{2\alpha_i(\eta_{i,0}/\eta_i)^2k_i\sigma_{i,0}(\partial \mathrm{FR}_i/\partial \mathrm{DP}_l)_{\mathrm{DP}_l=\mu_l}}{\eta_l^2}$$

$$\forall i=1,\ldots,m;\quad l\in W \tag{9A.7}$$

Equation (A.1) can be written as

$$\left(\frac{\partial \mathrm{FR}_i}{\partial \mathrm{DP}_l}\right)^2_{\mathrm{DP}_l=\mu_l}\sigma_l^2+\sum_{\substack{j=1\\ j\notin W}}^{p}\left(\frac{\partial \mathrm{FR}_i}{\partial \mathrm{DP}_j}\right)^2_{\mathrm{DP}_j=\mu_j}\sigma_j^2\le\left(\frac{\eta_{i,0}}{\eta_i}\right)^2\sigma_{i,0}^2\qquad \forall i=1,\ldots,m \tag{9A.8}$$

Substituting (A.5) and (A.6), we get and (A.6)

$$\sum_{\substack{j=1\\ j\notin W}}^{p}\frac{(\partial \mathrm{FR}_i/\partial \mathrm{DP}_j)^2_{\mathrm{DP}_j=\mu_j}}{4\left[\sum_{i=1}^{m}(\partial \mathrm{FR}_i/\partial \mathrm{DP}_j)^2_{\mathrm{DP}_j=\mu_j}\right]^2}\left(\frac{dd_j(\sigma_j)}{d(\sigma_j)}\right)^2$$

$$\le\left(\frac{\eta_{i,0}}{\eta_i}\right)^2\sigma_{i,0}^2\left[1-\left(\frac{\alpha_j(\eta_{i,0}/\eta_i)}{\eta_l}\right)^2\right] \tag{9A.9}$$

with $\alpha_i<\eta_l/(\eta_{i,0}/\eta_i)$.

CHAPTER 10

CASE STUDY: LOW-PASS FILTER AXIOMATIC QUALITY PROCESS

10.1 INTRODUCTION

In previous chapters we described the axiomatic quality process in terms of three sequential phases: the customer attributes (CAs)-to-FRs mapping phase, the conceptual design for capability (CDFC) phase, and the optimization phase. The objective of the CDFC phase is to assure conceptual robustness by eliminating or reducing the coupling vulnerability. Execution of a CDFC project phase will give a design team the maximum design controllability and adjustability leverage. The objective of the optimization phase is to address operational vulnerability through several techniques to release the design at the six-sigma level in all of its requirements. In particular, this case study addresses operational vulnerability optimization using analytical design tolerances techniques presented in Section 9.6.

Both design and manufacturing engineers should be concerned with the magnitude of the tolerances specified on the design blueprints, as tolerances will affect, in addition to the optimum control cost, design manufacturability. For example, design engineers know that tolerance stacking in assemblies controls critical clearances and interfaces in the system designed. Process engineers know that tight tolerances increase the cost of manufacturing. Tolerances also greatly influence the selection of manufacturing processes by process planners and affect the manufacturability of the physical structure released. In the axiomatic quality process, tolerance specification is an important link between design engineering

Axiomatic Quality: Integrating Axiomatic Design with Six-Sigma, Reliability, and Quality Engineering, by Basem Said El-Haik
ISBN 0-471-68273-X

and processing. The optimization phase offers a ground on which to build a continuous interface between the two functions, to open a dialogue based on common interests and competing requirements.

In this chapter we demonstrate the axiomatic quality process through a case study.[1] Specifically, in this chapter we walk through the key steps of axiomatic quality discussed earlier.

10.2 PROBLEM STATEMENT

An electric passive filter network manufacturer learned about the axiomatic quality method and wanted to apply it to filter design to improve customer satisfaction. In the past the company experienced many quality problems in passive network functional performance. A passive filter is a simple instrument used to obtain a record of the displacement of a mechanical system. The signal of the filter is processed through a transducer and a demodulator. First, the displacement signal is passed through an ac excited transducer to produce an amplitude-modulated displacement signal (Figure 10.1*a*). The transducer output is passed through a full-wave, phase-sensitive demodulator, producing the output shown in Figure 10.1*b*. The filter suppresses the high-frequency carrier signal and passes only the desired low-frequency displacement signal (Figure 10.1*c*). The signal is to be recorded using a light-beam oscillograph equipped with a mechanically damped galvanometer. The displacement signal to be recorded has a spectral content in the range [0,2] hertz. The displacement transducers are excited at 60 Hz. The demodulated output of the transducer consists of the signal desired, the rectified carrier near 120 Hz, and higher harmonics of the carrier. The task is to manufacture a filter network to match the demodulated transducer output to the galvanometer in the oscillograph. The passive filter is formed from the following three DPs: capacitor C and resistors R_1 and R_2.

The filter must suppress the carrier frequency while passing an undistorted signal of interest (Figure 10.1*d*). As depicted in Figure 10.2, the network also attenuates the signal so that the deflection record is scaled properly. The baseline network settings currently used are $R_g = 98\ \Omega$, $R_s = 120\ \Omega$, $G = 0.00065758$ A/in., $V_s = 0.015$V, $C = 282\ \mu$F, $R_1 = 550\ \Omega$, and $R_2 = 415\ \Omega$.

The FRs are:

FR_1 = attenuate the signal to obtain dc gain with full-scale deflection ± 3 in.
FR_2 = suppress the carrier without distorting the displacement signal with a filter pole at 6.84 Hz.

We would like to design and manufacture the network to minimize the total quality loss incurred by the customer as well as the precision control cost incurred by the company and release the two functional requirements at six-sigma quality levels, subject to coupling constraints with $\alpha = 70\%$ by following the axiomatic quality process. For the precision control cost, Spotts's equation (9.37) will be used.

[1]This case study is constructed based on Example 9.2 in Suh (1990, p. 106).

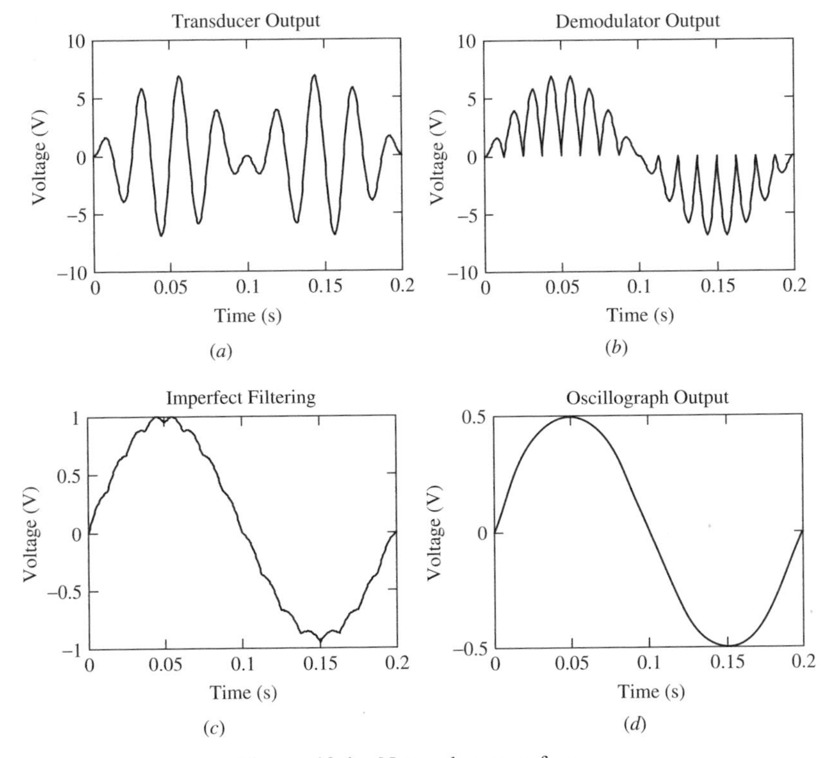

Figure 10.1 Network output forms.

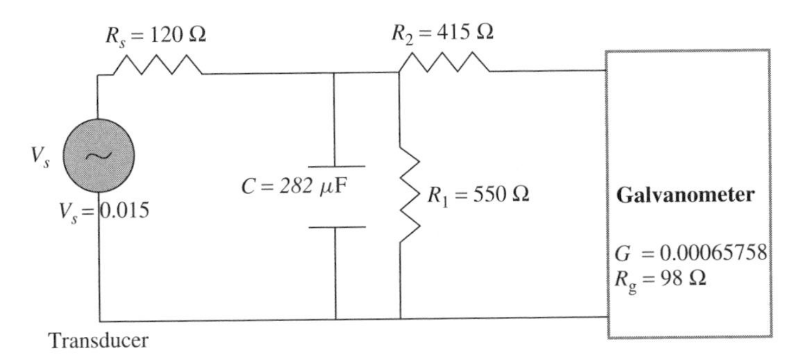

Figure 10.2 Low-pass filter circuit.

10.3 PASSIVE FILTER CONCEPTUAL DESIGN FOR THE CAPABILITY PHASE

Following the CDFC phase steps presented in Chapters 6 and 8, we have the FRs given by the following two mathematical transfer function formulas for the

FRs[2] by derivation:

$$\mathrm{FR}_1 = \frac{R_g R_1}{G[R_s R_1 + (R_s + R_1)(R_g + R_2)]} V_s \tag{10.1}$$

$$\mathrm{FR}_2 = \frac{1}{2\pi C}\left(\frac{1}{R_1} + \frac{1}{R_2 + R_g} + \frac{1}{R_s}\right) \tag{10.2}$$

Notice that the parameter design study is not required in this case because transfer functions can be derived. Substituting network current settings in equations (10.1) and (10.2), we have

$$\begin{aligned}\mathrm{FR}_1 &= \frac{(98)(0.015R_1)}{(0.00065758)120R_1 + (0.00065758)(120 + R_1)(98 + R_2)} \\ &= \frac{1.47R_1}{0.143353R_1 + (0.00065758R_1 + 0.07891)R_2 + 7.733141}\end{aligned} \tag{10.3}$$

$$\mathrm{FR}_2 = \frac{1}{2\pi C}\left(\frac{1}{R_1} + \frac{1}{R_2 + 98} + \frac{1}{120}\right) \tag{10.4}$$

In addition, the zigzagging step is not necessary here because the case study is at the LOI level of the design hierarchy, the component level. The design is redundant, since the design has two requirements ($m = 2$) and three design parameters ($p = 3$). The mapping equation $\{\mathrm{FRs}\}_{2\times1} = [\mathrm{A}]_{2\times3}\{\mathrm{DPs}\}_{3\times1}$ can be written as

$$\begin{Bmatrix}\mathrm{FR}_1\\ \mathrm{FR}_2\end{Bmatrix} = \begin{bmatrix} X & X & 0\\ X & X & X\end{bmatrix}\begin{Bmatrix}R_1\\ R_2\\ C\end{Bmatrix} \tag{10.5}$$

where X is a nonzero matrix element. The design matrix is coupled in both resistors but decoupled in every capacitor–resistor combination. No matrix reordering is necessary (Chapter 8), due to the small size of the matrix.

Decoupling Design Matrix We need to maintain the independence of the two functional requirements at the filter component level by uncoupling or decoupling the matrix (10.5). We would like to have an ideal design in which the number of FRs and DPs are equal. This means that we need to *fix* one DP. A design parameter fixing activity includes treating one DP as a constant (with minimal variation) in the axiomatic quality process by using techniques such as parameter design DOE with static nominal-the-best classification as well as tolerance design with upgraded material if necessary. Design parameter fixing might also include outsourcing to a capable high-quality supplier.

The fact that functional requirement FR_2 cannot be delivered without capacitor C reduces the fixing options to one of the two resistors, R_1 or R_2. Which one should we choose? The answer is the one that has the higher potential to achieve

[2]Equations (10.1) and (10.2) are from Suh (1990, Example 9.2).

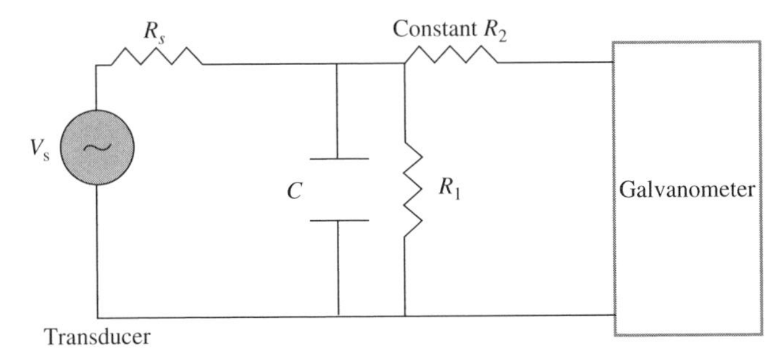

Figure 10.3 Passive filter network a.

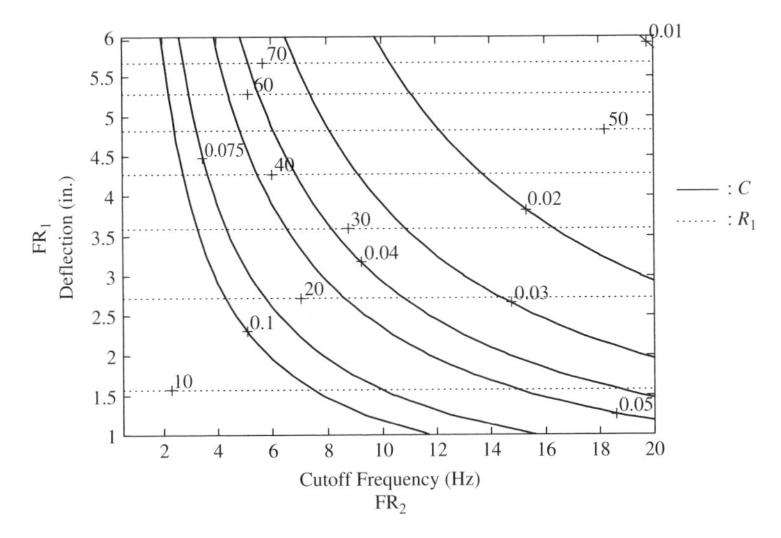

Figure 10.4 Isograms of network a.

a coupling-free design. A good answer can be obtained from (1) isogram plots or (2) an assessment of the reangularity and semangularity coupling measures. Figures 10.3 and 10.4 depict the network at constant R_2 (network a); Figures 10.5 and 10.6 depict the network at constant R_1 (network b). It is obvious that network b has better orthogonality among the DPs than network a, leading to a decision to fix resistance R_1. Based on this analysis, we fix R_1 at 550 Ω, the current setting, by very high quality resistance. The new design mapping is decoupled with the following design matrix:

$$\begin{Bmatrix} FR_1 \\ FR_2 \end{Bmatrix} = \begin{bmatrix} X & 0 \\ X & X \end{bmatrix} \begin{Bmatrix} DP_1 \\ DP_2 \end{Bmatrix} \tag{10.6}$$

where $DP_1 = R_2$ and $DP_2 = C$. In addition, we can conclude that DP_1 is the diagonal design parameter for FR_1 and DP_2 is the diagonal parameter for FR_2.

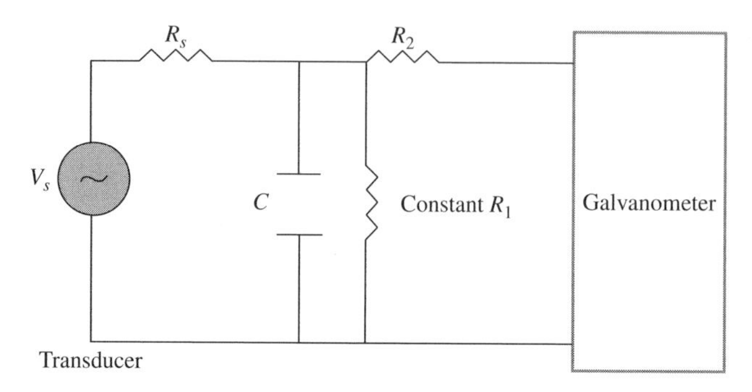

Figure 10.5 Passive filter network b.

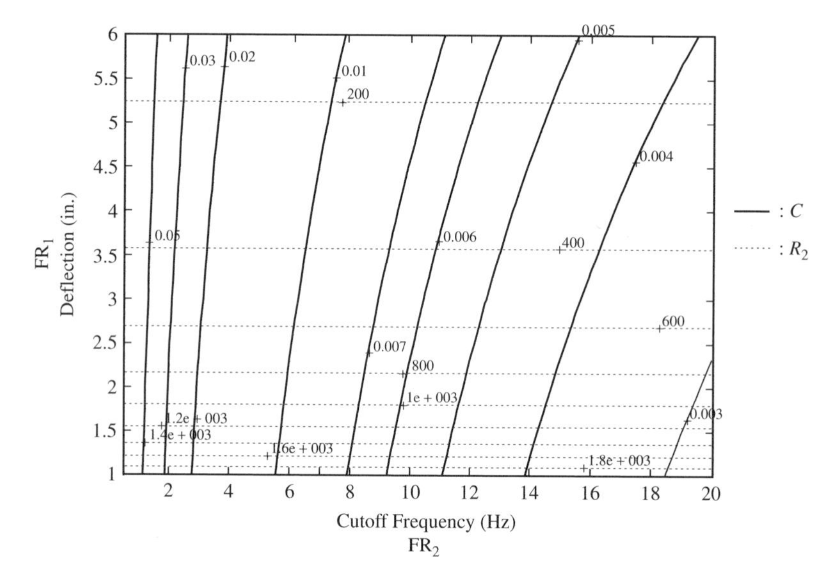

Figure 10.6 Isograms of network b.

Upon fixing R_1, the design is reduced to a decoupled design. Due to nonlinearity and the availability of transfer functions, the tolerance vulnerability optimization technique in Chapter 9 will be employed further in this case study, to reduce coupling.

10.4 PASSIVE FILTER TOLERANCE OPTIMIZATION PHASE

With respect to the optimization phase, we need the following prerequisites:

1. The ability to center the functional requirements $\{\mathrm{FR}_1, \mathrm{FR}_2, \ldots, \mathrm{FR}_m\}$ to customer targets $\{T_1, T_2, \ldots, T_m\}$ exists by correct selection of the nominal settings

of the design. The decoupling step in the CDFC phase results in the following DP nominal settings based on the decoupling sequence:

a. Substitute $R_1 = 550\ \Omega$ and $\text{FR}_1 = 3$ in. in (10.3) to obtain R_2 (or DP_1):

$$\text{DP}_1 = R_2 = \frac{1.47R_1/\text{FR}_1 - 0.14335244R_1 - 7.7331408}{0.00065758R_1 + 0.0789096}$$
$$= 415.1875\ \Omega \tag{10.7}$$

b. Substitute $R_1 = 550\ \Omega$, $R_2 = 415.1875\ \Omega$, and $\text{FR}_2 = 6.84$ Hz to obtain C (or DP_2):

$$\text{DP}_2 = C = \frac{1}{2\pi\text{FR}_2}\left(\frac{1}{R_1} + \frac{1}{R_2 + 98} + \frac{1}{120}\right)$$
$$= 0.000281\ \text{F} \tag{10.8}$$

These values of the DPs represent the targeted means (i.e., $\mu_{R_2} = 415.1875\ \Omega$ and $\mu_C = 0.000281$ F), as discussed in Chapter 9.

2. Using the respective transfer function equations in (10.3) and (10.4), we have the following derivatives (sensitivities):

$$\frac{\partial\text{FR}_1}{\partial R_2} = \frac{-1.47R_1(0.00065758R_1 + 0.0789096)}{[0.143353R_1 + (0.00065758R_1 + 0.0789096)R_2 + 7.733141]^2} \tag{10.9}$$
$$= -1.819821 \times 10^{-5}\ \text{in.}/\Omega$$

$$\frac{\partial\text{FR}_1}{\partial C} = 0 \tag{10.10}$$

$$\frac{\partial\text{FR}_2}{\partial R_2} = \frac{-1}{2\pi C(R_2 + 98)^2} = -0.0021457\ \text{Hz}/\Omega \tag{10.11}$$

$$\frac{\partial\text{FR}_2}{\partial C} = \left(\frac{-1}{2\pi C^2}\right)\left(\frac{1}{R_1} + \frac{1}{R_2 + 98} + \frac{1}{120}\right) = -23,467.606\ \text{Hz/F} \tag{10.12}$$

3. In the CDFC phase we concluded that DP_1 (resistance R_2) is the diagonal design parameter for FR_1 (the deflection), and DP_2 (capacitor C) is the diagonal parameter for FR_2 (6.84-Hz filter pole).

4. To assess the current mean and variance of the design, we will assume 10% of the nominal settings as the current design parameter half-tolerance width and equate this range to three-sigma specification limits for the DPs (i.e., $\eta_{R_2,0} = 3$ and $\eta_{C,0} = 3$), which results in $\sigma_{R_2} = 13.83958193\ \Omega$ and $\sigma_C = 9.38119 \times 10^{-6}$ F. Since the passive filter is an incremental design (i.e., we are seeking improvement of the current baseline), we can use the DP mean and variance to estimate the FR mean and variance using Taylor series expansion. These settings produce the following averages and variances of the FRs: $\mu_{\text{FR}_1} \cong 3.00$ in., $\mu_{\text{FR}_2} \cong 6.84$ Hz, $\sigma_{\text{FR}_1,0} \cong 0.045161799$ in., and $\sigma_{\text{FR}_2,0} \cong 0.22497823$ Hz.

TABLE 10.1 Capabilities Before Axiomatic Quality

		Specification			FR Capability Predicted				6σ Score
FR	Units	Target	LSL	USL	μ	σ	η	Shift	DPMO
FR_1	in.	3	2.999	3.001	3	0.05	0.02	1.50	1,017,605.3
FR_2	Hz	6.84	6.76	6.92	6.8372	0.22	0.53	1.50	1,406,232.9

5. To calculate the total quality loss of the network at the current settings, we need to make assumption about the relative costs incurred by the customer as being off target in all the FRs. We will assume that the average quality loss incurred by being off-target by 1 Hz is the same as that incurred by being off-target by 0.01 in., then $k_1 = \$100/\text{Hz}^2$ and $k_2 = \$0.01/\text{Hz}^2$.

$$
\begin{aligned}
L &\cong \sum_{i=1}^{2} k_i \sum_{j=1}^{2} \left(\frac{\partial \text{FR}_i}{\partial \text{DP}_j} \right)_{\text{DP}_j=\mu_j} \sigma_j^2 \\
&\cong 100 \frac{-1.47(550)[0.00065758(550) + 0.0789096]}{\{(0.143353)(550) + [(0.00065758)(550) + 0.0789096]R_2 + 7.733141\}^2} \sigma_{R_2}^2 + 0 \\
&\quad \cdot \frac{0.01}{2\pi C^2} \left[\frac{1}{(R_2+98)^2} \sigma_{R_2}^2 + \left(\frac{1}{R_1} + \frac{1}{R_2+98} + \frac{1}{120} \right)^2 \sigma_C^2 \right] \\
&\cong k_1 \sigma_{\text{FR}_1}^2 + k_2 \sigma_{\text{FR}_2}^2 \\
&\cong \$0.204464965/\text{Hz}^2
\end{aligned}
\tag{10.13}
$$

The current settings produce the FR capabilities shown in Table 10.1. Note that the sigma capability of both FRs as calculated using equation (1.1) is very problematic, and the specifications on FR_1 are very stringent.

With respect to precision control cost $d_j(\sigma_j)$, $j = 1, 2$, of the design parameters, we will assume that the resistance R_2 production technology can be upgraded to have a precise polycrystalline material deposition[3] device to control variation. Equivalently, we will assume that the capacitor C production technology can be upgraded to have a precise aluminum oxide[4] (Al_2O_3) deposition device to control variation within a wide range. In addition, we assume the precision control costs $d_j(\sigma_j)$ of both devices to be proportional to the reciprocal of the variance (Spotts, 1973), as depicted in equation (9.37). We estimate that the resistor deposition machine costs \$0.5 k (fixed cost) and \$0.005 k (variable cost) for 1 unit of resistance-controlled variance. It is estimated that the deposition cost

[3]The material used to gauge the resistance setting.

[4]The material used in the capacitor.

of the capacitor-deposition machine costs \$1.00 k (fixed cost) and \$0.007 k per 10^{-6} F (variable cost) for 1 unit of capacitor-controlled variance. That is, the precision control costs (in thousands of dollars) is given by

$$d_1(\sigma_{R_2}) = 0.5 + \frac{0.005}{\sigma_{R_2}^2} \tag{10.14}$$

$$d_2(\sigma_C) = 1 + \frac{0.007}{\sigma_C^2} \tag{10.15}$$

6. We assume the existence of a quality control function in the passive filter manufacturing company to control and monitor the design parameters continuously at the optimum solution level.

Theorem 9.1 tolerance optimization steps start from the most mapped-to requirement in the design matrix, beginning with the diagonal element of that requirement. In this example, the most mapped requirement is FR_2, the 6.84-Hz pole frequency, and its diagonal element is DP_2, the capacitor C.

The problem is solved by substituting the arguments presented above in the optimization program solution in the system of equations of Theorem 9.1. The capacitor control cost using Spotts's equation is

$$d(\sigma_C) = 1 + \frac{0.007}{0.204068^2} = \$1.17\ \text{k} \tag{10.16}$$

Substituting in equation (9.42), we get

$$\begin{aligned}\sigma_{R_2} &= \frac{(0.7)\,(0.53/6)\,(0.22497823)}{(3)(23{,}469.962)} \\ &= 2.04068 \times 10^{-7}\ \text{F}\end{aligned} \tag{10.17}$$

This controlled standard value represents a 97.82% improvement over the old capacitor standard deviation.

The variance of the resistance R_2, the off-diagonal element with respect to FR_2, can be found using equation (9.43):

$$\begin{aligned}\sigma_{R_2} &= \frac{\left|\partial d(\sigma_{R_2})/\partial \sigma_{R_2}\right|}{6.627 \times 10^{-6}} \\ &= \frac{c_{R_2}/\sigma_{R_2}^2}{6.627 \times 10^{-6}}\end{aligned} \tag{10.18}$$

It is estimated that a computerized electric resistance-producing machine costs $d(\sigma_{R_2}) = 0.5 + (0.005/\sigma_{R_2}^2)$\$k using Spotts's equation. Therefore,

$$\sigma_{R_2} = \left(\frac{0.005}{6.627 \times 10^{-6}}\right)^{1/3} = 9.10\ \Omega \tag{10.19}$$

TABLE 10.2 Capabilities After Use of Axiomatic Quality

		Specification			FR Capability predicted				6σ Score
FR	Units	Target	LSL	USL	μ	σ	η	Shift	DPMO
FR_1	Ω	3	2.999	3.001	3	0.000139785	6.02	1.50	0
FR_2	F	6.84	6.76	6.92	6.8372	0.017160657	5.97	1.50	0

This controlled standard value represents a 34.25% improvement over the old R_2 standard deviation. The control cost is given by

$$d(\sigma_{R_2}) = 0.5 + \frac{0.005}{0.0091^2} = \$0.56\text{ k} \tag{10.20}$$

The new capabilities are given in Table 10.2.

The total cost of control for both DPs equals \$1730. The controlled variance settings result in $\sigma_{FR_1} \cong 0.000165604$ in. and $\sigma_{FR_2} \cong 0.020111999$ Hz, with $L = \$6.78738 \times 10^{-6}/\text{Hz}^2$. The improvement in the total loss function equals $\$0.2044582/\text{Hz}^2$. The contribution of each FR to the loss function improvement of the FRs can be calculated using equation (9.38), which shows that a 99.76% improvement is coming from FR_1, the scale deflection requirement. This FR was the most problematic FR.

To assess the improvements in the robustness measures based on the sigma levels, we need to focus our attention on the degradation effect of requirement quality loss improvement over time. The assumption discussed in Section 9.4 is still valid.

$$\begin{aligned}
\Delta\text{QLF}_{\text{FR}_1} &= \text{QLF}_{1,\text{new}} - \text{QLF}_{1,\text{baseline}} \\
&= k_1\left[\left(\frac{\eta_{1,0}}{\eta_1}\right)^2 - 1\right]\sigma_{1,0}^2 \\
&= (100)\left[\left(\frac{0.02}{6.02}\right)^2 - 1\right](0.045161799^2) \\
&= -\$0.20396/\text{Hz}^2
\end{aligned} \tag{10.21}$$

$$\begin{aligned}
\Delta\text{QLF}_{\text{FR}_2} &= \text{QLF}_{2,\text{new}} - \text{QLF}_{2,\text{Baseline}} \\
&= k_2\left[\left(\frac{\eta_{2,0}}{\eta_2}\right)^2 - 1\right]\sigma_{2,0}^2 \\
&= (0.01)\left[\left(\frac{0.53}{5.97}\right)^2 - 1\right](0.22497823^2) \\
&= -\$5.02163 \times 10^{-4}/\text{Hz}^2
\end{aligned} \tag{10.22}$$

Calculation of signal-to-noise factor improvement is left as an exercise.

CHAPTER 11

AXIOMATIC RELIABILITY

11.1 INTRODUCTION

In the reliability literature, much research has been carried out on the functional design aspect of products from a physical composition perspective. Unfortunately, the design theory relation to reliability engineering has not advanced much beyond the state of reliability testing. Contemporary reliability engineering focuses on evaluation and assessment of design reliability based on analysis parametric techniques derived from actual test data from models or actual field data. Such analysis techniques have numerous associated testing costs, are after the fact, and are inconsistent with the axiomatic quality philosophy of bringing quality improvements upstream. Further, classical reliability evaluation of systems is based on several assumptions that do not always hold true: for example, the assumption that the modules comprising the physical structure work independently and that their inherent or designed failure time distributions have no mutual interdependence. Component failures are often assumed to be statistically mutually independent events. In reality, this assumption does not hold true in many product designs. There is a need for a more generalized reliability design theory that (1) strengthens the link to design theory with an appreciation of assumption validity (the design theory used in this book is axiomatic design), and (2) enables design teams to make a leading upstream reliability assessment prior to testing using available development of the axiomatic quality process.

Axiomatic Quality: Integrating Axiomatic Design with Six-Sigma, Reliability, and Quality Engineering, by Basem Said El-Haik
ISBN 0-471-68273-X

System reliability is defined as the probability that a system will perform its intended function for a specified interval of time under the conditions stated. Several reliability methods focus on the evaluation and assessment of the reliability of modules (components, subsystems, systems, and products). Even the definition and the measure of reliability may be quite different from the customer's perception. Indeed, most reliability science is based on probability-oriented analysis techniques. These techniques are used to analyze actual test data or field data in order to evaluate the performance of products. This approach is characterized by build–test–evaluate–fix cycles. Clearly, these methods increase the product development cycle time and life-cycle cost significantly, eventually undermining a company's competitive edge. Hence, improving reliability in the concept and product design phase is an imperative of the axiomatic quality and reliability process.

A study in the automotive industry (Palady, 1995) showed that in some situations, component failure caused only 15% of system failures. The remaining 85% of the causes of system failure are due to subsystem interactions, poor subsystem interface, improper part installation, and others. These failures, which are caused by coupling or interdependencies contributing to the severity of the damage, add to the cost of repair and make failures difficult to detect.

There is a need to develop a reliability design tool that will guide design teams to assess reliability in conceptual entities. This need hinges on the requirement to extend traditional reliability design methods to a conceptual reliability design method, which we call *axiomatic reliability*. This shift in the traditional reliability design paradigm should focus on improving reliability at the concept design stage prior to testing. Axiomatic reliability utilizes axiomatic quality concepts to analyze and assess design reliability as a number of related, possibly coupled entities with the goal of studying how these entities interact to influence the reliability of the overall system upstream of the development cycle. However, to move reliability activities upstream within the paradigm of operational vulnerability, there are several deficiencies in current reliability methods, including how to build reliability into the design process and how to be proactive to customer attributes from a reliability perspective. These are some of the issues dealt with in this chapter. The primary focus will be to enable reliability assessment at the axiomatic quality optimization phase of design modules and to develop relevant axiomatically driven measures to evaluate the inherent reliability in competing alternatives. Specifically, axiomatic reliability has the following objectives:

- To explore conceptual interfaces between design theory and reliability engineering
- To derive possible mathematical links that enable design teams to have firsthand assessment of design reliability in early developmental stages (Figure 1.7) and prior to testing
- To identify physical structure links to FR reliability and to analyze structure component reliability

11.2 WHY AXIOMATIC RELIABILITY?

In recent years a number of researchers have ventured into exploring new avenues for the advancement of reliability engineering. Some notable discussions and suggestions emanating from these activities include the reliability modeling effort, which addresses failure particulars that relate to the physics of failure, such as failure time, failure environment, failure mechanism, and failure mode (Pecht et al., 1994). Reliability prediction based on failure randomness needs to be changed since field failures are not generally random events. Cook (1990) promoted the idea that reliability should be appraised based on documented control of input parameters of manufacturing processes that affect the reliability of the design. Reliability models should be related to design parameters so that designers can trade off between the cost and the reliability of products (Fragole, 1993).

Reliability in design should be driven by the customer's perspective (Brunelle and Kapur, 1997; Yang and Kapur, 1997) and its techniques should be tied into efforts to deploy the customer's voice during the concept design stage. That is, deployment of the customer's voice into reliability design should be undertaken at an early design stage and therefore tie into design theory.

Higher losses due to poor reliability are faced when reliability activities are farther from design activities (i.e., when reliability is deployed farther downstream), since more effort and a higher cost are required to fix the problem. For example, reliability problems encountered in the field could lead to product recalls, which is more expensive than modifying and improving a design in the concept design stage. Hence, it is imperative to move reliability activities upstream, preferably to the optimization phase of the axiomatic quality process.

The development of concurrent design team practice has further compounded the problem of shifting reliability activities upstream. Fragole (1993) has commented: "If reliability technology is to survive in the concurrent engineering era, the technical tools used must be modified and the analytical thrust redirected so as to focus on the identification and characterization of the uncertainties involved in the developing design. Without this redirection, post hoc reliability analysis becomes largely irrelevant. If, however, this transition is successfully accomplished, reliability technology will not only be useful, but will also satisfy key design development needs."

The reliability theory evaluation of systems is based on the assumption that the components comprising a system work independently, whereas their inherent or designed failure-time distributions have no mutual interdependence (Bhattacharya, 1996). This assumption may be appropriate in situations where the operating modules are functionally uncoupled. Yet, in reality, the trade-offs between performance (functionality) and reliability are often subtle and increase design complexity (Lewis, 1987). Reliability is often improved at the expense of increased complexity (complexity is classically measured by the number of subsystems/components that comprise a system or product).

11.3 AXIOMATIC RELIABILITY IN THE DEVELOPMENT CYCLE

Concurrent engineering is presently the most popular method for designing services, products, and processes. The axiomatic quality process is no exception, whether deployed within a design for six-sigma (DFSS) program or stand-alone. This design philosophy is centered on the simultaneous design of a product and its manufacturing system. Teng and Ho (1995) emphasize that a successful product design must reach certain specified goals, such as minimum product life-cycle cost, good product quality, cost-sensitive manufacturing processes, and satisfactory product functionality delivered to customers. Garrett (1990) has discussed eight steps to simultaneous engineering, of which product quality and price are the two most important indexes that a company must have in order to compete. Reactive reliability techniques for screening defective products in a manufacturing process (acceptance testing: increased scrap and reworking costs) and correcting the defects through the use of repair/service stations and warranty coverage (increased field/postmanufacturing costs) are costly exercises, and they seriously undermine a company's competitive edge. Hence, improving reliability in the concept and product design phase is imperative, and this requires a more realistic reliability design philosophy than traditional philosophy based on the test–fix–test approach.

With the advent of OFD in the late 1980s, product design has undergone a shift in paradigm toward a customer-centered design approach. This shift in design paradigm has resulted in product designs getting more complex as design engineers struggle to satisfy all customer needs (or as many technically and economically feasible needs). Hence, design complexity is on the increase, resulting in reduced reliability due to dependent failures. Today, it is imperative for a design team to have a knowledge of the conceptual vulnerabilities (coupling) that allows single failures to combine and create a dependent failure. "Dependent failures are characteristic of most modern engineered systems, including structures, power distribution networks, electrical and electronic circuits, and complex process systems such as nuclear power plants. Failure to properly account for dependence among component failures can lead to significant overestimation of system reliability" (Greig, 1993).

Robert Lusser is one of the pioneers of reliability. He stated that the strength of a chain consisting of n rings is weaker than the strength of its weakest ring. Then he came to the following reliability predictive model for a series system consisting of n modules: $R_{\text{system}} = \prod_{k=1}^{n} R_k$. This predictive formula, called *Lusser's law*, serves as the basis for traditional reliability theory and techniques. Implicit in Lusser's law is the assumption of random failure. If a module cannot perform any of its functions, a failure has occurred. Generally, there are two types of failures, catastrophic (hard) and performance degradation (soft). *Performance degradation* means that the system FRs will degrade continuously over time, affecting customer satisfaction, but will not cause the product to cease to function. There are many discussions about specific forms of catastrophic failure and degradation, which can be categorized in the following four major conceptual models for failure (Yang and Xue, 1996):

1. *Stress–strength*. The system fails if and only if the stress (load) exceeds the strength. This model depends on critical events.
2. *Damage–endurance*. Stress causes damage that accumulates irreversibly; the item fails when and only when the damage exceeds the endurance. Accumulated damage does not disappear when the stresses are removed. This model depends on time or cycles.
3. *Challenge–response*. A design entity is bad, but only when the entity is challenged (needed) does it fail to response, reveal itself as bad, and cause the system to fail. The model depends on critical events.
4. *Tolerance–requirement*. A module's FRs are satisfactory if and only if its performance remains within the tolerances desired. The tolerance–requirement model can represent any kind of gradual degradation of an FR in terms of its mapped-to DPs.

As a natural extension of the axiomatic quality optimization phase, axiomatic reliability demands use of the category 4 tolerance-based approach.

In this chapter the objective is to explore, establish, and derive the mathematical relationship between axiomatic design and reliability theory, hence the term axiomatic reliability. The derivation presented here allows an assessment of design reliability in the initial developmental stages as reliability is addressed from the perspective of design axioms. In our development of axiomatic reliability (Figure 11.1), we explore two tracks: (1) functional reliability to assess design reliability (Section 11.4) using available transfer functions of a given mapping, and (2) physical structure importance assessment track (Section 11.7), in lieu of reliability, when transfer functions and structural component failure probability are difficult to obtain.

11.4 AXIOMATIC RELIABILITY IN THE DESIGN STAGES

Considering the four basic reliability models outlined in Section 11.3, models 1 to 3 usually represent catastrophic failures and model 4 represents performance degradation beginning at time 0, representing commencement of prototype testing or product launch. The premise is that treatment of soft failure will delay or may prevent catastrophic failure. The objective of non-time-dependent axiomatic reliability techniques is to assess and improve the design reliability at $t = 0$ by adopting the tolerance-requirement model of failure (category 4 in Section 11.3). The understanding is that FR degradation is being attributed to degradation in the DPs, which in turn can be attributed to (PV) degradation in the fashion depicted in Figure 11.2. From the axiomatic quality perspective and in the context of Figure 11.2, the variation at $t = 0$ should be reduced with optimized FR mean settings, as discussed in relation to the optimization phase (Chapter 9). The optimization phase effectiveness can be assessed and improved by axiomatic reliability techniques by identifying and then improving the reliability-sensitive design modules. Axiomatic reliability provides techniques to evaluate design

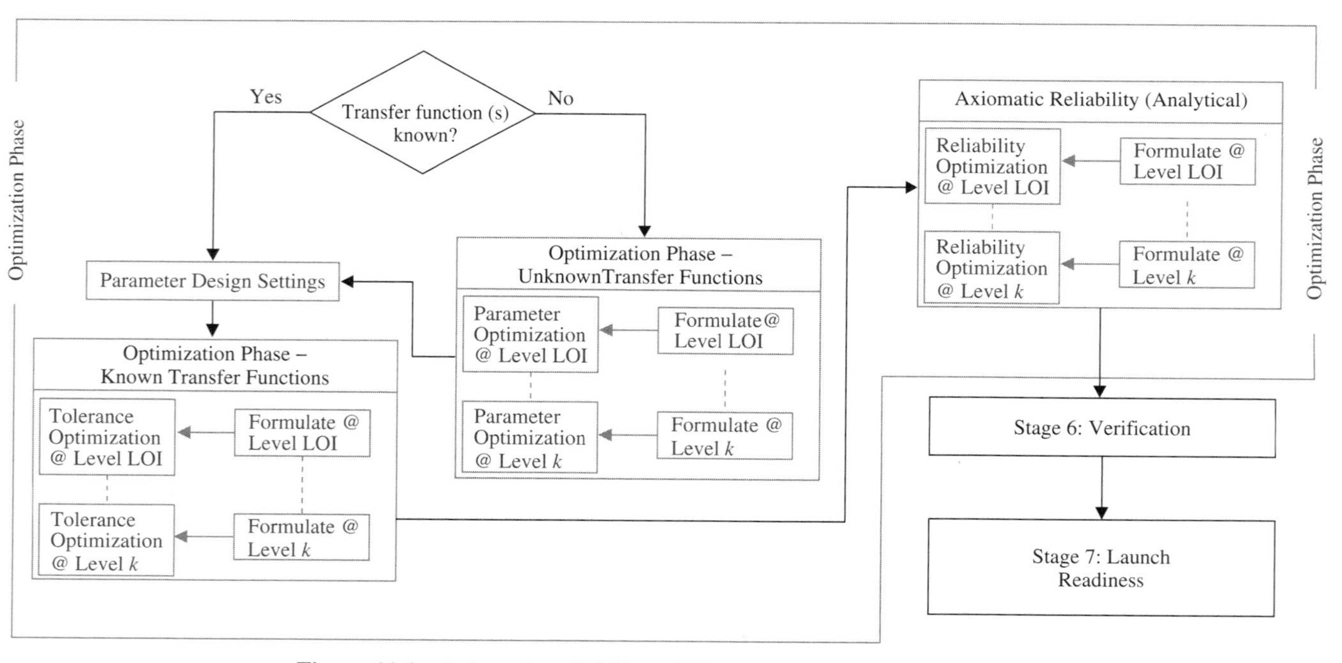

Figure 11.1 Axiomatic reliability within the axiomatic quality process.

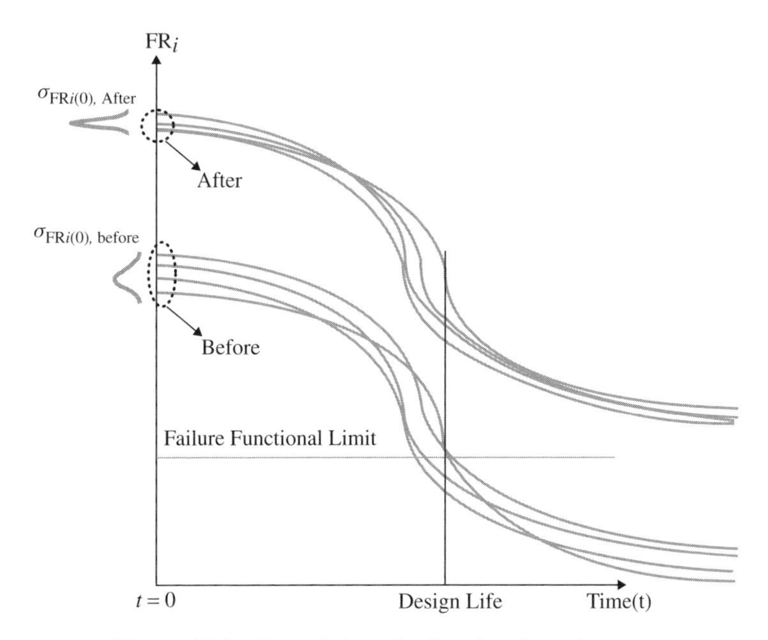

Figure 11.2 Degradation of a functional requirement.

reliability one mapping (or module) at a time and one hierarchical level at a time prior to testing based on baseline or surrogate distributions. In doing so, it provides the design team with insights to where reliability improvements (by operational vulnerability means similar to what is suggested in Section 11.8) need to be pursued and where resources need to be allocated.

The acceptable functional limits are defined by the customer for each requirement using the format $T_i \pm \Delta\text{FR}_i, i = 1, \ldots, m$ (Figure 11.3). We will assume that distributions of the DPs and PVs in the mapping of interest accounts for, in

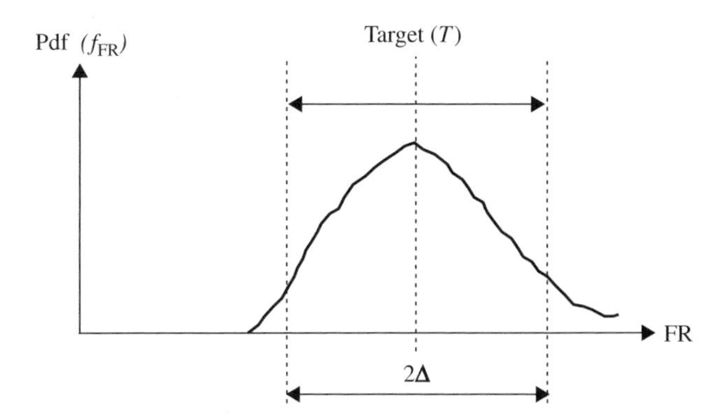

Figure 11.3 Hypothetical functional limits.

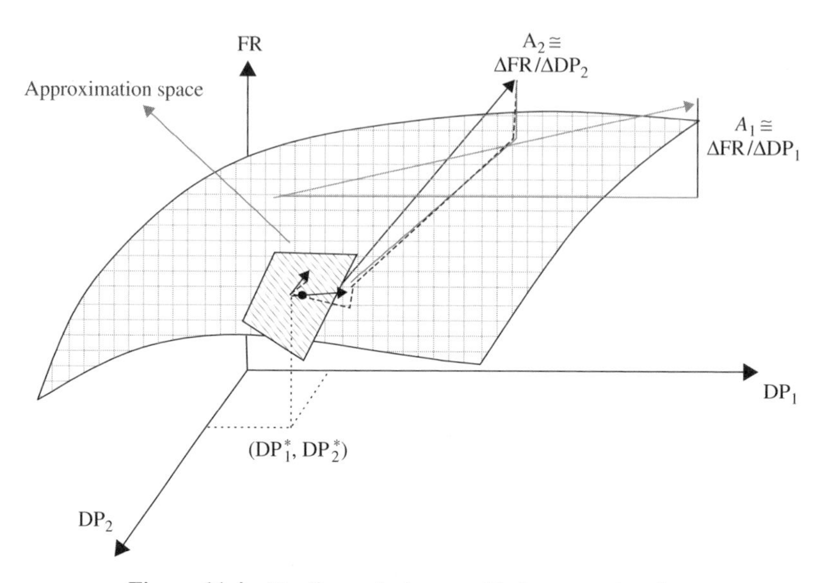

Figure 11.4 Nonlinear design sensitivity approximation.

addition to manufacturing variability, the effect of other noise factors, similar to the formulation discussed in Section 9.4. We also assume that the ideal design case (Theorem 2.4) holds (i.e., $m = p$ in the physical and $p = n$ in the process mappings, or at least in the overall matrix [**C**] with $m = n$).

Most of the derivation presented here is simplified by linearity assumption (i.e., the entries of the design matrices are constant or can be approximated by constants in nonlinear design mappings with some average representation in design subspaces of interest) (Figure 11.4). A design space is a set of points in a design parameter or process variable space that lies within the tolerance space of the FRs simultaneously. In an m-dimensional space of real numbers, a tolerance-constrained design space forms a convex polyhedron.[1] If the design is approximately linear within the system range at or below the LOI, (3.6) can be modeled as a linear set of transfer function equations in the mapping of interest. By adopting category 4 and employing (3.6), design reliability can assume any one of the definitions in (11.1)–(11.4). The axiomatic reliability at this point is in effect a function of the volume (probability) of a constrained space, the polyhedron pieced from design tolerance ranges. The design is reliable as long as its DPs stay within the design ranges defined. A violation of such constraint (tolerance limits) induced failure in one or more (depending on the coupling situation) of its functional requirements. This violation may happen at design launch ($t = 0$) or later, due to degradation (at $t > 0$) because of the noise factor

[1]In algebraic topology, a "polyhedron" is defined as a space that can be formed from elements such as line segments, triangles, tetrahedral, and their higher dimensional analogs by piecing them together along their faces.

effect. In this section we handle non-time-dependent axiomatic reliability. Time-dependent axiomatic reliability is beyond the scope of this chapter.

$$R_{\mathrm{FRs}} = \mathrm{Prob}\left(\begin{Bmatrix} T_1 - \Delta\mathrm{FR}_1 \\ T_2 - \Delta\mathrm{FR}_2 \\ \vdots \\ T_m - \Delta\mathrm{FR}_m \end{Bmatrix} \le \begin{bmatrix} A_{11} & A_{12} & \cdots & A_{1p} \\ A_{21} & A_{22} & \cdots & \cdot \\ \cdot & \cdot & \cdots & \cdot \\ \cdot & \cdot & \cdots & \cdot \\ A_{m1} & \cdot & \cdots & A_{mp} \end{bmatrix} \times \begin{Bmatrix} \mathrm{DP}_1 \\ \mathrm{DP}_2 \\ \vdots \\ \mathrm{DP}_p \end{Bmatrix} \le \begin{Bmatrix} T_1 + \Delta\mathrm{FR}_1 \\ T_2 + \Delta\mathrm{FR}_2 \\ \vdots \\ T_m + \Delta\mathrm{FR}_m \end{Bmatrix}\right) \tag{11.1}$$

$$R_{\mathrm{FRs}} = \mathrm{Prob}\left(\begin{matrix} \begin{Bmatrix} T_1 - \Delta\mathrm{FR}_1 \\ T_2 - \Delta\mathrm{FR}_2 \\ \vdots \\ T_m - \Delta\mathrm{FR}_m \end{Bmatrix} \\ \le \begin{bmatrix} A_{11} & A_{12} & \cdots & A_{1p} \\ A_{21} & A_{22} & \cdots & \cdot \\ \cdot & \cdot & \cdots & \cdot \\ \cdot & \cdot & \cdots & \cdot \\ A_{m1} & \cdot & \cdots & A_{mp} \end{bmatrix} \begin{bmatrix} B_{11} & B_{12} & \cdots & B_{1n} \\ B_{21} & B_{22} & \cdots & \cdot \\ \cdot & \cdot & \cdots & \cdot \\ \cdot & \cdot & \cdots & \cdot \\ B_{p1} & \cdot & \cdots & B_{pn} \end{bmatrix} \begin{Bmatrix} \mathrm{PV}_1 \\ \mathrm{PV}_2 \\ \vdots \\ \mathrm{PV}_n \end{Bmatrix} \\ \le \begin{Bmatrix} T_1 + \Delta\mathrm{FR}_1 \\ T_2 + \Delta\mathrm{FR}_2 \\ \vdots \\ T_m + \Delta\mathrm{FR}_m \end{Bmatrix} \end{matrix}\right) \tag{11.2}$$

$$= \mathrm{Prob}\left(\begin{matrix} \begin{bmatrix} C_{11} & C_{12} & \cdots & C_{1n} \\ C_{21} & C_{22} & \cdots & \cdot \\ \cdot & \cdot & \cdots & \cdot \\ \cdot & \cdot & \cdots & \cdot \\ C_{m1} & \cdot & \cdots & C_{mn} \end{bmatrix}^{-1} \begin{Bmatrix} T_1 - \Delta\mathrm{FR}_1 \\ T_2 - \Delta\mathrm{FR}_2 \\ \vdots \\ T_m - \Delta\mathrm{FR}_m \end{Bmatrix} \\ \le \begin{Bmatrix} \mathrm{PV}_1 \\ \mathrm{PV}_2 \\ \vdots \\ \mathrm{PV}_n \end{Bmatrix} \\ \le \begin{bmatrix} C_{11} & C_{12} & \cdots & C_{1n} \\ C_{21} & C_{22} & \cdots & \cdot \\ \cdot & \cdot & \cdots & \cdot \\ \cdot & \cdot & \cdots & \cdot \\ C_{m1} & \cdot & \cdots & C_{mn} \end{bmatrix}^{-1} \begin{Bmatrix} T_1 + \Delta\mathrm{FR}_1 \\ T_2 + \Delta\mathrm{FR}_2 \\ \vdots \\ T_m + \Delta\mathrm{FR}_m \end{Bmatrix} \end{matrix}\right) \tag{11.3}$$

$$= \text{Prob}\left(\frac{1}{|[C]|}\text{Adj}([C])\begin{Bmatrix} T_1 - \Delta\text{FR}_1 \\ T_2 - \Delta\text{FR}_2 \\ \vdots \\ T_m - \Delta\text{FR}_m \end{Bmatrix} \leq \begin{Bmatrix} \text{PV}_1 \\ \text{PV}_2 \\ \vdots \\ \text{PV}_n \end{Bmatrix} \leq \frac{1}{|[C]|}\text{Adj}([C])\begin{Bmatrix} T_1 + \Delta\text{FR}_1 \\ T_2 + \Delta\text{FR}_2 \\ \vdots \\ T_m + \Delta\text{FR}_m \end{Bmatrix}\right) \tag{11.4}$$

where Adj($[C]$) is the adjoined matrix $\mathbf{C}$. Let the PVs be continuous random variables and $\mathbf{C}$ be square; then $\{\mathbf{FR}\}$ is a vector of functions of n jointly distributed continuous random variables $\text{PV}_1, \text{PV}_2, \ldots, \text{PV}_n$, which can be expressed as $\{\mathbf{FR}\} = [\mathbf{C}]\{\mathbf{PV}\}$, or in transfer function format as[2]

$$\begin{aligned} \text{FR}_1 &= q_1(\text{PV}_1, \text{PV}_2, \ldots, \text{PV}_n) \\ \text{FR}_2 &= q_2(\text{PV}_1, \text{PV}_2, \ldots, \text{PV}_n) \\ &\vdots \\ \text{FR}_n &= q_n(\text{PV}_1, \text{PV}_2, \ldots, \text{PV}_n) \end{aligned} \tag{11.5}$$

We are interested in finding the distributions of the FRs when the DP or PV distributions are given. There are two requirements that transfer functions $\{q_1, q_2, \ldots, q_n\}$ need to satisfy here: First, they have to be continuous first-order partial derivatives at all points in the design and process domains, and second, the Jacobian (the sensitivity matrix determinant), $|[C']|$, is nonzero. The latter condition can be expressed mathematically as

$$|[C']| = \begin{vmatrix} \frac{\partial q_1}{\partial PV_1} & \frac{\partial q_1}{\partial PV_2} & \cdots & \frac{\partial q_1}{\partial PV_n} \\ \frac{\partial q_2}{\partial PV_1} & \frac{\partial q_2}{\partial PV_2} & \cdots & \cdot \\ \cdot & \cdot & \cdots & \cdot \\ \cdot & \cdot & \cdots & \cdot \\ \cdot & \cdot & \cdots & \cdot \\ \frac{\partial q_n}{\partial PV_1} & \frac{\partial q_n}{\partial PV_2} & \cdots & \frac{\partial q_n}{\partial PV_n} \end{vmatrix} \neq 0 \tag{11.6}$$

at all points $(\text{pv}_1, \text{pv}_2, \ldots, \text{pv}_n)$ in the design space defined by the tolerance ranges of the PVs.

[2]For $n \geq 4$ it may be safe to assume that the central limit theorem holds and the FR is normally distributed. However, statistical testing is required.

Theorem 11.1 Let $\mathbf{\{FR\}]}_{m\times 1}=\mathbf{[A]}_{m\times p}\mathbf{\{DP\}}_{p\times 1}$, $\mathbf{\{DP\}}_{p\times 1}=\mathbf{[B]}_{p\times n}\mathbf{\{PV\}}_{n\times 1}$, and $\mathbf{\{FR\}}_{m\times 1}=\mathbf{[C]}_{m\times n}\mathbf{\{PV\}}_{n\times 1}$ be given as design mappings, where $\mathbf{\{FR\}}_{m\times 1}$ is the vector of independent functional requirements with m components, $\mathbf{\{DP\}}_{p\times 1}$ is the vector of design parameters with p components, $\mathbf{\{PV\}}_{n\times 1}$ is the vector of process variables with n components, **A** is the physical design matrix, **B** is the process design matrix, and $\mathbf{[C]=[A][B]}$ is the overall design matrix. Let $\mathrm{PV}_1, \mathrm{PV}_2, \ldots, \mathrm{PV}_n$ be jointly distributed continuous random variables with continuous pdf's; the random variables $\mathrm{FR}_1, \mathrm{FR}_2, \ldots, \mathrm{FR}_n$ are jointly distributed with the following continuous pdf:

$$
\begin{aligned}
f_{\mathrm{FR}_1,\mathrm{FR}_2,\ldots,\mathrm{FR}_n}(\mathrm{fr}_1, \mathrm{fr}_2, \ldots, \mathrm{fr}_n) &= f_{\mathrm{DP}_1,\mathrm{DP}_2,\ldots,\mathrm{DP}_n}(d\mathrm{p}_1, d\mathrm{p}_2, \ldots, d\mathrm{p}_n)|[\mathrm{A}']|^{-1} \\
&= f_{\mathrm{PV}_1,\mathrm{PV}_2,\ldots,\mathrm{PV}_n}(\mathrm{pv}_1, \mathrm{pv}_2, \ldots, \mathrm{pv}_n)|[\mathrm{B}']|^{-1}|[\mathrm{A}']|^{-1} \\
&= f_{\mathrm{PV}_1,\mathrm{PV}_2,\ldots,\mathrm{PV}_n}(\mathrm{pv}_1, \mathrm{pv}_2, \ldots, \mathrm{pv}_n)|[C']|^{-1} \quad (11.7)
\end{aligned}
$$

where $\mathrm{fr}_1, \mathrm{fr}_2, \ldots, \mathrm{fr}_n; d\mathrm{p}_1, d\mathrm{p}_2, \ldots, d\mathrm{p}_n; \mathrm{pv}_1, \mathrm{pv}_2, \ldots, \mathrm{pv}_n$ are the generic continuous variables of the respective design attributes.

Proof Let Ω be the set of points $(\mathrm{fr}_1, \mathrm{fr}_2, \ldots, \mathrm{fr}_n)$ such that the n equations in (11.5) possess at least one solution given by

$$
\begin{aligned}
\mathrm{pv}_1 &= q_1^{-1}(\mathrm{fr}_1, \mathrm{fr}_2, \ldots, \mathrm{fr}_n) \\
\mathrm{pv}_2 &= q_2^{-1}(\mathrm{fr}_1, \mathrm{fr}_2, \ldots, \mathrm{fr}_n) \\
&\vdots \qquad\qquad\qquad\qquad (11.8)\\
\mathrm{pv}_n &= q_n^{-1}(\mathrm{fr}_1, \mathrm{fr}_2, \ldots, \mathrm{fr}_n)
\end{aligned}
$$

If $\mathrm{fr}_1, \mathrm{fr}_2, \ldots, \mathrm{fr}_n \in \Omega$, then $\mathrm{pv}_1, \mathrm{pv}_2, \ldots, \mathrm{pv}_n$ are given by (11.9). For $\mathrm{fr}_1, \mathrm{fr}_2, \ldots, \mathrm{fr}_n \notin \Omega$, we need to prove, in addition to (11.7), the equation

$$f_{\mathrm{FR}_1,\mathrm{FR}_2,\ldots,\mathrm{FR}_n}(\mathrm{fr}_1, \mathrm{fr}_2, \ldots, \mathrm{fr}_n) = 0 \quad (11.9)$$

Proof of (11.9) For any real numbers $(\Delta\mathrm{FR}_1, \Delta\mathrm{FR}_2, \ldots, \Delta\mathrm{FR}_n)$, we have

$$
\begin{aligned}
&f_{\mathrm{FR}_1,\mathrm{FR}_2,\ldots,\mathrm{FR}_n}(\Delta\mathrm{FR}_1, \Delta\mathrm{FR}_2, \ldots, \Delta\mathrm{FR}_n) \\
&\quad = \lim_{\Delta\mathrm{FR}_1\to o, \Delta\mathrm{FR}_2\to 0,\ldots,\Delta\mathrm{FR}_n\to 0} \frac{1}{\Delta\mathrm{FR}_1\Delta\mathrm{FR}_2\cdots\Delta\mathrm{FR}_n} \cdot \mathrm{Prob}(T_1 - \Delta\mathrm{FR}_1 \le \mathrm{FR}_1 \\
&\quad \le T_1 + \Delta\mathrm{FR}_1, T_2 - \Delta\mathrm{FR}_2 \le \mathrm{FR}_2 \le T_2 + \Delta\mathrm{FR}_2, \ldots, T_n - \Delta\mathrm{FR}_n \le \mathrm{FR}_n \\
&\quad \le T_n + \Delta\mathrm{FR}_n) \qquad (11.10)
\end{aligned}
$$

However, the probability in the RHS of (11.10) is given by

$$\begin{aligned}
&\text{Prob}(T_1 - \Delta\text{FR}_1 \le \text{FR}_1 \le T_1 + \Delta\text{FR}_1, T_2 - \Delta\text{FR}_2 \\
&\quad \le \text{FR}_2 \le T_2 + \Delta\text{FR}_2, \ldots, T_n - \Delta\text{FR}_n \le \text{FR}_n \le T_n + \Delta\text{FR}_n) \\
&= \iiint\limits_{\Omega_n} \cdots \int f_{\text{PV}_1,\text{PV}_2,\ldots,\text{PV}_n}(\text{pv}_1, \text{pv}_2, \ldots, \text{pv}_n)\, d\text{pv}_1\, d\text{pv}_2 \cdots d\text{pv}_n \qquad (11.11)
\end{aligned}$$

in which

$$\begin{aligned}
\Omega_n = T_1 - \Delta\text{FR}_1 &\le \text{FR}_1 \le T_1 + \Delta\text{FR}_1, T_2 - \Delta\text{FR}_2 \le \text{FR}_2 \le T_2 \\
&+ \Delta\text{FR}_2, \ldots, T_n - \Delta\text{FR}_n \le \text{FR}_n \le T_n + \Delta\text{FR}_n \qquad (11.12)
\end{aligned}$$

If $(T_1 - \Delta\text{FR}_1, T_2 - \Delta\text{FR}_2, \ldots, T_n - \Delta\text{FR}_n) \notin \Omega$, then for sufficiently small values of $(\Delta\text{FR}_1, \Delta\text{FR}_2, \ldots, \Delta\text{FR}_n)$, there are no points $(\text{pv}_1, \text{pv}_2, \ldots, \text{pv}_n)$ in Ω_n, and the probability in (11.10) is zero. It follows that $f_{\text{FR}_1,\text{FR}_2,\ldots,\text{FR}_n}(T_1 - \Delta\text{FR}_1, T_2 - \Delta\text{FR}_2, \ldots, T_n - \Delta\text{FR}_n) = 0$ for $(T_1 - \Delta\text{FR}_1, T_2 - \Delta\text{FR}_2, \ldots, T_n - \Delta\text{FR}_n) \notin \Omega$ [proof of (11.9)].

Proof of (11.7) By changing the variables in multiple integrals, we transform the integral in the RHS of (11.11) to

$$\begin{aligned}
&\text{Prob}(T_1 - \Delta\text{FR}_1 \le \text{FR}_1 \le T_1 + \Delta\text{FR}_1, T_2 - \Delta\text{FR}_2 \le \text{FR}_2 \le T_2 \\
&\quad + \Delta\text{FR}_2, \ldots, T_n - \Delta\text{FR}_n \le \text{FR}_n \le T_n + \Delta\text{FR}_n) \\
&= \left(\int_{T_1-\Delta\text{FR}_1}^{T_1+\Delta\text{FR}_1} d\text{fr}_1 \int_{T_2-\Delta\text{FR}_2}^{T_2+\Delta\text{FR}_2} d\text{fr}_2 \ldots \int_{T_n-\Delta\text{FR}_n}^{T_n+\Delta\text{FR}_n} d\text{fr}_n f_{\text{PV}_1,\text{PV}_2,\ldots,\text{PV}_n}\right. \\
&\quad \left.\times (\text{pv}_1, \text{pv}_2, \ldots, \text{pv}_n)\, |[C^{'}]|^{-1}\, d\text{pv}_1\, d\text{pv}_2 \cdots d\text{pv}_n\right)
\end{aligned}$$

Substituting in (11.10) and taking the limits, we have

$$\begin{aligned}
&f_{\text{FR}_1,\text{FR}_2,\ldots,\text{FR}_n}(\Delta\text{FR}_1, \Delta\text{FR}_2, \ldots, \Delta\text{FR}_n) \\
&= \lim_{\Delta\text{FR}_1 \to o, \Delta\text{FR}_2 \to 0, \ldots, \Delta\text{FR}_n \to 0} \frac{1}{\Delta\text{FR}_1 \Delta\text{FR}_2 \cdots \Delta\text{FR}_n} \\
&\quad \cdot \left(\int_{T_1-\Delta\text{FR}_1}^{T_1+\Delta\text{FR}_1} d\text{fr}_1 \int_{T_2-\Delta\text{FR}_2}^{T_2+\Delta\text{FR}_2} d\text{fr}_2 \ldots \int_{T_n-\Delta\text{FR}_n}^{T_n+\Delta\text{FR}_n} d\text{fr}_n f_{\text{PV}_1,\text{PV}_2,\ldots,\text{PV}_n}\right. \\
&\quad \left.\times (\text{pv}_1, \text{pv}_2, \ldots, \text{pv}_n)|[C^{'}]|^{-1}\, d\text{pv}_1\, d\text{pv}_2 \cdots d\text{pv}_n\right) \\
&= f_{\text{PV}_1,\text{PV}_2,\ldots,\text{PV}_n}(\text{pv}_1, \text{pv}_2, \ldots, \text{pv}_n)|[C^{'}]|^{-1}
\end{aligned}$$

Theorem 11.1 is very significant. It allows the computation of reliability in the respective mapping under different coupling categories. By utilizing Theorem 11.1, equations (11.1)–(11.3) can be expressed in one of the following forms:

$$\int_{T_n-\Delta\mathrm{FR}_n}^{T_n+\Delta\mathrm{FR}_n}\int_{T_{n-1}-\Delta\mathrm{FR}_{n-1}}^{T_{n-1}+\Delta\mathrm{FR}_{n-1}}\cdots\int_{T_1-\Delta\mathrm{FR}_1}^{T_1+\Delta\mathrm{FR}_1} f_{\mathrm{PV}_1,\mathrm{PV}_2,\ldots,\mathrm{PV}_n}$$
$$\times\,(\mathrm{pv}_1,\mathrm{pv}_2,\ldots,\mathrm{pv}_n)|[C']|^{-1}\,d\mathrm{fr}_1\,d\mathrm{fr}_2\cdots d\mathrm{fr}_n \tag{11.13}$$

$$\int_{(T_n-\Delta\mathrm{FR}_n+\mathrm{other})/A_{nn}}^{(T_n+\Delta\mathrm{FR}_n+\mathrm{other})/A_{nn}}\int_{(T_{n-1}-\Delta\mathrm{FR}_{n-1}+\mathrm{other})/A_{n-1,n-1}}^{(T_{n-1}+\Delta\mathrm{FR}_{n-1}+\mathrm{other})/A_{n-1,n-1}}\cdots\int_{(T_1-\Delta\mathrm{FR}_1+\mathrm{other})/A_{11}}^{(T_1+\Delta\mathrm{FR}_1+\mathrm{other})/A_{11}}$$
$$\times\, f_{\mathrm{DP}_1,\mathrm{DP}_2,\ldots,\mathrm{DP}_n}(d\mathrm{p}_1,d\mathrm{p}_2,\ldots,d\mathrm{p}_n)d\,d\mathrm{p}_1 d\,d\mathrm{p}_2\cdots d\,d\mathrm{p}_n \tag{11.14}$$

$$\int_{T_n-\Delta\mathrm{FR}_n}^{T_n+\Delta\mathrm{FR}_n}\int_{T_{n-1}-\Delta\mathrm{FR}_{n-1}}^{T_{n-1}+\Delta\mathrm{FR}_{n-1}}\cdots\int_{T_1-\Delta\mathrm{FR}_1}^{T_1+\Delta\mathrm{FR}_1} f_{\mathrm{DP}_1,\mathrm{DP}_2,\ldots,\mathrm{DP}_n}$$
$$\times\,(d\mathrm{p}_1,d\mathrm{p}_2,\ldots,d\mathrm{p}_n)|[\mathrm{A}']|^{-1}d\mathrm{fr}_1 d\mathrm{fr}_2\cdots d\mathrm{fr}_n \tag{11.15}$$

$$\int_{(T_n-\Delta\mathrm{FR}_n+\mathrm{other})/A_{nn}}^{(T_n+\Delta\mathrm{FR}_n+\mathrm{other})/A_{nn}}\int_{(T_{n-1}-\Delta\mathrm{FR}_{n-1}+\mathrm{other})/A_{n-1,n-1}}^{(T_{n-1}+\Delta\mathrm{FR}_{n-1}+\mathrm{other})/A_{n-1,n-1}}\cdots\int_{(T_1-\Delta\mathrm{FR}_1+\mathrm{other})/A_{11}}^{(T_1+\Delta\mathrm{FR}_1+\mathrm{other})/A_{11}}$$
$$\times\, f_{\mathrm{PV}_1,\mathrm{PV}_2,\ldots,\mathrm{PV}_n}(\mathrm{pv}_1,\mathrm{pv}_2,\ldots,\mathrm{pv}_n)|[B']|^{-1}d\,d\mathrm{p}_1 d\,d\mathrm{p}_2\cdots d\,d\mathrm{p}_n \tag{11.16}$$

where the term "other" in the integration limits refers to other terms.

For coupled and decoupled design classifications, the integration limits are very dependent on the mapping classification from a coupling standpoint. To evaluate the integral of these equations numerically, Simpson's rule or Gaussian quadrature can be used. A nested integral order can be forced by the coupling classification of the design matrix. The solution computational complexity of (11.13)–(11.16) is dependent on the following three design aspects: coupling, linearity, and the statistical independence of the random variables (integration argument), each of which has its own unique effect on the reliability assessment equations. Coupling will affect the integration limits and sequence, and linearity will affect the Jacobian calculation. For the linear design case (or for a nonlinear design in the spirit of Figure 11.4), the design matrix Jacobian is constant and can be factored out, resulting in significant computational burden relief. The statistical independence will affect the mathematical form of the joint distributions. We realize that the number of design combinations in these aspects is infinite, considering the continuous coupling measures in (3.4) and (3.5). However, the number of combinations is reduced to 12 when the coupling space of Figure 9.2 is considered. The taxonomy is shown in Figure 11.5. Note that the computational burden increases as we move away from the origin[3] in any taxonomy dimension. In this chapter we cover only the cases in Sections 11.5 and 11.6. Other cases are forthcoming in future publications by the author.

[3]The origin is a design that represents the triplet linear, independent, uncoupled.

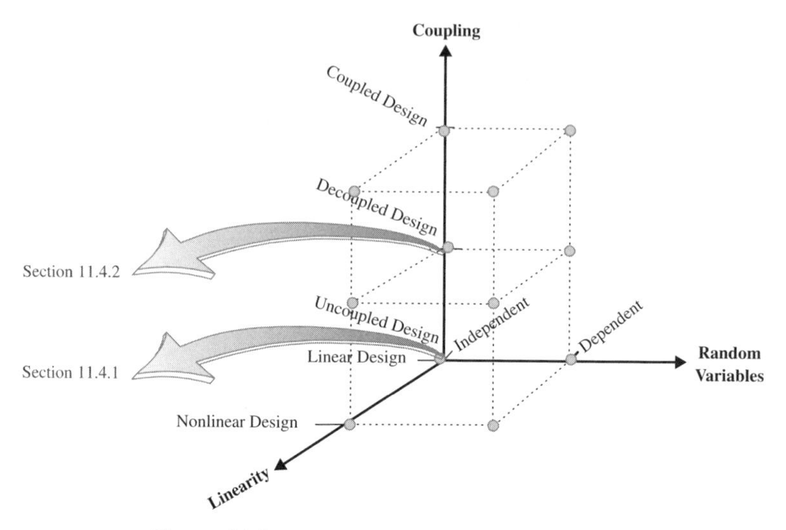

Figure 11.5 Taxonomy of axiomatic reliability.

11.4.1 Linear, Independent, Uncoupled Design

We learned in Chapter 8 that the design is uncoupled when both design matrixes in the physical and process mappings are uncoupled for a given hierarchical level. In this case we have

$$[C'] = \begin{bmatrix} C'_{11} & 0 & \cdots & 0 \\ 0 & C'_{22} & \cdots & . \\ . & . & \cdots & . \\ . & . & \cdots & . \\ 0 & . & \cdots & C'_{nn} \end{bmatrix} \qquad [C']^{-1} = \begin{bmatrix} \frac{1}{C'_{11}} & 0 & \cdots & 0 \\ 0 & \frac{1}{C'_{22}} & \cdots & . \\ . & . & \cdots & . \\ . & . & \cdots & . \\ 0 & . & \cdots & \frac{1}{C'_{nn}} \end{bmatrix} \tag{11.17}$$

where $C_{ii} = A_{ii}B_{ii}$. The inverse Jacobian determinant is given by

$$|[C']|^{-1} = \prod_{i=1}^{n} \frac{1}{C'_{ii}} \tag{11.18}$$

The Jacobian can be factored out when the design is linear. Since the design matrix is uncoupled, the sequence of integration is arbitrary while the integration limits are related only to one random variable, a DP or PV in the mapping of interest. In addition, the integration arguments (random variables) are independent, which results in the following multiplicative form of the joint distribution:

$$f_{\mathrm{PV}_1,\mathrm{PV}_2,\ldots,\mathrm{PV}_n}(\mathrm{pv}_1, \mathrm{pv}_2, \ldots, \mathrm{pv}_n) = \prod_{i=1}^{n} f_{\mathrm{PV}_i}(\mathrm{pv}_i) \tag{11.19}$$

where $f_{\mathrm{PV}_i}(\mathrm{pv}_i)$ is the distribution of the random variable pv_i. In this case the reliability statement in (11.14), for example, becomes

$$R_{\mathrm{FRs}} = \int_{(T_n-\Delta \mathrm{FR}_n)/C_{nn}}^{(T_n+\Delta \mathrm{FR}_n)/C_{nn}} \int_{(T_{n-1}-\Delta \mathrm{FR}_{n-1})/C_{n-1,n-1}}^{(T_{n-1}+\Delta \mathrm{FR}_{n-1})/C_{n-1,n-1}} \cdots \int_{(T_1-\Delta \mathrm{FR}_1)/C_{11}}^{(T_1+\Delta \mathrm{FR}_1)/C_{11}} f_{\mathrm{PV}_1,\mathrm{PV}_2,\ldots,\mathrm{PV}_n}$$
$$\times (\mathrm{pv}_1, \mathrm{pv}_2, \ldots, \mathrm{pv}_n)\, d\mathrm{pv}_1\, d\mathrm{pv}_2 \cdots d\mathrm{pv}_n$$
$$= \int_{T_n-\Delta \mathrm{FR}_n}^{T_n+\Delta \mathrm{FR}_n} \int_{T_{n-1}-\Delta \mathrm{FR}_{n-1}}^{T_{n-1}+\Delta \mathrm{FR}_{n-1}} \cdots \int_{T_1-\Delta \mathrm{FR}_1}^{T_1+\Delta \mathrm{FR}_1} f_{\mathrm{PV}_1,\mathrm{PV}_2,\ldots,\mathrm{PV}_n}$$
$$\times (\mathrm{pv}_1, \mathrm{pv}_2, \ldots, \mathrm{pv}_n) |[C']|^{-1}\, d\mathrm{pv}_1\, d\mathrm{pv}_2 \cdots d\mathrm{pv}_n$$
$$= \left(\prod_{i=1}^{n} \frac{1}{C'_{ii}}\right) \int_{T_1-\Delta \mathrm{FR}_1}^{T_1+\Delta \mathrm{FR}_1} f_{\mathrm{PV}}(\mathrm{pv}_1)\, d\mathrm{pv}_1 \int_{T_2-\Delta \mathrm{FR}_2}^{T_2+\Delta \mathrm{FR}_2} \cdots \int_{T_n-\Delta \mathrm{FR}_n}^{T_n+\Delta \mathrm{FR}_n} f_{\mathrm{PV}_n}(\mathrm{pv}_n)\, d\mathrm{pv}_n \tag{11.20}$$

Let R_{FR_i} be the reliability of meeting the design ith functional requirement, FR_i, given by

$$R_{\mathrm{FR}_i} = \frac{1}{C'_{ii}} \int_{T_i-\Delta \mathrm{FR}_i}^{T_i+\Delta \mathrm{FR}_i} f_{\mathrm{PV}_i}(\mathrm{pv}_i)\, d\mathrm{pv}_i \tag{11.21}$$

Hence, (11.20) becomes

$$R_{\mathrm{FRs}} = \prod_{i=1}^{n} R_{\mathrm{FR}_i} \tag{11.22}$$

which is Lusser's reliability law (i.e., $R = R_1 \times R_2 \times \cdots \times R_n$). Graphically, this formulation implies a series physical structure arrangement as depicted in Figure 11.6 with the P-diagram building blocks. Notice that the referenced "series" arrangement indicates the domino effect of the signal factors with no DP linkages or coupling. The conclusion is that an uncoupled design can be arranged in a modular structure with series arrangement, provided that coupling is not created in the process due to physical constraints such as packaging, size, or weight.

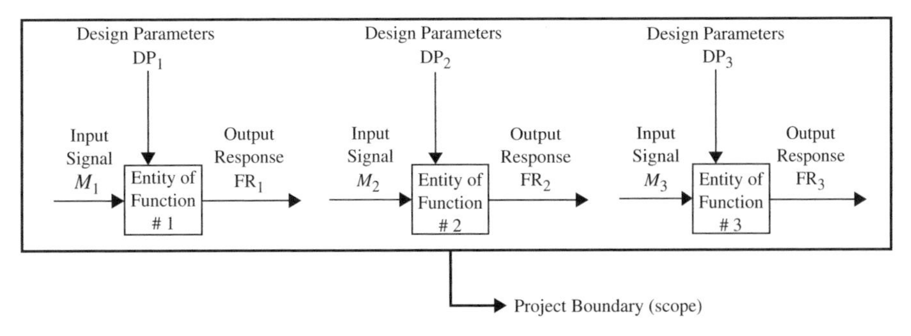

Figure 11.6 Series physical structure arrangement.

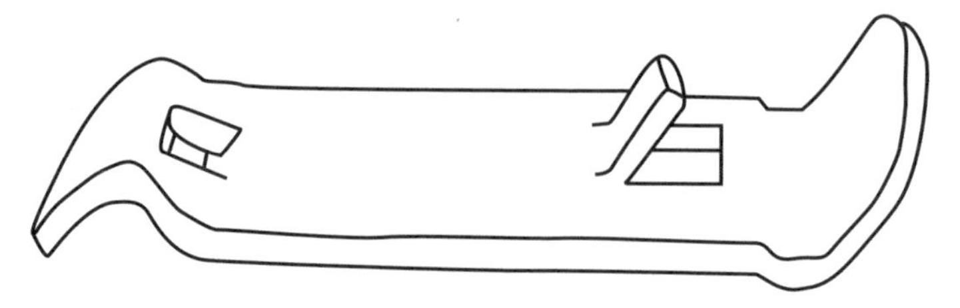

Figure 11.7 Bottle–can opener.

We would like to design for a minimum number of parts by using the idea of physical integration, not functional coupling (i.e., multi-FR parts with multiple DPs uncoupled in time or space). For example, consider the bottle–can opener in Figure 11.7.

The FRs are:

$$\mathrm{FR}_1 = \text{open beverage bottle}$$

$$\mathrm{FR}_2 = \text{open beverage can}$$

The DPs are:

$$\mathrm{DP}_1 = \text{beverage opener side}$$

$$\mathrm{DP}_2 = \text{can opener side}$$

The design mapping is depicted as

$$\begin{Bmatrix} \mathrm{FR}_1 \\ \mathrm{FR}_2 \end{Bmatrix} = \begin{bmatrix} X & 0 \\ 0 & X \end{bmatrix} \begin{Bmatrix} \mathrm{DP}_1 \\ \mathrm{DP}_2 \end{Bmatrix}$$

an uncoupled design.

A simple device that satisfies the FRs above can be made by stamping a sheet metal as shown in Figure 11.7. Notice that a single modular part can be made without functional coupling and hosted, physically, in the same device. Functional coupling should not be confused with physical integration.

11.4.2 Linear, Independent, Decoupled Design

The convention in the decoupled design matrix given in Chapter 3 is adopted. In the decoupled case, the inverse Jacobian is the given by (11.18). The reliability in this case can is given as

$$R_{\mathrm{FRs}} = \int_{T_n-\Delta\mathrm{FR}_n}^{T_n+\Delta\mathrm{FR}_n} \int_{T_{n-1}-\Delta\mathrm{FR}_{n-1}}^{T_{n-1}+\Delta\mathrm{FR}_{n-1}} \cdots \int_{T_1-\Delta\mathrm{FR}_1}^{T_1+\Delta\mathrm{FR}_1} f_{\mathrm{PV}_1,\mathrm{PV}_2,\ldots,\mathrm{PV}_n}$$
$$\times (\mathrm{pv}_1, \mathrm{pv}_2, \ldots, \mathrm{pv}_n)|[C']|^{-1}\, d\mathrm{fr}_1\, d\mathrm{fr}_2 \cdots d\mathrm{fr}_n$$

$$= \left(\prod_{i=1}^{n} \frac{1}{C_{ii}}\right) \left(\int_{(-T_n+\Delta FR_n+\sum_{i=1}^{n-1} A_{n,i}PV_i)/C_{nn}}^{(T_n+\Delta FR_n-\sum_{i=1}^{n-1} A_{n,i}PV_i)/C_{nn}} \right.$$

$$\times f_{PV_n}(pv_n) \cdots \left(\int_{(-T_2+\Delta FR_2+C_{21}PV_1)/C_{22}}^{(T_2+\Delta FR_2-C_{21}PV_1)/C_{22}} \right.$$

$$\left.\left.\times f_{PV_2}(pv_2) \left(\int_{(-T_1+\Delta FR_1)/C_{11}}^{(T_1+\Delta FR)/C_{11}} f_{PV_1}(pv_1)\, dpv_1\right) dpv_2\right) \cdots dpv_n\right) \quad (11.23)$$

As was discussed in Chapter 8, a decoupled design may be expressed with a lower or upper triangular design matrix. In an ideal overall design, with m FRs and n PVs, $m = n$; then a lower triangular design matrix is characterized by $C_{ij} = 0$ for $i < k, i = 1, \ldots, m, k = 1, \ldots, n$. Similarly, a decoupled design is upper triangular if $C_{ij} = 0$ for $i > k$. In a lower triangular, design matrix we have the following recursive relationships because of the decoupling structure:

$$pv_k = \frac{T_i - \sum_{k=1}^{i-1} C_{ik} pv_k}{C_{kk}} \qquad i = 1, \ldots, n; \quad k = 1, \ldots, n; \quad k < i \quad (11.24)$$

That is, the next random process variable can be expressed as a function of the preceding random variable in the decoupling sequence as revealed by (3.2). This fact facilitates the employment of Theorem 11.1. For example, $f_{PV_2}(pv_2)$ can be expressed as $f_{PV_1}(pv_1)(C_{22}/C_{21})$, $f_{PV_3}(pv_3) = f_{PV_1}(pv_1)(C_{22}C_{33}/C_{31} - C_{32}C_{21})$, and so on. We recognize the bracket terms as the Jacobian inverse between PV_2 and PV_3 with PV_1, respectively. Let the paired inverse Jacobian be denoted as J_{1i}^{-1}; then

$$R_{FRs} = \left(\prod_{i=1}^{n} \frac{1}{C_{ii}}\right) \left(\int_{\left[(1-C_{nn})T_n - C_{nn}\Delta FR_n - \sum_{j=2}^{n-1} C_{nj}(T_j - \Delta FR_j)\right]/C_{n1}}^{\left[(1-C_{nn})T_n + C_{nn}\Delta FR_n - \sum_{j=2}^{n-1} C_{nj}(T_j + \Delta FR_j)\right]/C_{n1}} \right.$$

$$\times f_{PV_1}(pv_1) \cdots \left(\int_{(1-C_{22})T_2 - \Delta FR_2/C_{21}}^{(1-C_{22})T_2 + \Delta FR_2/C_{21}} f_{PV_1}(pv_1) \right.$$

$$\left.\left.\times \left(\int_{T_1-\Delta FR_1}^{T_1+\Delta FR_1} f_{PV_1}(pv_1)\, dpv_1\right) dpv_1\right) \cdots dpv_1\right) \quad (11.25)$$

When the design is linear, the Jacobian is constant and (11.4) is still valid. Otherwise, the Jacobian cannot be factored out from the integration and a closed form will be case dependent. Equation (11.25) computes the reliability of the design, R_{FRs}. The computational burden may grow with the number of PVs. In such cases, nondeterministic integration methods such as Monte Carlo can be used.

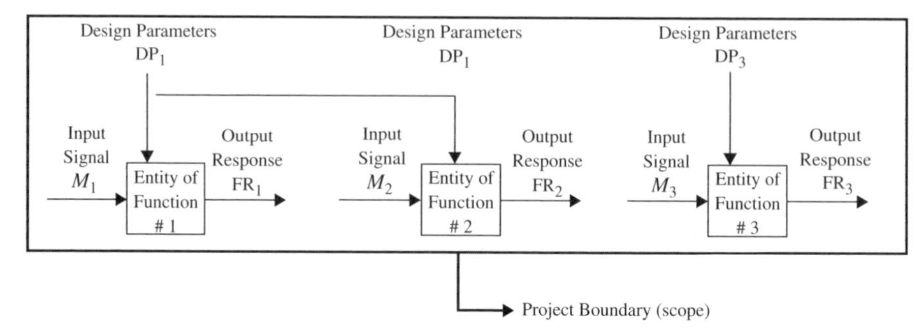

Figure 11.8 Coupled/decoupled physical structure arrangement.

In addition, for a decoupled lower triangular design matrix with positive diagonal entries, (11.4) becomes[4]

$$
R_{\mathrm{FRs}} = \mathrm{Prob}\left(\left\{\begin{array}{c} \dfrac{T_1 - \Delta\mathrm{FR}_1}{C_{11}} \\ \dfrac{T_2 - \Delta\mathrm{FR}_2 - \Delta\mathrm{FR}_1}{C_{22}} \\ \vdots \\ \dfrac{T_n - \Delta\mathrm{FR}_n - \sum_{k=1}^{n-1} C_{kn}\Delta\mathrm{FR}_k}{C_{nn}} \end{array}\right\} \le \left\{\begin{array}{c} \mathrm{PV}_1 \\ \mathrm{PV}_2 \\ \vdots \\ \mathrm{PV}_n \end{array}\right\} \le \left\{\begin{array}{c} \dfrac{T_1 + \Delta\mathrm{FR}_1}{C_{11}} \\ \dfrac{T_2 + \Delta\mathrm{FR}_2 - \Delta\mathrm{FR}_1}{C_{22}} \\ \vdots \\ \dfrac{T_n + \Delta\mathrm{FR}_n - \sum_{k=1}^{n-1} C_{kn}\Delta\mathrm{FR}_k}{C_{nn}} \end{array}\right\}\right) \tag{11.26}
$$

Graphically, this formulation implies a series physical structure arrangement as depicted in Figure 11.8 with the P-diagram format. Notice that the structure is not pure series as was the case in Section 11.5.1. Both FR_1 and FR_2 are coupled in DP_1 in the design mapping.

The decoupled nonlinear case is discussed next through the passive filter design case study discussed in Chapter 10.

11.5 CASE STUDY: PASSIVE FILTER DESIGN

This application is adopted from the Chapter 10 case study, which involves the design of an electrical passive filter. The two circuit designs proposed are given in

[4]For a lower triangular matrix with negative diagonal elements, switch the constraints from $\le$ to $\ge$.

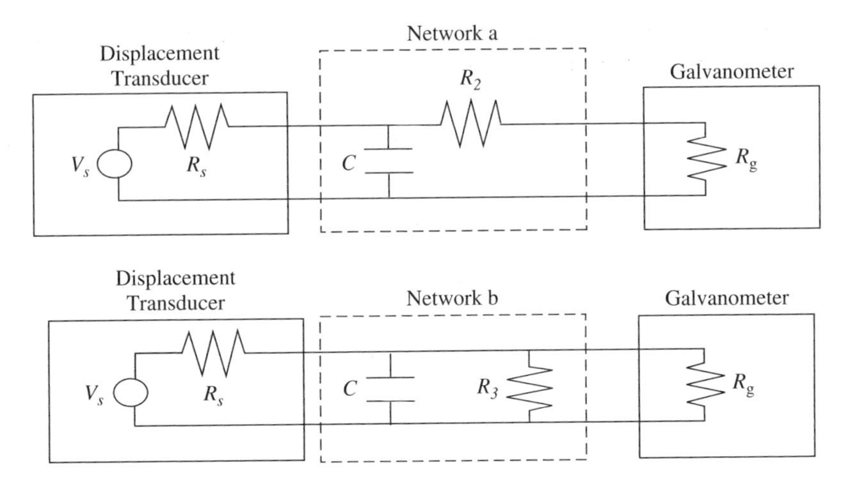

Figure 11.9 Passive filter design networks.

Figure 11.9 as network a and network b. We chose to analyze the design options that employ (as a transducer) a strain gauge bridge rather than a linear variable differential transformer (LVDT[5]). The functional requirements of the design have been specified as:

$\mathrm{FR}_l = \omega_c =$ design a low-pass filter with a filter pole at 6.84 Hz or 42.98 rad/s

$\mathrm{FR} = D =$ obtain dc gain such that the full-scale deflection results in ± 3 in. light beam deflection

The two design parameters are:

$$\mathrm{DP}_1 = C = \text{capacitance}$$

$$\mathrm{DP}_2 = R = \text{resistance}$$

The design parameter for network a is R_2. The design parameter for network b is R_3.

The transfer function expressions for the FR_1 and FR_2 terms of the design parameters, the transducer and galvanometer characteristics, are adopted from Suh (1990). The filter must suppress the carrier frequency while passing an undistorted signal of interest. The displacement transducer is replaced by an output resistance R_s and a demodulated source signal V_s, and the galvanometer is represented by impedance R_g. The network attenuates the signal so that the deflection record is scaled properly. The network settings currently used are $R_g = 98\ \Omega$, $R_s = 120\ \Omega$, $G = 0.00065758$, and $V_s = 0.015$ V.

[5]An inductance element that produces an electrical output proportional to the displacement of a separate movable core; used to measure position.

The passive filter network can be analyzed using Kirchhoff's current law to obtain the following transfer functions. For network a:

$$\omega_c = \frac{1}{C}\frac{R_s + R_g + R_2}{R_g + R_2} \quad \text{rad/s} \tag{11.27}$$

$$D = \frac{R_g}{R_s + R_g + R_2}\frac{V_s}{G} \quad \text{in.} \tag{11.28}$$

For network b:

$$\omega_c = \frac{1}{C}\frac{(R_g + R_3)R_s + R_3 R_g}{R_3 R_g R_s} \quad \text{rad/s} \tag{11.29}$$

$$D = \frac{R_3 R_g}{(R_g + R_3)R_s + R_3 R_g}\frac{V_s}{G} \quad \text{in.} \tag{11.30}$$

One may solve for the nominal DP values that place the FRs precisely on their target values. Using the formulas and values above, the design parameters that target values that satisfy the FRs are given as:

Network a: $C = 231\ \mu\text{F}$, $R_2 = 527\ \Omega$

Network b: $C = 1474\ \mu\text{F}$, $R_3 = 22.3\ \Omega$

However, noise factors will cause these values to drift, producing a variation and, over time, degradation of the FRs from their target values. Hence, we will be interested in having some tolerance around the DP target (nominal) values. Now, let us consider the reliability assessment of the design at the design stages, prior to launch ($t = 0$) for network a.

Case 1: Uniform Distribution The uniform distribution for both capacitor (DP_1) and resistor (DP_2) will be used for illustration as

$$\begin{aligned} f_{DP_1,DP_2}(dp_1, dp_2) &= f_{C,R_2}(c, r_2) \\ &= f_C(c) f_{R_2}(r_2) \\ &= \frac{1}{2\Delta C}\frac{1}{2\Delta R_2} \end{aligned} \tag{11.31}$$

Assuming that C and R_2 are statistically independent and have probability densities, their product is the combined random variable (C, R_2) joint density. Then

$$\frac{\partial\omega_c}{\partial C} = \left(-\frac{1}{C^2}\right)\frac{R_s + R_g + R_2}{R_g + R_2} \tag{11.32}$$

$$\frac{\partial\omega_c}{\partial R_2} = -\frac{1}{C(R_g + R_2)^2} \tag{11.33}$$

$$\frac{\partial D}{\partial C} = 0 \tag{11.34}$$

$$\frac{\partial D}{\partial R_2} = \frac{1}{(R_s + R_g + R_2)^2}\left(-\frac{R_g V_s}{G}\right) \tag{11.35}$$

$$[A'] = \begin{bmatrix} \dfrac{\partial \omega_c}{\partial C} & \dfrac{\partial \omega_c}{\partial R_2} \\ \dfrac{\partial D}{\partial C} & \dfrac{\partial D}{\partial R_2} \end{bmatrix} \tag{11.36}$$

and the Jacobian determinant inverse is

$$\begin{aligned} \|[A']\|^{-1} &= K[C^2((R_g + R_s)R_g + R_2^2 + R_2(2R_g + R_s))] \\ &\quad \text{with } K = GR_s/R_g V_s \end{aligned} \tag{11.37}$$

$$\begin{aligned} R_{\text{FRs}} &= \int_{T_n-\Delta\text{FR}_n}^{T_n+\Delta\text{FR}_n} \int_{T_{n-1}-\Delta\text{FR}_{n-1}}^{T_{n-1}+\Delta\text{FR}_{n-1}} \cdots \int_{T_1-\Delta\text{FR}_1}^{T_1+\Delta\text{FR}_1} \frac{1}{2\Delta C}\frac{1}{2\Delta R_2}\|[A']\|^{-1}\, d\text{fr}_1\, d\text{fr}_2 \\ &= K\frac{1}{2\Delta C}\frac{1}{2\Delta R_2}\int_{(T1-\Delta\text{FR}_1-A_{12}R_2)/A_{11}}^{(T_1+\Delta\text{FR}_1-A_{12}R_2)/A_{11}} \int_{(T_2-\Delta\text{FR}_2)/A_{22}}^{(T_2+\Delta\text{FR}_2)/A_{22}} \\ &\quad \times [c^2((R_g + R_s)R_g + r_2^2 + r_2(2R_g + R_s))]\, dc\, dr_2 \end{aligned} \tag{11.38}$$

where r_2 and c are dummy variables for R_2 and C, respectively.

$$\underbrace{\left[K\int_{(T_1-\Delta\text{FR}_1-A_{12}R_2)/A_{11}}^{(T_1+\Delta\text{FR}_1-A_{12}R_2)/A_{11}} \frac{1}{2\Delta C}c^2\, dc\right]}_{R_{\text{FR}_1}} \underbrace{\left[\int_{(T_2-\Delta\text{FR}_2)/A_{22}}^{(T_2+\Delta\text{FR}_2)/A_{22}} \frac{1}{2\Delta R_2}((R_g + R_s)R_g + r_2^2 + r_2(2R_g + R_s))\, dr_2\right]}_{R_{\text{FR}_2}} \tag{11.39}$$

$$\frac{K}{2\Delta C}\underbrace{\left[\frac{c^3}{3}\right]_{(T_1-\Delta\text{FR}_1-A_{12}R_2)/A_{11}}^{(T_1+\Delta\text{FR}_1-A_{12}R_2)/A_{11}}}_{R_{\text{FR}_1}} \underbrace{\left[\frac{1}{2\Delta R_2}\left((R_g + R_s)R_g r_2 + \frac{r_2^3}{3} + (2R_g + R_s)\frac{r_2^2}{2}\right)\right]_{(T_2-\Delta\text{FR}_2)/A_{22}}^{(T_2+\Delta\text{FR}_2)/A_{22}}}_{R_{\text{FR}_2}} \tag{11.40}$$

Let $T_1 \pm \Delta\text{FR}_1 = 42.98 \pm 2.149$ rad/s, $T_2 \pm \Delta\text{FR}_2 = 3 \pm 0.15$ in., $C^* \pm \Delta C = 31 \pm 15$ μF, and $R_2^* \pm \Delta R_2 = 527 \pm 26.35\ \Omega$. By substitution, we get $R_{\text{FRs}} = 95.97\%$.

Case 2: Exponential Distribution The exponential distribution (although an odd assumption to balance the computational complexity) for both capacitor

(DP_1) and resistor (DP_2) will be used:

$$\begin{aligned} f_{\mathrm{DP}_1,\mathrm{DP}_2}(d\mathrm{p}_1, d\mathrm{p}_2) &= f_{C,R_2}(c, r_2) \\ &= f_C(c) f_{R_2}(r_2) \\ &= \left(\frac{1}{\mu_c} e^{-c/\mu_c}\right)\left(\frac{1}{\mu_{r_2}} e^{-r_2/\mu_{r_2}}\right) \end{aligned} \tag{11.41}$$

where r_2 and c are dummy variables for R_2 and C, respectively. Assuming that C and R_2 are statistically independent and have probability densities, the product is the combined random variable (C, R_2) joint density:

$$\begin{aligned} R_{\mathrm{FRs}} &= \int_{T_n-\Delta\mathrm{FR}_n}^{T_n+\Delta\mathrm{FR}_n} \int_{T_{n-1}-\Delta\mathrm{FR}_{n-1}}^{T_{n-1}+\Delta\mathrm{FR}_{n-1}} \cdots\cdot \int_{T_1-\Delta\mathrm{FR}_1}^{T_1+\Delta\mathrm{FR}_1} \\ &\quad \times \left(\frac{1}{\mu_c} e^{-c/\mu_c}\right)\left(\frac{1}{\mu_{r_2}} e^{-r_2/\mu_{r_2}}\right) |[A']|^{-1}\, d\mathrm{fr}_1\, d\mathrm{fr}_2 \\ &= \int_{(T_1-\Delta\mathrm{FR}_1-A_{12}R_2)/A_{11}}^{(T_1+\Delta\mathrm{FR}_1-A_{12}R_2)/A_{11}} \int_{(T_2-\Delta\mathrm{FR}_2)/A_{22}}^{(T_2+\Delta\mathrm{FR}_2)/A_{22}} \left(\frac{1}{\mu_c} e^{-c/\mu_c}\right)\left(\frac{1}{\mu_{r_2}} e^{-r_2/\mu_{r_2}}\right) \\ &\quad \times K[c^2((R_g + R_s)R_g + r_2^2 + r_2(2R_g + R_s))]\, dc\, dr_2 \\ &= \underbrace{\left[K \int_{(T_2-\Delta\mathrm{FR}_2-A_{12}R_2)/A_{11}}^{(T_1+\Delta\mathrm{FR}_1-A_{12}R_2)/A_{11}} \left(\frac{1}{\mu_c} e^{-c/\mu_c}\right) c^2\, dc\right]}_{R_{\mathrm{FR}_1}} \\ &\quad \times \underbrace{\left[\int_{(T_2-\Delta\mathrm{FR}_2)/A_{22}}^{(T_2+\Delta\mathrm{FR}_2)/A_{22}} \left(\frac{1}{\mu_{r_2}} e^{-r_2/\mu_{r_2}}\right)[(R_g + R_s)R_g + r_2^2 + r_2(2R_g + R_s)]\, dr_2\right]}_{R_{\mathrm{FR}_2}} \\ &= \underbrace{K\left[e^{-c/\mu_c}(-\mu_c c^2 - 2\mu_c^2 - 2\mu_c^3)\right]\Big|_{(T_1-\Delta\mathrm{FR}_1-A_{12}R_2)/A_{11}}^{(T_1+\Delta\mathrm{FR}_1-A_{12}R_2)/A_{11}}}_{R_{\mathrm{FR}_1}} \\ &\quad \times \underbrace{\left[\int_{T_2-\Delta\mathrm{FR}_2/A_{22}}^{T_2+\Delta\mathrm{FR}_2/A_{22}} \left(\frac{1}{\mu_{r_2}} e^{-r_2/\mu_{r_2}}\right)[(R_g + R_s)R_g + r_2^2 + r_2(2R_g + R_s)]\, dr_2\right]}_{R_{\mathrm{FR}_2}} \end{aligned} \tag{11.42}$$

$$\begin{aligned} R_{\mathrm{FR}_2} &= \int_{(T_2-\Delta\mathrm{FR}_2)/A_{22}}^{(T_2+\Delta\mathrm{FR}_2)/A_{22}} \left(\frac{1}{\mu_{r_2}} e^{-r_2/\mu_{r_2}}\right) \\ &\quad \times [(R_g + R_s)R_g + r_2^2 + r_2(2R_g + R_s)]\, dr_2 \end{aligned}$$

$$
\begin{aligned}
&= (R_g + R_s)R_g \int_{(T_2-\Delta\mathrm{FR}_2)/A_{22}}^{(T_2+\Delta\mathrm{FR}_2)/A_{22}} \left(\frac{1}{\mu_{r_2}} e^{-r_2/\mu_{r_2}}\right) dr_2 + \int_{(T_2-\Delta\mathrm{FR}_2)/A_{22}}^{(T_2+\Delta\mathrm{FR}_2)/A_{22}} \\
&\quad \times \left(\frac{1}{\mu_{r_2}} e^{-r_2/\mu_{r_2}}\right) r_2^2\, dr_2 + \int_{(T_2-\Delta\mathrm{FR}_2)/A_{22}}^{(T_2+\Delta\mathrm{FR}_2)/A_{22}} \left(\frac{1}{\mu_{r_2}} e^{-r_2/\mu_{r_2}}\right) r_2(2R_g + R_s)\, dr_2 \\
&= \underbrace{K \left(e^{-c/\mu_c}(-\mu_c c^2 - 2\mu_c^2 - 2\mu_c^3)\right)\Big|_{(T_1-\Delta\mathrm{FR}_1-A_{12}R_2)/A_{11}}^{(T_1+\Delta\mathrm{FR}_1-A_{12}R_2)/A_{11}}} \\
&\quad + (R_g + R_s)R_g \left(-\mu_{r_2} e^{-r_2/\mu_{r_2}}\right)\Big|_{(T_2-\Delta\mathrm{FR}_2)/A_{22}}^{(T_2+\Delta\mathrm{FR}_2)/A_{22}} \\
&\quad + \left(e^{-r_2/\mu_{r_2}}(-\mu_{r_2} r_2^2 - 2\mu_{r_2}^2 - 2\mu_{r_2}^3)\right)\Big|_{(T_2-\Delta\mathrm{FR}_2)/A_{22}}^{(T_2+\Delta\mathrm{FR}_2)/A_{22}} \\
&\quad + (2R_g + R_s)\left(\mu_{r_2}^2 e^{-r_2/\mu_{r_2}}(-\mu_{r_2} r_2 - 1)\right)\Big|_{(T_2-\Delta\mathrm{FR}_2)/A_{22}}^{(T_2+\Delta\mathrm{FR}_2)/A_{22}}
\end{aligned}
\tag{11.43}
$$

11.6 PHYSICAL STRUCTURE AXIOMATIC RELIABILITY FORMULATION

The physical structure formulation assumes the availability of structural component failure probabilities. Let the functional requirement be delivered by a structure built off the morphological matrix (Chapter 6) with some modular components. A typical structure example is shown in Figure 11.10 for an automobile engine. It is obvious that functional requirements are split among the various subsystems and components in a modular structure.

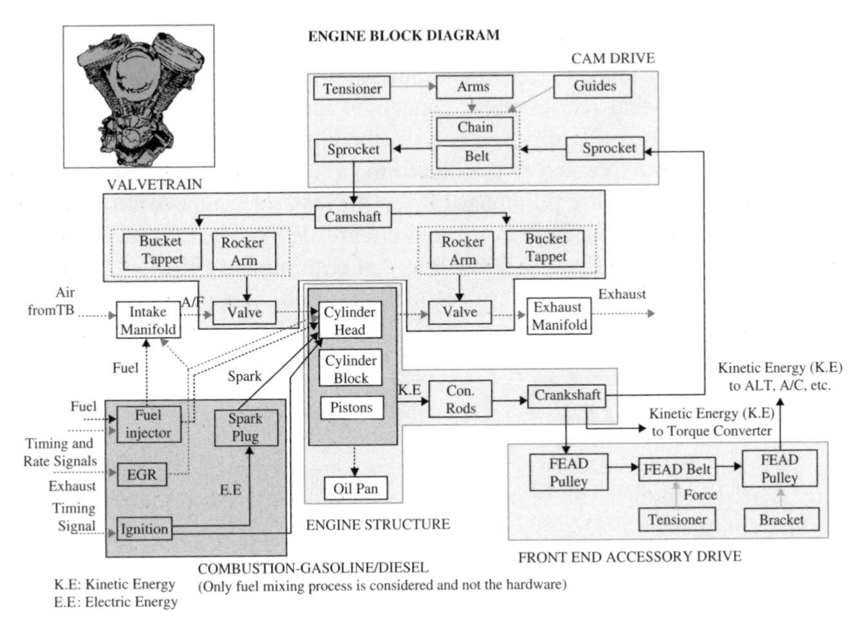

Figure 11.10 Engine physical structure.

In effect, the physical structure built is the synthesis axiomatic quality step (step A.2 of Figure 6.3) in preparation for concept selection. This step can be modeled mathematically as **{DP}=[S]{SC}**, where SC is a structure component array and [**S**] is the synthesis matrix formed from the binary variable $S_{ik} = 1$ if $\mathrm{DP}_i \rightarrow \mathrm{SC}_k$ and 0 otherwise. Therefore, **{FR}=[T]** $\otimes$ **[S]{SC}=[TS]{SC}**, where **[TS]=[T]** $\otimes$ **{S]** representing the relationship between the functional requirements and the physical structure. The matrix [**T**] was introduced in Chapter 7 as the technology matrix. The symbol $\otimes$ is a composite binary matrix operator with entry $\mathrm{SC}_{ik} = \cup_{i=1}^{m} (T_{ij} \cap S_{jk})$.

For example, let

$$[T] = \begin{bmatrix} 1 & 0 & 1 & 1 \\ 1 & 1 & 1 & 0 \\ 0 & 1 & 1 & 0 \end{bmatrix} \qquad [S] = \begin{bmatrix} 1 & 1 & 0 & 0 \\ 0 & 1 & 0 & 1 \\ 1 & 1 & 1 & 0 \\ 1 & 0 & 0 & 0 \end{bmatrix}$$

which is a redundant design matrix in four components; then

$$\mathrm{SC}_{11} = [(1 \cap 1) \cup (0 \cap 0) \cup (1 \cap 1) \cup (1 \cap 1)] = 1$$
$$\mathrm{SC}_{12} = [(1 \cap 1) \cup (0 \cap 1) \cup (1 \cap 1) \cup (1 \cap 0)] = 1$$
$$\mathrm{SC}_{13} = [(1 \cap 0) \cup (0 \cap 0) \cup (1 \cap 1) \cup (1 \cap 0)] = 1$$
$$\mathrm{SC}_{14} = [(1 \cap 0) \cup (0 \cap 1) \cup (1 \cap 0) \cup (1 \cap 0)] = 0$$
$$\vdots$$

In this formulation, the assumption is that every structural component, SC, has a binary state of existence (successor failure) and this assumption produces two states for the functional requirements concerned (delivered or not delivered). In addition, the failure of a component causes the simultaneous failure of all FRs that mapped to it as expressed in the binary mapping [**TS**]. The matrix maps the dependency of functional requirement FR_i on the state of component k, and TS_{ik} indicates the binary product of row T_i and column S_k. When $\mathrm{TS}_{ik} = 1$, the state of component k has an effect on the delivery of requirement FR_i. Let $\mathrm{Prob}_i(\mathrm{FR}_i)$ be the probability of loosing the ith requirement in the mapping. Therefore, the reliability of this requirement R_{FR_i} is given as $R_{\mathrm{FR}_i} = 1 - \mathrm{Prob}(\mathrm{FR}_i)$ and

$$R_{\mathrm{FR}_i} = \prod_{k=1}^{K} (1 - \mathrm{Prob}(\mathrm{SC}_k))^{\mathrm{TS}_{ik}} \tag{11.44}$$

where K is the number of elements of array {**SC**} (i.e., the number of design structural components). That is,

$$R_{\mathrm{FR}_i} = [1 - \mathrm{Prob}(\mathrm{SC}_1)]^{\mathrm{TS}_{11}} [1 - \mathrm{Prob}(\mathrm{SC}_2)]^{\mathrm{TS}_{12}}$$
$$\times [1 - \mathrm{Prob}(\mathrm{SC}_3)]^{\mathrm{TS}_{13}} [1 - \mathrm{Prob}(\mathrm{SC}_4)]^{\mathrm{TS}_{14}}$$

The reliability of a design is the probability of delivering all its m FRs successfully, which can be expressed mathematically through $\cup$, the logical "or" operators, as

$$\begin{aligned} R_{\mathrm{FRs}} &= (R_{\mathrm{FR1}} \cap R_{\mathrm{FR}_2} \cdots \cap R_{\mathrm{FR}_{m-1}} \cap R_{\mathrm{FR}_m}) \\ &= \prod_{i=1}^{m} R_{\mathrm{FR}_i} \\ &= \prod_{i=1}^{m} \prod_{k=1}^{K} [1 - \mathrm{Prob}(\mathrm{SC}_k)]^{\mathrm{TS}_{ik}} \end{aligned} \qquad (11.45)$$

another form of Lusser's law.

Equation (11.48) implies that the reliability of a function is dependent on the reliability of components that are related to it. The logical (binary) indicator TS_{ik} establishes this relationship. It also evaluates the reliability of the system from the reliability of delivering all the functions it is designed to deliver. Hence, we have been successful in building the component reliabilities and relationship between components and functions into the overall design reliability.

11.6.1 Time-Dependent Physical Structure Axiomatic Reliability Assessment

Reliability is the probability that a system will perform its intended function for a specified interval of time under stated conditions (Ramakumar, 1993). That is, the most important argument of reliability as a function is time, and the probability of failure is a monotone-increasing time function which implies that there is a threshold beyond which the design entity is considered failed and not functioning. Also implied are the binary-state status (failure, success), independence of component failures, and the fact that component failure results in all FRs that are mapped to it by the matrix [**TS**].

The reliability of a physical structure using the formulation above when linked to the time variable enables an assessment of the life-cycle cost and replacement analysis by the prediction of failure occurrence. The goal of reliability engineering is to evaluate the inherent reliability of a product or process and pinpoint potential areas for reliability improvement. Realistically, all failures cannot be eliminated from a design, so another goal of reliability engineering is to identify the most likely failures and then identify appropriate actions to mitigate the effects of those failures.

The reliability evaluation of a product or process can include a number of different axiomatic reliability analyses. Depending on the phase of the product life cycle, certain types of analyses are appropriate. As the reliability analyses are being performed, it is possible to anticipate the reliability effects of design changes and corrections (Section 11.5). The different axiomatic reliability analyses are all related and examine the reliability of the product or system from different perspectives, in order to determine possible problems and assist in analyzing corrections and improvements.

Reliability engineering activity should be an ongoing process starting at the conceptual phase of a product design and continuing throughout all phases of a product life cycle. The goal always needs to be to identify potential reliability problems as early as possible in the product life cycle. Although it may never be too late to improve the reliability of a product, changes in a design are orders of magnitude less expensive in the early part of a design phase rather than once the product is manufactured and in service.

The probability of failure as a function of time can be expressed as $\mathrm{Prob}(\tau \leq t) = F(t)$, where τ is a random variable denoting the time of failure. $F(t)$ is the probability that the system will fail by time $\tau = t$ (i.e., the failure distribution function). Reliability, on the other hand, can be defined as

$$\begin{aligned} R(t) &= 1 - F(t) \\ &= 1 - \int_0^t f(\tau)\,d\tau \\ &= \int_t^\infty f(\tau)\,d\tau \end{aligned} \tag{11.46}$$

Upon data availability, the time to failure τ can be modeled using several probability density function formats (e.g., normal, exponential, Weibull). Inducing the time reliability into (11.48) and (11.49) yields

$$R_{\mathrm{FR}_i} = \prod_{k=1}^{K} [R_{\mathrm{SC}_k}(t)]^{\mathrm{TS}_{ik}} \tag{11.47}$$

$$\begin{aligned} R_{\mathrm{FRs}} &= \prod_{i=1}^{m} \prod_{k=1}^{K} [R_{\mathrm{SC}_k}(t)]^{\mathrm{TS}_{ik}} \\ &= \prod_{k=1}^{K} [R_{\mathrm{SC}_k}(t)]^{\cup_{i=1}^{m} \mathrm{TS}_{ik}} \\ &= \prod_{k=1}^{K} [1 - F_{\mathrm{SC}_k}(t)]^{\cup_{i=1}^{m} \mathrm{TS}_{ik}} \end{aligned} \tag{11.48}$$

This formulation is a critical departure from traditional methods. This treatment of axiomatic reliability estimation can be applied at an early stage of development (Figure 1.8), as at this stage the functions of the system are identified, and these functions lead to the identification of modules that are needed to deliver these functions (step A.2 of axiomatic quality, Figure 6.3). This is more pertinent for incremental design situations. Thus, this methodology includes the functional performance of systems and products along with the traditional means of estimating reliability, which is based on component failures. Since this method is applied at the product concept design stage, it can be suitably applied to evaluate the ability of a number of competing concept designs that provide the functions required of the system or product (see, e.g., Cekecek and Yang, 2002).

11.7 PHYSICAL STRUCTURE AXIOMATIC IMPORTANCE FORMULATION

There are situations, in particular creative design projects, where no information is available about component reliabilities, thus making the assessment of axiomatic reliability derivation in Section 11.7 irrelevant. For reliability assessment in this case, it is wise to consider the relative importance of various components to the structure, which is called *structural importance* (Birnbaum, 1969). This will help allocate special attention to the most important components by making them more robust to noise factors. If the component (or other modules forming the structure) reliabilities are known, the concept of structural importance is expressed by reliability assessment using (11.48)–(11.49), indicating the relative importance of components to the physical structure based on test or fleet data.

The physical structure often proceeds in steps (by design hierarchy). Structuring a system of lower hierarchical levels starts by synthesizing the components into subsystem modules and the subsystems into systems. The reliability of a module structure, denoted as ψ, is discussed below. The synthesis of several modules forms the overall design structure within the design project scope.

11.7.1 Structured Modules

When a module structure is formed from K components, we can describe each of its states by a vector of component stats. The module structure can only assume any one of the 2^K represented by vertexes of the unit cube in K-dimensional space: $(0, 0, \ldots, 0), (1, 0, \ldots, 0), \ldots, (1, 1, \ldots, 1)$ with each entry $u_{\mathrm{SC}_k} = 1$ when SC_k is functioning and 0 when SC_k fails, producing K-tuples of 1's and 0's, U. It is obvious that the state of some components may cause the entire module structure to function or fail depending on whether the vertex U has its coordinate u_{SC_k} equal to 1 or 0.

The component SC_k is critical for the module physical structure ψ at some state vector U (vertex of unit cube) when

$$\theta_k(U) = \psi(u_{\mathrm{SC}_k} = 1, U) - \psi(u_{\mathrm{SC}_k} = 0, U) = 1 \tag{11.49}$$

The module structural components SC_k is critical for the success of the structure ψ, that is, delivering the FRs if

$$(1 - u_{\mathrm{SC}_k})\theta_k(U) = 1 \tag{11.50}$$

and the module structural components SC_k is critical for the failure of the structure ψ, that is, failing to deliver the FRs if

$$(u_{\mathrm{SC}_k})\theta_k(U) = 1 \tag{11.51}$$

We define the module structural importance of component SC_k for success of ψ as

$$X_{SC_k}(\psi, u_{SC_k} = 1) = 2^{-K} \sum_{\{(0,0,\ldots,0),(0,1,\ldots,0),(1,1,\ldots,1)\}} (1 - u_{SC_k})\theta_k(U) \qquad (11.52)$$

where the sum is over all vertexes of the unit cube state vectors. By the same argument we define the module structural importance of component SC_k for failure of ψ as

$$X_{SC_k}(\psi, u_{SC_k} = 0) = 2^{-K} \sum_{\{(0,0,\ldots,0),(0,1,\ldots,0),(1,1,\ldots,1)\}} (u_{SC_k})\theta_k(U) \qquad (11.53)$$

The structural importance of component SC_k in the physical structure ψ is

$$X_{SC_k} = X_{SC_k}(\psi, u_{SC_k} = 1) + X_{SC_k}(\psi, u_{SC_k} = 0) \qquad (11.54)$$

If SC_k is critical at U for the successful delivery of ψ, then SC_k is critical for structure failure, and if SC_k is critical at U for structure failure, it is critical for structure success. There is a 1-to-1 relationship between cube vertexes at which SC_k is critical for functioning and those at which it is critical for module structure failure; therefore, the number of either kind of vertexes is the same, and hence

$$\tfrac{1}{2} X_{SC_k} = X_{SC_k}(\psi, u_{SC_k} = 1) = X_{SC_k}(\psi, u_{SC_k} = 0) \qquad (11.55)$$

Example 11.1 Assume that a module structure ψ is formed using the P-diagram format from at least l components out of K total components; that is, at least l components have to function for the structure to deliver its FRs successfully:

$$X_{SC_k} = 2^{-K} 2 \binom{K-1}{l-1} \qquad k = 1, 2, \ldots, K \qquad (11.56)$$

All components have the same structural importance, and this importance is largest for

$$k = \begin{cases} \tfrac{1}{2}K & K \text{ is even} \\ \tfrac{1}{2}K + 1 & K \text{ is odd} \end{cases} \qquad (11.57)$$

The importance of every component is lowest in the case of K-out-of-K and of K in parallel 1-out-of-K structure when $X_{SC_k} = 2^{-K} \cdot 2$.

11.8 DESIGN FOR RELIABILITY

Previously, reliability has been defined as the probability that a physical entity delivers its FRs for an intended period under defined operating conditions. The time can be measured in several ways. For example, time in-service and mileage are both acceptable for automotive cases, whereas for switches and circuit breakers it is the number of open–close cycles. The design team should use the design for reliability (DFR) approach while limiting the life-cycle cost of the design.

A reliable design should anticipate all that can wrong. We view DFR as a means to maintain and sustain six-sigma capability over time. The adoption of the axiomatic quality process produces designs with high-level reliability and robustness levels in a proactive setup. In the context of this book, the DFR spirit is a core value of the axiomatic quality process. Ii is deployed by adopting:

- Conceptual measures to reduce coupling in the CDFC phase by employing design axioms and reliability science concurrently (Chapters 6 and 8)
- Axiomatic reliability techniques to calculate the reliability of key parts and to design ways to reduce or eliminate functional coupling and other design vulnerabilities
- Derating: using modules below their specified nominal values
- Design failure mode and effect analysis (DFEMA): alternative methods for failures[6]
- Operational vulnerability practices by making the design insensitive to all uncontrollable sources of variation (noise factors; see Chapter 9)
- Redundancy, where necessary, which calls for a parallel system to back up an important part or subsystem in case it fails

Reliability has to deal with a wide spectrum of issues, including human errors, technical malfunctions, environmental factors, inadequate design practices, and material variability. The design team can improve the reliability of a design by:

- Minimizing damage caused by shipping, service, and repair
- Counteracting environmental and degradation factors
- Using the information axiom and reducing design complexity
- Maximizing the use of standard components
- Determining all underlying causes of defects (not simply symptoms) by using DFMEA
- Controlling the significant and critical factors using statistical process control (SPC) where applicable
- Tracking all yield and defect rates from both in-house and external suppliers and developing strategies to address them

To minimize the probability of failure, it is first necessary to identify all possible modes of failure and the mechanism by which these failures occur. The detailed examination of DFR is developed after structure development, followed

[6]A *failure* is an unplanned occurrence that causes a system or component not to meet its FRs under the operating conditions specified.

by prototyping (Figure 1.8). Considerations regarding reliability should be taken into account in the CDFC phase when the independence axiom is employed. The team should take advantage of existing knowledge and experience of similar entities and any axiomatic reliability techniques that are applicable.

Failure avoidance, in particular when related to safety, is key. Various hazard analysis approaches are available. In general, these approaches start by highlighting hazardous elements and then proceed to identify all events that may transform these elements into hazardous conditions and their symptoms. The team then has to identify corrective actions that will eliminate or reduce these conditions. One of these approaches is *fault tree analysis* (FTA), which uses deductive logic gates to combine events that can produce the failure or fault of interest. Other tools that can be used in conjunction with FTA include DFMEA and PFMEA, as well as the fishbone diagram.

11.9 SUMMARY

In this chapter we introduced axiomatic reliability as a new subject in design theory. Axiomatic reliability provides a set of techniques based on design theory (axiomatic design in this case) that enable design teams to assess and improve the reliability of their designs. Axiomatic reliability is a proactive approach to design that utilizes the data available prior to testing. The axiomatic reliability portfolio of techniques is very suitable to a wide spectrum of applications that fit both incremental (redesign) situations and creative (white sheet) designs. Axiomatic reliability suggests a robust strategy to prevent both performance degradation and hard failures.

REFERENCES

Alabano, L. D., Conner J. J., and Suh, N. P. (1993), A framework for performance-based design, *Journal of Research in Engineering Design*, Vol. 5, pp. 105–119.

Altshuller, G. S. (1988), *Creativity as Exact Science*, Gordon and Breach, New York.

Altshuller, G. S. (1990), On the theory of solving inventive problems, *Design Methods and Theories*, Vol. 24, No. 2, pp. 1216–1222.

Arcidiacono, G., Campatelli, G., and Citti, P. (2002), Axiomatic design for six sigma, *Proceedings of the 2nd International Conference on Axiomatic Design*, MIT, Cambridge, MA, June 2002.

Arciszewsky, T. (1988), ARIZ 77: an innovative design method, *Design Methods and Theories*, Vol. 22, No. 2, pp. 796–820.

Ashby, W. R. (1973), Some peculiarities of complex systems, *Cybernetic Medicine Journal*, No. 9, pp. 1–7.

Babic, B. (1999), Axiomatic design of flexible manufacturing systems, *International Journal of Production Research*, Vol. 37, No. 5, pp. 1159–1173.

Bhattacharya, A. (1996), Reliability evaluation of systems with dependent failures, *International Journal of Systems Science*, Vol. 27, No. 9, pp. 881–885.

Birnbaum, Z. W. (1969), On the importance of different components in multicomponent system, *Proceedings of the 2nd International Symposium on Multivariate Analysis*, in *Multivariate Analysis II*, Krishnaiah, P. (ed.), Academic Press, New York.

Bowker, A. H., and Lieberman, G. J. (1959), *Engineering Statistics*, Prentice Hall, Upper Saddle River, NJ.

Axiomatic Quality: Integrating Axiomatic Design with Six-Sigma, Reliability, and Quality Engineering, by Basem Said El-Haik
ISBN 0-471-68273-X

Brejcha, M. F. (1982), *Automatic Transmission*, 2nd ed., Prentice Hall, Upper Saddle River, NJ.

Breyfogle, F. W. (1999), *Implementing Six Sigma: Smarter Solutions Using Statistical Methods*, Wiley, New York.

Brunnelle, R. D., and Kapur, K. C. (1997), Customer-centered reliability methodology, *Proceedings of the Annual Reliability and Maintainability Symposium*, pp. 286–292.

Carnap, R. (1977), *Two Essays on Entropy*, University of California Press, Berkeley, CA.

Cekecek, E., and Yang, K. (2004), Design vulnerability analysis and design improvement by using warranty data, *Quality and Reliability Engineering International*, Vol. 20, pp. 121–133.

Chase, K. W., and Greenwood, W. H. (1988), Design issues in mechanical tolerance analysis, *Manufacturing Review*, Vol. 1, No. 1, pp. 50–59.

Chen, S. J., and Hwang, C. L. (1992), *Fuzzy Multiple Attributes Decision Making*, Springer-Verlag, New York.

Chen, G., and Kapur, K. C. (1989), Quality evaluation using loss function, *International Industrial Engineering Conference and Societies' Manufacturing and Productivity Symposium Proceedings.*

Clausing, D. P. (1994), *Total Quality Development: A Step by Step Guide to World-Class Concurrent Engineering*, ASME Press, New York.

Cohen, L. (1988), Quality function deployment and application perspective from Digital Equipment Corporation, *National Productivity Review Journal*, Vol. 7, No. 3, pp. 197–208.

Cohen, L. (1995), *Quality Function Deployment: How to Make QFD Work for You*, Addison-Wesley, Reading, MA.

Cook, D. L. (1990), Evolution of VLSI reliability engineering, *Proceedings of the International Reliability Physics Symposium*, pp. 2–11.

Creveling, C. M. (1997), *Tolerance Design: A Handbook for Developing Optimal Specifications*, Addison-Wesley, Reading, MA.

Dovoino, I. (1993), Forecasting additional functions in technical systems, *Proceeding of ICED-93*, The Hague, The Netherlands, Vol. 1, pp. 247–277.

Dubois, D., and Prade, H. (1979), Fuzzy real algebra: some results, *Fuzzy Sets and Systems Journal*, No. 2, pp. 327–348.

Dubois, D., and Prade, H. (1982), On several representations of an uncertainty of evidence, in *Fuzzy Information and Decision Processes*, Gupta, M. M., and Sanches, E. (eds.), North-Holland, Amsterdam, pp. 167–182.

Dubois, D., and Prade, H. (1988), *Possibility Theory*, Wiley, New York.

Durmusoglu, M. B., Kulak, O., and Tufecki, S. (2002), An implementation methodology for transition from traditional manufacturing to cellular manufacturing using axiomatic design, *Proceedings of the 2000 International Conference on Axiomatic Design*, MIT, Cambridge, MA, June.

Dutta, A. (1985), Reasoning with imprecise knowledge in expert systems, *Information Sciences Journal*, No. 37, pp. 2–24.

El-Haik, B. S., and Yang, K. (1999), The components of complexity in engineering design, *IIE Transactions*, Vol. 31, No. 10, pp. 925–934.

El-Haik, B. S., Johnston, P., and Mohsen, H. (1995). A framework to integrate QFD, Pugh concept selection, and value engineering in concept design, *Proceedings of the*

Total Product Development Conference, American Supplier Institute, Dearborn, MI, November 1–3.

El-Haik, B. S., Johnston, P., Slater, L., and Nolf, J. (1997), A robust design optimization study of room temperature vulcanizing (RTV) silicon seal for a generic oil pan system, *Proceedings of the American Supplier Institute Symposium*, Dearborn, MI, November 5–6.

Fey, V. R., Rivin, E. I., and Verkin, I. M. (1994), Application of the theory of inventive problem solving to design and manufacturing systems, *CIRP Annals*, Vol. 43, No. 1, pp. 107–110.

Fowlkes, W. Y., and Creveling, C. M. (1995), *Engineering Methods for Robust Product Design: Using Taguchi Methods in Technology and Product Development*, Addison-Wesley, Reading, MA.

Fragole, J. R. (1993), Designing for success: reliability technology in the current design era, *Proceedings of the Annual Reliability and Maintainability Symposium*, pp. 77–82.

Fredrikson, B. (1994), Holistic systems engineering in product development, *The Saab-Scania Griffin*, Saab-Scania, AB, Linkoping, Sweden, November.

Garrett, R. (1990), Eight steps to simultaneous engineering, *Manufacturing Engineering*, November, pp. 41–47.

Gebala, D. A., and Suh, N. P. (1992), An application of axiomatic design, *Journal of Research in Engineering Design*, Vol. 3, pp. 149–162.

Greig, G. L. (1993), Second moment reliability analysis of redundant systems with dependent failures, *Reliability Engineering and Systems Safety*, Vol. 41, No. 1, pp. 57–70.

Harry, M. J. (1994), *The Vision of Six-Sigma: A Roadmap for Breakthrough*, Sigma Publishing Company, Phoenix, AZ.

Harry, M. J. (1998), Six sigma: a breakthrough strategy for profitability, *Quality Progress*, May, pp. 60–64.

Hartley, R. V. (1928), Transmission of information, *Bell System Technical Journal*, No. 7, pp. 535–563.

Hauser, J. R., and Clausing, D. (1988), The house of quality, *Harvard Business Review*, Vol. 66, No. 3, pp. 63–73.

Hillstrom, F. (1994), On axiomatic design in modular product development, licentiate thesis, Machine and Vehicle Design, Chalmers University of Technology, Goteborg, Sweden.

Hines, W. W., and Montgomery, D. C. (1980), *Probability and Statistics for Engineering and Management Science*, 2nd ed., Wiley, New York.

Hintersteiner, J. D. (1999), A fractal representation for systems, presented at the International CIRP Design Seminar, Enschede, The Netherlands, March 24–26.

Hintersteiner, J. D., and Nain, A. S. (1999), Integrating software into systems: an axiomatic design approach, *Proceedings of the 3rd International Conference on Engineering Design and Automation*, Vancouver, BC, Canada, August 1–4.

Hintersteiner, J. D., and Tate, D. (1998), Command and control in axiomatic design theory: its role and placement in system architecture, *Proceedings of the 2nd International Conference on Engineering Design and Automation*, Maui, HI, August, pp. 9–12.

Hubka, V. (1980), *Principles of Engineering Design*, Butterworth Scientific Publishing, London.

Igata H. (1996), Application of axiomatic design to rapid-prototyping support for real-time control software, S.M. thesis, Department of Mechanical Engineering, MIT, Cambridge, MA, May.

Jacobs, B. (2002), Were the ancient Egyptians system engineers? How the building of Khufu's great pyramid satisfies systems engineering axioms, University of Maryland, College Park, MD; download from *http://www.bandisoftware.com/Incose2002.pdf.*

Jaynes, E. T. (1957a), Information theory and statistical mechanics: I, *Physical Review Journal*, No. 106, pp. 620–630.

Jaynes, E. T. (1957b), Information theory and statistical mechanics: II, *Physical Review Journal*, No. 108, pp. 171–190.

Johnson, R. A., and Wichem, D. W. (1982), *Applied Multivariate Statistical Analysis*, Prentice Hall, Englewood Cliffs, NJ.

Kacker, R. N. (1985), Off-line quality control, parameter design, and the Taguchi method, *Journal of Quality Technology*, No. 17, pp. 176–188.

Kapur, K. C. (1988), An approach for the development for specifications for quality improvement, *Quality Engineering*, Vol. 1, No. 1, pp. 63–77.

Kapur, K. C. (1991a), Quality engineering and tolerance design, in *Concurrent Engineering: Automation, Tools and Techniques*, pp. 287–306.

Kapur, K. C. (1991b), Quality improvement through robust design, *International Institute of Industrial Engineering Conference Proceedings*.

Killander, A. J. (1995), Concurrent development requires uncoupled concepts and projects, presented at the International Conference on Concurrent Engineering, Reston, VA.

Kim, S. J., and Suh, N. P. (1987), Knowledge-based synthesis system for injection molding, *Robotics and Computer Integrated Manufacturing Journal*, Vol. 3, No. 2, p. 181.

Kim, S. J., Suh, N. P., and Kim, S. (1991), Design of software systems based on AD, *Annals of CIRP*, Vol. 40, pp. 165–170.

Klir, J. G., and Folger, T. A. (1988), *Fuzzy Sets, Uncertainty, and Information*, Prentice Hall, Upper Saddle River, NJ.

Ku, H. H. (1966), Notes on the use of propagation of error formulas, *Journal of Research of the National Bureau of Standards: C. Engineering and Instrumentation*, Vol. 70, No. 4, pp. 263–273.

Lake, J. (1994), *Axioms for Systems Engineering*, Systems, System Management International, VA.

Lee, T. S. (1999), The system architecture concept in axiomatic design theory: hypotheses generation and case-study validation, S.M. thesis, Department of Mechanical Engineering, MIT, Cambridge, MA.

Lehner, M. (1997), *The Complete Pyramids*, Thames & Hudson, London.

Lentz, V. A., Lerner, B., and Whitecomb, C. (2002), The validation of a modular commercial product architecture, *Proceedings of the 2nd International Conference on Axiomatic Design,* MIT, Cambridge, MA, June.

Leung, Y. (1980), Maximum entropy estimation with inexact information, in *Fuzzy Set Possibility Theory: Recent Developments*, Yager, R. R. (ed.), Pergamon Press, New York, pp. 32–37.

Lewis, E. E. (1987), *Introduction to Reliability Engineering*, Wiley, New York.

Luenberger, D. G. (1989), *Linear and Non-linear Programming*, 2nd ed., Addison-Wesley, Reading MA.

Murty, K. G. (1983), *Linear Programming*, Wiley, New York.

Nair, V. N. (1992), Taguchi's parameter design: a panel discussion, *Econometrics*, Vol. 34, No. 2.

Nakazawa, H., and Suh, N. P. (1984), Process planning based on information concept, *Journal of Robotics and Computer Integrated Manufacturing*, Vol. 1, pp. 115–123.

National Academy of Engineering (NAE), (2002), *Approaches to Improve Design*, available on the Web at *http://www.nap.edu/NI000469/html/*

Nordlund, M. (1996), An information framework for engineering design based on axiomatic design, doctoral dissertation, Department of Manufacturing Systems, Royal Institute of Technology (KTH), Stockholm, Sweden.

Nordlund, M., Tate, D., and Suh, N. P. (1996), Growth of axiomatic design through industrial practice, *Proceedings of the 3rd CIRP Workshop on Design and Implementation of Intelligent Manufacturing Systems*, Tokyo, June 19–21, pp. 77–84.

Pahl, G., and Beitz, W. (1988), *Engineering Design: A Systematic Approach*, 2nd ed., Springer-Verlag, New York.

Pal, N. R., and Pal, S. K. (1992), Higher order fuzzy entropy and hybrid entropy of a set, *Information Science Journal*, Vol. 61, No. 3, pp. 211–231.

Palady, P. (1995), *Failure Mode and Effect Analysis*, PT Publications, West Palm Beach, FL.

Pecht, M., et. al. (1994), Reliability predictions: their use and misuse, *Proceedings of the Annual Reliability and Maintainability Symposium*, pp. 386–389.

Phadke, M. S. (1989), *Quality Engineering Using Robust Design*, Prentice Hall, Upper Saddle River, NJ.

Prazen, E. (1960), *Modern Probability Theory and Its Applications*, Wiley, New York.

Pugh, S. (1991), *Total Design*, Addison-Wesley, Reading, MA.

Pugh, S. (1996), in *Creating Innovative Products Using Total Design*, Clausing, D. P., and Andrade, (eds.), Addison-Wesley, Reading, MA.

Ramakumar, R. (1993), *Engineering Reliability: Fundamentals and Applications*, Prentice Hall, Upper Saddle River, NJ.

Rantanen, K. (1988), Altshuller's methodology in solving inventive problems, presented at ICED-88, Budapest, August 23–25.

Rinderle, J. R. (1982), Measures of functional coupling in design, Ph.D. dissertation, MIT, Cambridge, MA.

Salkin, H. M., and Mathur, K. (1989), *Foundations of Integer Programming*, Elsevier Science Publishing, New York.

Shannon, C. E. (1948), The mathematical theory of communication, *Bell System Technical Journal*, No. 27, pp. 379–423, 623–656.

Shiba, S., Graham, A., Walden, D., and Asay, D. (1993), *A New American TQM: Four Practical Revolutions in Management*, Productivity Press, Portland, OR.

Simon, H. A. (1981), *The Science of the Artificial*, 2nd ed., MIT Press, Cambridge, MA.

Sohlenius, G. (1998), *The Productivity of Manufacturing through Manufacturing System Design*, Royal Institute of Technology, Stockholm, Sweden, working paper.

Sohlenius, G., Kjellberg, A., and Holmstedt, P. (1999), Productivity system design and competence management, presented at the 12th World Productivity Congress, Edinburgh.

Spotts, M. F. (1973), Allocation of tolerance to minimize cost of assembly, *Transactions of the ASME*, pp. 762–764.

Srinivasan, R. S., and Wood, K. L. (1992). A computational investigation into the structure of form and size error based on machining mechanics, *Advances in Design Automation*, Phoenix, AZ, pp. 161–171.

Stark, H., and Woods, W. W. (1986), *Probability, Random Processes, and Estimation Theory for Engineers*, Prentice Hall, Upper Saddle River, NJ.

Suh, N. P. (1984), Development of the science base for the manufacturing field through the axiomatic approach, *Robotics and Computer Integrated Manufacturing Journal*, Vol. 1, No. 3/4.

Suh, N. P. (1990), *The Principles of Design*, Oxford University Press, New York.

Suh, N. P. (1995), Design and operation of large systems, *Journal of Manufacturing Systems*, Vol. 14, No. 3.

Annals of CIRP, Vol. 14, No. 3, pp. 203–213.

Suh, N. P. (1996), *Axiomatic design course*, Ford Motor Co., Dearborn, MI, April.

Suh, N. P. (1997), Design of systems, *Annals of CIRP*, Vol. 46, No. 1, pp. 75–80.

Suh, N. P. (2001), *Axiomatic Design: Advances and Applications*, Oxford University Press, New York.

Suh, N. P., and Rinderle, J. R. (1982), Qualitative and quantitative use of design and manufacturing axiom, *CIRP Annals*, Vol. 31, No. 1, pp. 333–338.

Suh, N. P., Cochran, D. S., and Lima, P. C. (1998), Manufacturing system design, *CIRP Annals*, Vol. 47, No. 2, pp. 627–639.

Suskkov, V. V., Mars, N. J., and Wognum, P. M. (1995), Introduction to TIPS: a theory for creative design, *Artificial Intelligence in Engineering*, Vol. 9, pp. 177–189.

Swenson, A., and Nordlund, M. (1996), *Axiomatic Design of Water Faucet*, Saab Linkoping, Sweden.

Tadikamalia, P. (1994), The confusion about six sigma, *Quality Progress*, July.

Taguchi, G. (1986), *Introduction to Quality Engineering*, Nordica International, Hong Kong; UNIPUB/Kraus International Publications, White Plains, NY; American Supplier Institute, Dearborn, MI.

Taguchi, G. (1987), *System of Experimental Design*, Vols. 1 and 2, American Supplier Institute, Dearborn, MI.

Taguchi, G. (1993), *Taguchi on Robust Technology Development: Bring Quality Engineering Upstream*, ASME, New York.

Taguchi, G., and Wu, Y. (1980), *Introduction to Off-Line Quality Control*, Central Japan Quality Control Association, Tokyo.

Taguchi, G., Elsayed, E., and Hsiang, T. (1989), *Quality Engineering in Production Systems*, McGraw-Hill, New York.

Taguchi, G., Chowdhury, S., and Taguchi, S. (1999), *Robust Engineering: Learn How to Boost Quality While Reducing Costs and Time to Market*, McGraw-Hill, New York.

Tate, D., and Nordlund, M. (1998), A design process roadmap as a general tool for structuring and supporting design activities, *SDPS Journal of Integrated Design and Process Science*, Vol. 2, No. 3, pp. 11–19.

Teng, S., and Ho, S. (1995), Reliability analysis for the design of an inflator, *Quality and Reliability Engineering International*, Vol. 11, pp. 203–214.

Trewn, J. (1999), Functional reliability design and evaluation methodology: a systems approach, Ph.D. dissertation, Wayne State University, Detroit, MI.

Tribus, M. (1961), *Thermostatics and Thermodynamics*, D. Van Nostrand, Princeton, NJ.

Tsourikov, V. M. (1993), Inventive machine: second generation, *Artificial Intelligence and Society*, No. 7, pp. 62–77.

Ulman, D. G. (1992), *The Mechanical Design Process*, McGraw-Hill, New York.

Ulrich, K. T., and Eppinger, S. D. (1995), *Product Design and Development*, McGraw-Hill, New York.

Vasseur, H., Kurfess, T. and Cagan, J. (1993), Optimal tolerance allocation for improved productivity, *Proceedings of 1993 NSF Design & Manufacturing Systems Conference*, Charlotte, NC, pp. 715–719.

Verein Deutscher Ingenieure (1986), *Systematic Approach to the Design of Technical Systems and Products*, VDI-2221, translation of the German edition, VDI-Verlag, Dusseldorf, Germany.

Wasiloff, J., and El-Haik, B. S. (2004), Practical application of DFSS with a focus on axiomatic design: a transmission planetary case study, *Proceedings of the International Society of Automotive Engineers*, Detroit, MI, March.

Weaver, W. (1948), Science and complexity, *American Scientist*, No. 36, pp. 536–544.

Wilson, A. G. (1970), *Entropy in Urban and Regional Modeling*, Pion Publishing, London.

Wood, G. W., Srinivasan, R. S., Tumer, I. Y., and Càvin, R. (1993), Fractal-based tolerancing: Theory, dynamic process modeling, test bed development, and experiment, *Proceedings of 1993 NSF Design & Manufacturing Systems Conference*, Charlotte, NC, pp. 731–740.

Xie, W. X., and Bedrosian, S. D. (1984), An Information measure for fuzzy sets, *IEEE Transactions on Systems, Man, and Cybernetics*, Vol. 14, pp. 151–156.

Yang, K., and El-Haik, B. (2003), *Design for Six Sigma: A Roadmap for Product Development*, McGraw-Hill, New York.

Yang, K., and Kapur, K. C. (1997), Customer driven reliability integration of QFD and robust design, *Proceedings of the IEEE Annual Reliability and Maintainability Symposium*, pp. 251–257.

Yang, K., and Xue, J. (1996), Continuous state reliability analysis, *Proceedings of the IEEE Annual Reliability and Maintainability Symposium*, pp. 251–257.

Zadeh, L. A. (1965), Fuzzy sets, *Information and Control Journal*, No. 8, pp. 338–353.

Zadeh, L. A. (1968), Probability measures of fuzzy events, *Journal of Mathematical Analysis and Applications*, Vol. 23, pp. 421–427.

Zadeh, L. A. (1975), The concept of linguistic variable and its application to approximate reasoning, *Information Sciences Journal*, Vol. 8, Parts I (pp. 199–249), II (pp. 301–357), and III (pp. 43–96).

Zadeh, L. A. (1978), Fuzzy sets as a basis for theory of possibility, *Fuzzy Sets and Systems*, No. 1, pp. 3–28.

Zhang, H. C., and Huq, M. E. (1994), Tolerancing techniques: The state-of-the-art, *International Journal of Production Research*, Vol. 30, No. 9, pp. 2111–2135.

Zimmermann, H. J. (1985), *Fuzzy Set Theory and Its Applications*, Kluwer-Nijhoff, Hingham, MA.

Zoltin, B., et al. (1996), TRIZ/ideation methodology for customer driven innovation, *Proceedings of the 8th Symposium on Quality Function Deployment*, Novi, MI, June, pp. 448–510.

Zwicky, F. (1984), *Morphological Analysis and Construction*, Wiley-Interscience, New York.

On-Line Resources

AllRefer.com encyclopedia @ *http://www.reference.allrefer.com/encyclopedia.*

Mathworld Wolfram research @ *http://www.mathworld.wolfram.com.*

INDEX

Axiomatic Quality: Integrating Axiomatic Design with Six-Sigma, Reliability, and Quality Engineering, by Basem Said El-Haik
ISBN 0-471-68273-X